Mediterranean Protected Areas in the Era of Overtourism

Ante Mandić · Lidija Petrić
Editors

Mediterranean Protected Areas in the Era of Overtourism

Challenges and Solutions

Springer

Editors
Ante Mandić
Faculty of Economics
Business and Tourism
University of Split
Split, Croatia

Lidija Petrić
Faculty of Economics
Business and Tourism
University of Split
Split, Croatia

ISBN 978-3-030-69192-9 ISBN 978-3-030-69193-6 (eBook)
https://doi.org/10.1007/978-3-030-69193-6

This Springer imprint is published by the registered company Springer Nature Switzerland AG
The registered company address is: Gewerbestrasse 11, 6330 Cham, Switzerland

Contents

Editors and Contributors

About the Editors

Dr. Ante Mandić is an assistant professor at the University of Split, Faculty of Economics, Business and Tourism—Department of Tourism and Economy, Croatia. His research and teaching focus on sustainable tourism management, nature-based tourism and destination management. He has extensive experience in tourism-related projects and has worked as an expert on EU funded projects (ERDF MED, INTERREG). Most recently he has been enrolled as a researcher on Horizon 2020 SmartCulTour project. He is a member of the IUCN World Commission on Protected Areas (WCPA) and Tourism and Protected Areas Specialist Group (TAPAS Group, and Europarc Federation). He currently serves as an associate editor for the Journal of Ecotourism (Taylor & Francis). Personal web page: https://antemandic.com/.

Dr. Lidija Petrić is a full professor at the Faculty of Economics, Business and Tourism, University of Split. The primary areas of her scientific interest are the economics of tourism, the role of tourism in regional and urban development, destination strategic management and planning, visitor management strategies, cultural tourism. She wrote a number of scientific and professional papers, reviews, conference papers and scientific-professional studies. As a guest lecturer, she attended professional conferences and seminars, postgraduate studies, as well as international conferences and seminars. As an expert, she cooperated with international associations such as UNDP, GEF, WWF, UNEP, the public bodies from national to the local level as well as non-governmental institutions, participating in and coordinating many projects. She also participated as a leading researcher in several tourism-related projects funded by the EU and is currently leading a research team of her Faculty on the HORIZON 2020 Smart CulTour Project. She is a member of the editorial board of several scientific journals as well as of scientific committees of several international conferences. She is a member of the International Association of Regional Economists of Europe—the European Regional Scientists Association, the Scientific Committee for Tourism at the Croatian Academy of Sciences and Arts and the Croatian National Committee for UNESCO.

Contributors

Karl Agius Institute for Tourism, Travel and Culture, University of Malta, Msida, Malta

Ivo Beroš Institute for Tourism, Zagreb, Croatia

Charles Besancon Global Park Solutions, Missoula, Montana, USA

Jan van der Borg Department of Economics (IT), University Ca' Foscari of Venice, Venice, Italy;
Division of Geography and Tourism (B), KU Leuven University, Louvain, Belgium

Zélia Breda University of Aveiro, Aveiro, Portugal

Nicola Camatti Department of Economics (IT), University Ca' Foscari of Venice, Venice, Italy

Hrvoje Carić Institute for Tourism, Zagreb, Croatia

Emanuel Castro Associação Geopark Estrela (AGE), Guarda, Portugal

Samantha Chaperon Department of Marketing, Events and Tourism, University of Greenwich, London, UK

Carlos Costa University of Aveiro, Aveiro, Portugal

Ivana Damnjanović Faculty of Health and Business Studies, Singidunum University, Valjevo, Serbia

Tomaz Ponce Dentinho Centro de Estudos de Economia Aplicada do Atlântico (CEEAplA), University of the Azores, Ponta Delgada, Portugal

Paul F. J. Eagles University of Waterloo, Waterloo, Canada

Gonçalo Fernandes Instituto Politécnico da Guarda, CITUR, CICS.NOVA, Guarda, Portugal

Magda Fernandes Associação Geopark Estrela (AGE), Guarda, Portugal

Sonia Ferrari University of Calabria, Rende, Italy

Ulrike Gretzel University of Southern California, Los Angeles, CA, USA

Burcin Kalabay Hatipoglu School of Business, Canberra and IIRG, The University of New South of Wales, Sydney, Australia

Nicolaia Iaffaldano University of Bari, Bari, Italy

Neven Ivandić Institute for Tourism, Zagreb, Croatia

Ana Carolina Jordão University of Aveiro, Aveiro, Portugal

Damir Krešić Institute for Tourism Zagreb, Zagreb, Croatia

Zvonimir Kuliš Faculty of Economics, Business and Tourism, University of Split, Split, Croatia

Sina Kuzuoglu Department of Recreation and Leisure Studies, University of Waterloo, Waterloo, ON, Canada

Ante Mandić Faculty of Economics, Business and Tourism, University of Split, Split, Croatia

Izidora Marković Vukadin Institute for Tourism, Zagreb, Croatia

Stephen F. McCool University of Montana, Montana, USA

Josip Mikulić Faculty of Economics and Business, University of Zagreb, Zagreb, Croatia;
Institute for Tourism Zagreb, Zagreb, Croatia

Vinko Muštra Faculty of Economics, Business and Tourism, University of Split, Split, Croatia

Andjelko Novosel Geocustos, Zagreb, Croatia

Lidija Petrić Faculty of Economics, Business and Tourism, University of Split, Split, Croatia

Smiljana Pivčević Faculty of Economics, Business and Tourism, University of Split, Split, Croatia

João Romão Yasuda Women's University, Hiroshima, Japan

Hugues Seraphin University of Winchester Business School, Winchester, UK

Blanka Šimundić Faculty of Economics, Business and Tourism, University of Split, Split, Croatia

Ognjen Skunca Granulum Salis, d.o.o., Zagreb, Croatia

Carolyn Smith Department of Geography, University of Cambridge, Cambridge, UK

Simon Smith University of Winchester Business School, Winchester, UK

Ivana Stevic University of Aveiro, Aveiro, Portugal

Neda Telišman-Košuta Institute for Tourism, Zagreb, Croatia

Rosa Branca Tracana Escola Superior de Educação, Comunicação e Desporto, Ci&DEI, Instituto Politécnico da Guarda, Guarda, Portugal

Medéia Veríssimo University of Aveiro, Aveiro, Portugal

Vesna Vukadin Park Bureau, d.o.o., Zagreb, Croatia

Dorra Yahiaoui Kedge Business School, Talence, France

Chapter 1
Introduction to Mediterranean Protected Areas in the Era of Overtourism

Ante Mandić and Lidija Petrić

Protected areas (PAs) are clearly defined geographical spaces that are recognised as such and are dedicated to achieving long-term nature conservation, with the associated ecosystem services and cultural values. To be able to accomplish this, they are managed through legal or other effective means (Dudley 2008). They encompass a wide variety of natural and semi-natural environments, described and categorised in the IUCN's Protected Area Categories System (Dudley 2010). Well managed, PAs provide environmental, social and economic benefits to the society (Croy et al. 2020), which can be enjoyed at local, regional and international levels (Kettunen et al. 2017) and support sustainable development goals (SDGs) (Mackinnon et al. 2020). Academics and those involved practically have acknowledged that equitably and effectively managed PA is essential for halting biodiversity loss (Zafra-Calvo and Geldmann 2020). Although PAs are common and extensive conservation interventions, their effectiveness in reducing biodiversity loss is usually simply assumed, implicitly or explicitly (Ribas et al. 2020). But the global analysis of impacts associated with PA management interventions (Coad et al. 2015) suggests that an understanding of what constitutes good management remains an ongoing challenge. Successful area-based conservation requires better collaboration with the indigenous peoples, community groups and private initiatives that are central to its success (Maxwell et al. 2020). Within this process, a stronger focus is needed on the monitoring of biodiversity outcomes (Rodrigues and Cazalis 2020). Taylor et al. (2020) even call for more "ecocentrism" as a response to documented negative anthropogenic environmental changes, suggesting that the moral argument for biodiversity conservation should stem from convictions that ecosystems have value and interests that should be respected regardless of whether they serve human needs. High

A. Mandić (✉) · L. Petrić
Faculty of Economics, Business and Tourism, University of Split, Split, Croatia
e-mail: ante.mandic@efst.hr

L. Petrić
e-mail: lidija.petric@efst.hr

A. Mandić and L. Petrić (eds.), *Mediterranean Protected Areas in the Era of Overtourism*, https://doi.org/10.1007/978-3-030-69193-6_1

anthropocentrism in PA management has been spurred, among other things, by the introduction of the ecosystem services (ES) approach, where individuals focus on maximising profits and benefits from nature (Hummel et al. 2019), while policy-makers fail to ensure that such benefits can only be accrued while the natural system is being protected.

At the EU level, the development of a wise and sustainable protected area use paradigm has given rise to the Natura 2000 network and Emerald Network (European Environmental Agency (EEA) 2020). With more than 120 000 sites designated across 52 countries, today Europe accounts for more PAs than any other region of the world. The backbone of its policy regarding heritage sites is the United Nations Convention on Biological Diversity, EU directives, i.e. Birds Directive and Habitats Directive and the recently adopted EU Biodiversity Strategy for 2030 and an associated action plan. The 2030 Strategy is looking for recovery of ecosystems with benefits for people, the climate and the planet through the advancement of the EU network of PAs on land and at sea, policy responses, and more sophisticated governance frameworks to ensure their effectiveness. This policy entitled "*Bringing nature back into our lives*", in the first paragraph emphasises "*We need nature in our lives*", stressing that nature is vital for our mental and physical wellbeing and that a healthy and resilient society depends on giving nature the space it needs. These claims have been recently confirmed as the COVID-19 pandemic has demonstrated that the emergence of many of the new scourges of our time can be attributed to increased invasive human impacts on natural systems (Corlett et al. 2020).

Although the intrinsic value of PAs remains the fundamental reason for their protection and management, their support to local and national economies through the provision of ES, in particular, tourism and recreation (cultural ES) has gained greater weight and acceptance in recent years. Today, tourism is recognised as the most extensive use of PAs, with positive and negative influences (Leung et al. 2018; Spenceley 2015). The growing demand for nature-based tourism and the development of associated businesses has initiated the transformation of many PAs into tourism destinations (Mandić 2019, 2020; Mandić and Petrić 2020). The study of Gon et al. (2018) demonstrated how nature-based tourism is more frequent in PAs that are of higher biodiversity, older, larger, more accessible from urban areas and at a higher elevation. Ever-increasing visitation (residents, domestic and international tourists) induced a deepening crisis within these unique ecosystems associated with, among other effects, population growth, increasing consumption, climate change, increasing reliance on visitor-based revenues, and growth in demand for rural outdoor and nature-based recreation from increasingly urbanised societies (Weaver and Lawton 2017). All of this has facilitated the development of a complex decision-making environment characterised by pronounced anthropocentrism and a continuous quest for new managerial solutions and knowledge-sharing. Within the complex, contentious and changing environment, PA managers are challenged to harmonise biodiversity conservation, provision of leisure services and sustained local community development. However, can this goal be reached? A fruitful discussion on reconciling the demand for recreation and the need for biology conservation from a managerial perspective was recently delivered in *Reframing sustainable tourism*, edited by

McCool and Bosak (2016). In one of the chapters, McCool (2016) concluded that the limited capacity of PA agencies to manage tourism and visitation inevitably suggests that decision-makers must have a better understanding of tools and solutions and their suitability for addressing different issues. Consequently, their developing capability in these areas ultimately will lead to more efficient, effective and equitable decisions. Along with these non-disputable claims, we would like to add that any PA as a locally adapted socioecological system needs to be resilient on external driving forces, among which excessive tourism, recently discussed under the discourse of overtourism, seems to be one of the most hazardous. As global tourism continuously increases, we are becoming more and more aware that solutions adapted and transposed from urban settings into natural environments often fail. At the same time, limited capacities prevent PA managers from forestalling problems. This is particularly pronounced in those PAs where excessive visitor numbers, arising either by chance or deliberately, for the sake of profits, give cause for concern. All of this suggests there is a need to initiate a debate on the root causes of excessive seasonal visitation in PAs. Within this framework, special attention should be given to research into the spatial interrelations between nature-based (tourism) areas and their adjoining urban destinations.

1.1 Focus on the Mediterranean

Although travel restrictions due to the COVID-19 pandemic have suspended the upward trend, international tourism will probably recover the levels of 2019 in a period of two to four years (World Travel Organization 2020) and we will witness growth again. Europe, the world's most visited destination in 2019, accounted for 51.1% of overall international tourist arrivals, indicating a 3.9% annual growth rate. Within this region, the Mediterranean is the most prosperous subregion, assimilating 20.9% of tourism flows, and facing an annual growth of 5.4% (Fig. 1.1).

Many of the Mediterranean tourism destinations are currently experiencing high tourism-dependency, continuous increase in a number of visitors, overcrowding and most recently even anti-tourist and anti-tourism movements and social unrest. Within the recent literature (Capocchi et al. 2019; Cheung and Li 2019; Milano et al. 2019; UNWTO 2018), these symptoms are discussed under the discourse of overtourism—a phenomenon highly pronounced in the Mediterranean where it relates to loss of authenticity, concentration effects of visitor flows, the continuous growth of visitor arrivals and significant risks to the future attractiveness of the destination (Mandić 2021). Peeters et al. (2018) suggest that overtourism is still at the very beginning of the policy cycle, meaning that on the EU level, it has not entered the policy-making stage and is only fragmentarily addressed at the destination level. The diversity of tourism destinations and sites, as well as the lack of policies and tools at the EU level, suggest that there cannot be a straightforward top-down approach to tackling overtourism, but instead destinations are expected to inaugurate tailor-made actions according to their specific needs.

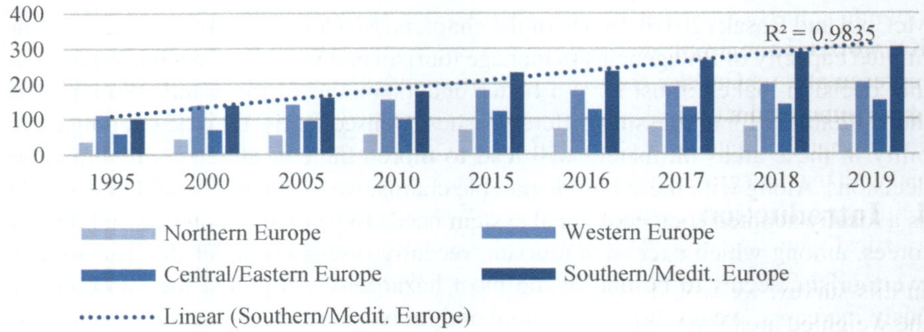

Fig. 1.1 International tourist arrivals—Europe by subregion (*Source* Adapted from UNWTO Tourism Highlights 2018; 2019, and UNWTO World Tourism Barometer 2020)

Considering that 32% of all European protected areas are located in Southern—Mediterranean Europe (EEA 2012), it is reasonable to expect they are influenced by the same tendencies. The Mediterranean region encompassing leading destinations, including France, Spain, Italy, Croatia, Greece and Turkey, accounts for 40.8% of the European international tourist arrivals and 38.6% of its international tourism receipts (World Travel Organization 2019). In most of these countries, tourism has a significant role in sustaining the national economy, considering that it accounts for a significant share of GDP, i.e. in France 7.5%; in Spain 11.1%; in Italy 6.0%; in Croatia 18.9%; in Greece 6.4% and in Turkey 4.3% (OECD 2018; World Tourism Organization 2018). Considering the increasing demand for nature-based tourism, as well as the growing importance of exceptional scenic areas and authenticity in the provision of mindful and transformative travel experience, their future tourism development and competitiveness are to a high degree dependent on a preserved environment and heritage. In the *Report for the EU Transport and Tourism Committee*, Peeters et al. (2018) particularly emphasise the severity of the effects associated with overtourism within the PAs, which are at risk of losing their appeal. The authors refer to Plitvice Lakes National Park in Croatia and Parc Naturel Régional des Monts d'Ardèche, in France, both facing increasing visitation, high tourism intensity, lack of management capacity, environmental issues, uncivilised behaviour and economic interest prevailing over the protection of the resources.

Although the discussions about the human pressure on resources have a long history pioneered in the PAs, the issues provoked by overtourism reinforce the need for careful planning and management of tourism, as well as for respect for wellbeing of the local community (Wall 2020). In light of the aforesaid, the need emerges for more systematic research on issues induced by extensive visitation, on the way nature-based and urban destinations interact as well as on sharing knowledge and acceptable practices associated with mitigation of overtourism. Along with that, there is also a need to discuss tourism impacts interdependently in a systematic way, taking into account resident and visitor perspectives and understandings of the phenomenon. Bearing that in mind, we decided to edit this book to extend recent writing on overtourism in the particular context of PAs and heritage sites and to investigate in what ways the management of fragile ecosystems can be improved taking into account critical concepts in both policy and practice. The book comprises studies that facilitate the understanding of the way these destinations adapt to and deal with the transgression of physical, ecological, social, economic, psychological or political capacity thresholds. The contributors discuss the causes and consequences holistically looking in detail at the emerging issues in the management and operation of PAs and providing the perspectives of academia, industry, NGOs, visitors and community.

The volume is organised into four parts. Following the introduction (this chapter), Part I, reflecting on *governance and management* includes three chapters. In Chap. 2, Petrić and Mandić deliver a theoretical and empirical analysis of the governance and management of PAs in an era of overtourism. To discuss both phenomena, the authors employ the governance-management assessment by consideration of output measures constituents of PAME. Following critical reflection on current research, the empirical analysis of Krka NP, a vibrant nature-based destination in Croatia, demonstrated the insufficient commitment of managing authority to aligning with good governance principles. Such deficiencies constrain the adaptation of ACM, which is primarily reflected in insufficient stakeholder involvement, lack of transparency and inefficiency. The authors reflect on the consequences to the sustainability of ecosystem, bearing in mind the need to reconcile conservation, visitation and development goals.

Mandić and Marković Vukadin (Chap. 3) discuss the impacts of overtourism on PAs. The authors critically reflect on the usability of the established tourism and visitor management frameworks (LAC, VIM, VERP, TOMM, ROS, PAP/RAC) to address the challenges associated with excessive visitation. The analysis involving expert opinion and evaluation of frameworks based on nineteen proposed criteria suggests there is no "one for all" solution in terms of visitor use frameworks. However, constituents of existing frameworks can be crucial for mitigating influences related to extensive visitation. The findings suggest that the conventional approaches to nature-based tourism and visitor management should be advanced by consideration of new theoretical and practical advances, employing the system approach in which PAs are seen in interrelation with other ecosystems. The authors suggest that managers should require tools and resources, goal-orientation and monitoring to obviate the pressures.

In Chap. 4, Pivčević et al. analyse the role and critical elements of a successful public participatory process (PP) in PAs. The authors present two case studies in which local communities have raised increasing concerns regarding the sustainability of the development of nature-based tourism. The critical discussion of models employed suggests that in both cases, PP remains a theoretical concept with modest, practical implementation, with a tendency to gain momentum in the next planning period. The authors suggest that the relationship between PP, PA management and tourism development is very dynamic and complex, as the analysis provides evidence that the practice of PP in PA management can have both positive and negative effects on the mitigation of pressures associated with tourism visitation. While the implementation of PP to satisfy the formal legal requirements will probably be ineffective, implementation to empower the local community and niche stakeholders can be beneficial to the entire ecosystem.

Part II containing six chapters focuses on the *local community and wellbeing*. In Chap. 5, Damnjanović critically reflects on overtourism-related impacts on local communities within a PA and wellbeing. The study emphasises the interdependencies of multiple stakeholder groups, the need for collaboration in a time of uncertainty as well as the lack of research on socio-cultural impacts of tourism development within PAs. The large number of studies dealing with overtourism in urban destinations and other settings has demonstrated that overuse by the tourism industry can disrupt the essence of the locale, which often takes on the responsibility for resource protection and stewardship. Unfortunately, such impacts are but little discussed as the mainstream research prioritises the analysis of the environmental and economic consequences of nature-based tourism development. While the conservation of resources remains the fundamental goal of a PA, Damnjanović points out that a holistic approach to nature-based tourism development also requires the prevention and mitigation of negative impacts of mass visitation on local communities and their wellbeing.

Agius and Chaperon (Chap. 6) suggest that overtourism has led to extensive challenges for island communities in the Mediterranean region. As a result, attention is shifting towards alternative forms of tourism, such as ecotourism, which embrace sustainability principles. This also involves the repositioning of PAs in terrestrial and marine environments as not solely conservation instruments but also as venues for ecotourism. The authors discuss the importance of stakeholder involvement in the process of creating and managing a PA as well as some of the challenges in doing this. The study applies qualitative interviews in six islands in the central Mediterranean region with different levels of protection to discuss stakeholder management and to supply recommendations on how genuine stakeholder involvement and a rebalancing of power could foster competitive ecotourism development and improvement of the wellbeing of local communities in an era of overtourism.

Tourism is relevant for development, as suggested in the literature, only if it is adequately managed if it contributes to the improvement of the local community quality of life, visitor experiences and the preservation of natural and cultural resources. Fernandes et al. (Chap. 7) demonstrate the potential of UNESCO Geoparks to stimulate the creation of innovative local enterprises, new jobs, high-quality training courses, as well as revenue through geo-tourism, assuming resources being preserved. The conclusions are drawn from the UNESCO Estrela Geopark, which, in a synergetic way, promotes in situ experiences and the transmission of knowledge and values about the territory, its heritage and sustainability. The authors emphasise that a relationship between geoconservation, science, education and tourism must be established to facilitate an integrated approach to the territory, in both natural and anthropocentric dimensions, fostering it as a pedagogical resource, and ensuring its sustainability.

In Chap. 8, Seraphin et al. focus on higher education students as future leaders who will be responsible for the sustainable use of finite resources. The context of this chapter is set within the subject of overtourism, which is, according to authors, irreversibly damaging the world's cultural and natural heritage. A case study focused on Marseille Calanques National Park in France, and Kedge Business School, which is part of the Principles for Responsible Management Education (PRME) network, provides a specific context for discussion. The findings support the conclusion that when it comes to engagement with overtourism in terms of helping to solve challenges, there is an apparent disconnect between values taught and what is being practised. Students reveal the gap between theory and practice and point out the necessity for universities to enable the practising of values, i.e. practical empowerment as fundamental in terms of achieving true sustainability in future.

Telišman Košuta and Ivandić (Chap. 9) reflect on collaborative destination governance in nature-based destinations adjoining vibrant PAs. They design the research so as to broaden the understanding of social carrying capacity as this may enable the improvement of overall destination management and governance. The threat of diverging perceptions between residents and visitors, which is limiting the sustainability of tourism development could be resolved by developing destination cohesion and advancing community values. The authors emphasise better monitoring of a destination's tourism carrying capacity, community capacity building, improvement of the quality of tourism planning and capacity for plan implementation, as well as the promotion of a collaborative culture among local stakeholders as preconditions for the advancement of the sustainability of tourism development on a local level.

The impacts of tourism development are felt more intensively by communities living in tourism destinations; thus, residents have to be involved in the tourism development planning process. Jordão et al. (Chap. 10) analyse the perception of local people regarding the impacts of tourism in a heritage destination in Portugal to define the limits of acceptable change (LAC) in this fragile ecosystem. The LAC approach enables integrated planning and adaptivity and thus is discussed as a solution to reconcile the resource conservation, community wellbeing and tourism development. The analysis encompassing focus groups with stakeholders suggests that the maintenance of a scenario of imbalance and tourism growth, without proper planning, monitoring and control will only lead to an aggravation of several emerging problems in Porto heritage site. The study reaffirms the necessity to set the limits of tourism growth, to empower residents and to develop infrastructure to meet the new needs of both residents and visitors.

Part III encompasses four chapters and focuses *on visitor experience design and management.*

Building a viable tourism management plan requires integrative, holistic approaches that provide a foundation for outstanding visitor experience opportunities. In Chap. 11, McCoole et al. illustrate how management in an era of rapidly growing visitation can respond in a holistic way to produce competitive experiences with opportunities to enjoy other nearby offers and foster the development of the regional economy. The conclusions are drawn from the Plitvice Lakes National Park in Croatia, where the recently introduced general management plan enabled the development of integrated programs of promotional changes, capacity building, visitor use modelling, trail construction, as well as the use of modern technology to manage visitors within dynamically changing visitor use patterns.

Kuzuoglu and Hatipoglu (Chap. 12) employ social-ecological system (SES) theory and qualitative study design to question tourism governance and mismanagement of visitor pressures in a vibrant World Heritage Site (WHS). The authors explore if co-management structures in tourism governance could facilitate adaptive communities that can manage change while maintaining the cultural uniqueness of living heritage sites. The findings suggest that WHS governance arrangements involve multiple stakeholders and often experience conflicts. Thus, formal and informal conflict resolution mechanisms are needed. Overcrowding and social and environmental disturbances are symptoms of a failing governance system. The public authorities should take the responsibility and be more accountable for developing tourism activity. The authors demonstrate that the adaptive co-management approach is necessary for the community to tackle social and physical disruption.

Inadequate management and planning of tourism and conservation often result in diminished attractiveness for visitors, as well as increased risk to the protected ecosystems. Thus optimisation of visitor flows is the priority in many Mediterranean destinations. Chap. 13 by Carić and Beroš investigates aspects of intensive short-term visits (excursions and daily tours) to a NATURA 2000 site. The study builds on recent experiences of the establishment of an action plan for the EU funded infrastructure on the territory of the site. The authors discuss how innovative use of established analytic tools such as radar tracking, environmental risk assessment and visitor crowding perception could be used to manage the risk of overcrowding and improve visitor experiences.

Iaffaldano and Ferrari in Chap. 14 introduce The Jonian Dolphin Conservation (JDC) association from the city of Taranto, Italy, which is dedicated to raising awareness of sustainability and nature conservation through the organisation of educational experiences. The authors employ a qualitative, inductive approach, and run a series of interviews with relevant stakeholders and visitors to demonstrate how a subject that works for environmental conservation could offer engaging educational activities, unique tourism experiences and inspire sustainable tourism development and the urban repositioning of a destination facing a deteriorating image. The study demonstrates the importance of the JDC initiative for the sustainability and resilience of the ecosystem and the development of pro-environmental attitudes and behaviour. The authors reveal the diversity of attitudes of stakeholders towards the future of tourism development and demonstrate the consensus as a prerequisite for a sustainable future.

Part IV reflecting on *intelligence in nature-based tourism development* includes five chapters and is followed by the conclusion section (Chap. 20).

Overtourism constitutes a significant challenge for tourism in the PAs of the Mediterranean, thus requiring appropriate solutions. Supported by a detailed analysis of the relations between natural resources, tourism, specialisation patterns, growth, resilience and sustainable development, within the perspective of Chap. 15, Romão emphasises how an excessive dependence on the tourism sector may lead to a strong regional vulnerability both to overtourism and to no-tourism. The potential over-specialisation in tourism brings long-term problems related to lack of regional resilience, low levels of technological development, low added value and degradation of the social capital. Supporting the development of clusters of activities not related to tourism, along with the promotion of the potential interactions between tourism and related sectors appear to be crucial challenges for smart specialisation and regional innovation strategies in the Mediterranean area.

Camatti et al. (Chap. 16) discuss how to change the tourism-growth focused mindset. The authors emphasise the need for tourism destinations to undertake complementary strategies that address sustainable development challenges in a holistic, inclusive and iterative manner. Governance frameworks, business models and socio-cultural values are vital instruments to achieve sustainable development but insufficient on their own. The authors claim that sustainable tourism is an ambitious but not unattainable goal, which should be treated as an iterative, reflexive process embedded within its locality and directed by its wide range of stakeholders. Furthermore, there is also a need to reframe the collective mindset both in reference to individuals' perception of what sustainability means and one's worldview and value systems more generally. The destination should employ measures capable of inducing a paradigm shift for long-term, resilient, sustainable tourism development.

Mitigating overtourism has become a growing concern for destinations that need to protect invaluable cultural and natural resources. In Chap. 17, Gretzel analyses the advantages and limitations of technological solutions to overtourism in the light of their long-term viability and their ability to create change instead of shifting the problem to other areas or postponing impacts. The study claims that technologies like social media are partially to be blamed for their role in increasing visitation; however, at the same time, they could be used as a means to combat massive visitor flows. Greater adaptation of smart technologies in the provision of tourism experiences, greater analytical capabilities of tourism service providers and destinations open up new opportunities for creating and managing tourist flows and influencing specific tourist behaviour.

The socioecological systems of PAs are influenced by and responsive to socio-cultural, economic and ecological elements all characterised by inherent complexity, dynamism, uncertainty and adaptation across multiple scales. Bearing that in mind, within Chap. 18, Šimundić et al. discuss the stakeholders' interactions within PAs. These interactions are highly dynamic and complex, influenced among others, with trade-offs between conservation and recreation, the economic dependency of surrounding communities and institutional arrangements. The authors introduce resilience thinking as a tool for understanding the complexity of PA systems and investigate the implementation of principles for building resilience in such complex surrounding, characterised with simultaneous but conflicting goals of nature conservation and tourism development.

Once established, mass tourism mobilises the specialisation of the place in the touristic activity and promotes economic growth. At the same time, other places remain protected due to the backwash effects of the touristic activity. In Chap. 19, Ponce Dentinho investigates the spacial interaction effect of sustainability, which is crucial for understanding the resilience and sustainability of people and places.

The author proposes a demographic growth model that includes spatial interaction sustainability to understand the relation between tourism and economic growth and to discuss the regional policy implications. The analysis demonstrates the existence of a tourism spatial-specialisation as well as low impacts on urban hierarchy. Furthermore, it demonstrates the relevance of backwash effects related to tourism and spread effects associated with the activities displaced by tourism to the surrounding areas, which is essential for nature conservation, as growing demand increases the value of natural resources. At the same time, the author concludes that backwash effects from nearby central places may prevent the degradation of exceptional natural areas.

References

Capocchi A, Vallone C, Amaduzzi A, Pierotti M (2019) Is 'overtourism' a new issue in tourism development or just a new term for an already known phenomenon? Curr Issues Tour 0(0):1–5. https://doi.org/10.1080/13683500.2019.1638353

Cheung KS, Li LH (2019) Understanding visitor–resident relations in overtourism: developing resilience for sustainable tourism. J Sustain Tour 27(8):1197–1216. https://doi.org/10.1080/09669582.2019.1606815

Coad L, Leverington F, Knights K, Geldmann J, Eassom A, Kapos V, Kingston N, De Lima M, Zamora C, Cuardros I, Nolte C, Burgess ND, Hockings M (2015) Measuring impact of protected area management interventions: current and future use of the global database of protected area management effectiveness. Philos Trans R Soc B: Biol Sci 370(1681). https://doi.org/10.1098/rstb.2014.0281

Corlett RT, Primack RB, Devictor V, Maas B, Goswami VR, Bates AE, Koh LP, Regan TJ, Loyola R, Pakeman RJ, Cumming GS, Pidgeon A, Johns D, Roth R (2020) Impacts of the coronavirus pandemic on biodiversity conservation. Biol Cons 246:108571. https://doi.org/10.1016/j.biocon.2020.108571

Croy WG, Moyle BD, Moyle CJ (2020) Perceived benefits of parks: The roles of information source exposure and park use. Journal of Sustainable Tourism 28(11): 1–20. https://doi.org/10.1080/09669582.2020.1757683

Dudley N (ed) (2008) Guidelines for applying protected area management categories. IUCN, Gland, Switzerland

Dudley N (2010) Guidelines for applying protected area management categories. In Guidelines for applying protected area management categories. https://doi.org/10.2305/iucn.ch.2008.paps.2.en

EEA (2012) Protected areas in Europe—an overview (Issue 5). European Environmental Agency

EEA (2020) Natura 2000 data—the European network of protected sites. Available at https://www.eea.europa.eu/data-and-maps/data/natura-11

Gon M, Dietz T, Liu J (2018) Global relationships between biodiversity and nature-based tourism in protected areas. Ecosyst Serv 34(November 2017):11–23. https://doi.org/10.1016/j.ecoser.2018.09.004

Hummel C, Poursanidis D, Orenstein D, Elliott M, Cristian M, Cazacu C, Ziv, G, Chrysoulakis N, Meer J, Van Der, Hummel H (2019) Science of the total environment protected area management: Fusion and confusion with the ecosystem services approach. Science of the Total Environme 651: 2432–2443. https://doi.org/10.1016/j.scitotenv.2018.10.033

Kettunen M, Illes A, Tucker G, Baldock D, Rayment M, Ebrahim N, Verstraeten Y, Primmer E, Rantala S, Rekola A, Ring I, Droste N, Santos R (2017) Integration approach to financing of biodiversity: evaluation of results and analysis of options for the future. January 2017.

Leung Y, Spenceley A, Hvenegaard G, Buckley R, Editors V, Groves C, Editor S (2018) Tourism and visitor management in protected areas: guidelines for sustainability. In Tourism and visitor

management in protected areas: guidelines for sustainability (Issue 27). https://doi.org/10.2305/iucn.ch.2018.pag.27.en
Mackinnon K, Pasha MKS, Phillips A (2020) Editorial essay: Covid-19 and protected and conserved areas. June. https://doi.org/10.2305/IUCN.CH.2020.PARKS-26-1MH.en
Mandić A (2019) Nature-based solutions for sustainable tourism development in protected natural areas: a review. Environ Syst Decis. https://doi.org/10.1007/s10669-019-09718-2
Mandić A (2020) Structuring challenges of sustainable tourism development in protected natural areas with driving force—pressure—state—impact—response (DPSIR) framework. Environ Syst Decis. https://doi.org/10.1007/s10669-020-09759-y
Mandić A (2021) Optimisation of tourism development in destinations: an approach used to alleviate the impacts of overtourism in the Mediterranean region. In: Spenceley A (ed) Handbook for Sustainable Tourism Practitioners: The Essential Toolbox. Edward Elgar Publishing Limited, UK.
Mandić A, Petrić L (2020) The impacts of location and attributes of protected natural areas on hotel prices: implications for sustainable tourism development. In Environment, development and sustainability (Issue 0123456789). Springer, The Netherlands. https://doi.org/10.1007/s10668-020-00611-6
Maxwell SL, Cazalis V, Dudley N, Hoffmann M, Rodrigues ASL, Stolton S, Visconti P, Woodley S, Kingston N, Lewis E, Maron M, Strassburg BBN, Wenger A, Jonas HD, Venter O, Watson JEM (2020) Area-based conservation in the twenty-first century. Nature 586(7828):217–227. https://doi.org/10.1038/s41586-020-2773-z
McCool S (2016) Tourism in protected areas: Frameworks for working through the challenges in an era of change, complexity and uncertainty. In: McCool S, Bosak K (eds) Reframing sustainable tourism. Springer
Milano C, Novelli M, Cheer JM (2019) Overtourism and tourismphobia: a journey through four decades of tourism development, planning and local concerns. Tour Plan Dev 16(4):353–357. https://doi.org/10.1080/21568316.2019.1599604
OECD (2018) OECD tourism trends and policies 2018 (vol 48, Issue 01)
Peeters P, Gössling S, Klijs J, Milano C, Novelli M, Dijkmans C, Eijgelaar E, Hartman S, Heslinga J, Isaac R, Mitas O, Moretti S, Nawijn J, Papp B, Postma A (2018) Research for TRAN Committee—Overtourism: impact and possible policy responses. October. http://www.europarl.europa.eu/RegData/etudes/STUD/2018/629184/IPOL_STU(2018)629184_EN.pdf
Ribas LG dos S, Pressey RL, Loyola R, Bini LM (2020) A global comparative analysis of impact evaluation methods in estimating the effectiveness of protected areas. Biol Conserv 246(April):108595. https://doi.org/10.1016/j.biocon.2020.108595
Rodrigues ASL, Cazalis V (2020) The multifaceted challenge of evaluating protected area effectiveness. Nat Commun, 1–4. https://doi.org/10.1038/s41467-020-18989-2
Spenceley A (2015) Tourism and the IUCN World Parks Congress 2014. J Sustain Tour 23(7):1114–1116. https://doi.org/10.1080/09669582.2015.1046704
Taylor B, Chapron G, Kopnina H, Orlikowska E, Gray J, Piccolo JJ (2020) The need for ecocentrism in biodiversity conservation. Conserv Biol 34(5):1089–1096. https://doi.org/10.1111/cobi.13541
UNWTO (2018) 'Overtourism'?—understanding and managing urban tourism growth beyond perceptions, executive summary. World Tourism Organization (UNWTO); Centre of Expertise Leisure, Tourism & Hospitality; NHTV Breda University of Applied Sciences; and NHL Stenden University of Applied Sciences. https://doi.org/10.18111/9789284420070
Wall G (2020) From carrying capacity to overtourism: a perspective article. Tour Rev 75(1):212–215. https://doi.org/10.1108/TR-08-2019-0356
Weaver DB, Lawton LJ (2017) A new visitation paradigm for protected areas. Tour Manag 60:140–146. https://doi.org/10.1016/j.tourman.2016.11.018
World Tourism Organization (2018) European Union Tourism Trends. In: European Union Tourism Trends. https://doi.org/10.18111/9789284419470
World Travel Organization (2019) International Tourism Highlights. Unwto, 1–24

World Travel Organization (2020) world tourism barometer N°18 January 2020. UNWTO 15 September 2020 18(5): 1–36. https://www.unwto.org/news/international-tourist-numbers-down-65-in-first-half-of-2020-unwto-reports
Zafra-Calvo N, Geldmann J (2020) Protected areas to deliver biodiversity need management effectiveness and equity. Glob Ecol Conserv 22:e01026. https://doi.org/10.1016/j.gecco.2020.e01026

Part I
Governance and Management

Chapter 2
Governance and Management of Protected Natural Areas in the Era of Overtourism

Lidija Petrić and Ante Mandić

Abstract In view of ever-growing visitation, proactive management is needed to reconcile the goals of both conservation and recreation. The literature suggests that one of the solutions might be a transition from the traditional top-down governance of protected area (PA) to adaptive co-management (ACM), which advocates participation, collaboration, and iterative learning. This chapter aims to analyse the key elements and structures of governance and management of PA while focusing on both conservation effectiveness and socio-economic sustainability goals. Attention is given to the discussion of PA governance and management assessment (PAME) methodologies and their relationships. Finally, the governance-management assessment method developed by Shields et al. (2016) is applied to the case of Krka NP in Croatia. The assessment is based on the governance indicators and related output measures that are a part of the PAME. The results of the analysis indicate insufficient commitment on the part of Krka National Park management authority to following United Nations Development Programme (UNDP) good governance principles. Although the last decade saw significant advances in both governance and management effectiveness, significant improvements should be made, particularly regarding the stakeholders' engagement as the fundamental aspect of ACM. The implications of the analysis are discussed with particular reference to excessive visitation.

Keywords Protected area governance · Management effectiveness · Overtourism · Adaptive co-management · Krka National Park, Croatia

L. Petrić (✉) · A. Mandić
Faculty of Economics, Business and Tourism, University of Split, Split, Croatia
e-mail: lipe@efst.hr

A. Mandić
e-mail: ante.mandic@efst.hr

A. Mandić and L. Petrić (eds.), *Mediterranean Protected Areas in the Era of Overtourism*, https://doi.org/10.1007/978-3-030-69193-6_2

2.1 Introduction

Global population growth, accompanied by industrial production, pollution and urbanisation, and most recently the COVID-19 pandemic have spurred the growth in the demand for nature-based experiences and ecotourism. Initially defined by the International Ecotourism Society (TIES), ecotourism reflects 'responsible travel to natural areas that conserves the environment, improves the well-being of local people' (Lopez Gutierrez et al. 2019), and emphasises learning and ethical planning (Fennel, 2014, in Lopez Gutierrez et al. 2019). Most ecotourism destinations are located in protected areas (PA) (Spenceley et al. 2015), attracting a large number of visitors. According to Balmford et al. (2015), approximately eight billion visits per year have been recorded in 556 terrestrial PA distributed over 51 countries. Mean annual visit rates for individual PA in this sample ranged from zero to over 10 million visits/y. Of these, 3.8 billion visits per year were estimated to have occurred in European terrestrial PAs. In all, the PA researched are estimated to have yielded global gross direct expenditure associated with visits (excluding indirect and induced expenditure) of US $600 billion/year worldwide (at 2014 prices).

The demand for ecotourism is high and continues to grow (CBI 2020). In the 2019 Booking.com survey,[1] almost three quarters (72%) of travellers believed that people need to act now and make sustainable travel choices to save the planet for future generations. Euronews[2] stressed in its *Travel and Tourism Trend Report* that after 2020, ecotourism could be the critical travel trend. The Centre for Responsible Travel (2018) reports that travellers are increasingly seeking opportunities to reconnect with nature, other people, and their meaning, most likely caused by an increasingly digitally connected, work-centric, and material world. In the European Commission (EC 2016) study on *Preferences of Europeans Towards Tourism*, respondents in the EU countries stressed 'nature' as a fundamental reason for the choice of a destination, ranging from, for example, 5% in Cyprus up to 26% of respondents in the Czech Republic, 25% in the Netherland (Fig. 2.1). Additionally, a substantial number of respondents from different countries pointed to nature as being either the primary or the secondary motivation for travelling (for example, nature is the reason to return to a destination for 25% of Germans, 36% of French, and 50% of Dutch).

Newsome's recent research (2020) has demonstrated that post-COVID-19 tourism demand is likely to shift towards sparsely populated, nature preserved destinations. This trend is most likely to be supported by the increasing attention of the tourism industry to business investment outside mass tourism, as a reaction to increasing public awareness of the adverse impact of overtourism in urban destinations. The expected demand shift towards nature-based destinations, some of which are already experiencing overtourism, could additionally aggravate established challenges, such as biological and socio-economic fragmentation. Eagles (2009) suggests that this

[1] https://globalnews.booking.com/bookingcom-reveals-key-findings-from-its-2019-sustainable-travel-report, Retrieved November 2, 2020.

[2] https://www.euronews.com/2020/10/26/travel-after-2020-the-new-reality, Retrieved November 2, 2020.

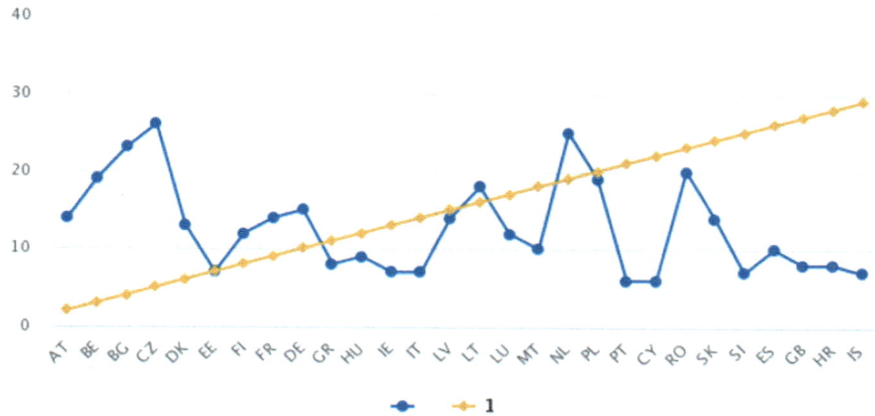

Fig. 2.1 Share of travellers in each EU country who saw nature as the main reason to go on holidays in 2016 (*Source* Flash Eurobarometer 432 study: Preferences of Europeans towards tourism, available at: https://data.europa.eu/euodp/data/dataset/S2065_432_ENG)

may be particularly an issue in countries such as Australia and New Zealand, which consider ecotourism a key component of export income. However, the challenges related to extensive visitation in PA are seen even in Mediterranean mass tourism destinations oriented dominantly to either 3S (sun, sea, sand) or urban (cultural) tourism. Most of the visits are paid to 129 national parks (NP), out of 14,690 PA registered in this region (or 14.11% of Europe's total) (EEA 2020). The Schägner et al. (2016) prediction model estimates that there are more than two billion recreational visits a year in 419 European existing and potential NPs. The estimated annual numbers of visitors to PA in specific Mediterranean countries range from 4,121,000 in Slovenia (which has an area of 1157 km^2), 121,666,000 in Spain (10,450 km^2), 71,408,000 in France (13,565 km^2), and up to 195,719,000 annual visitors in Italy (which has an area of 17,419 km^2).

Natural resources are the foundation of the appeal of tourism in PA. Any damage caused might have severe consequences for PA and the surrounding communities (Islam et al. 2018a). Despite potential risks, the reduction of government funding forces PA to rely increasingly on tourism revenues (Weaver and Lawton 2017). A reconciliation between recreation and conservation requires proactivity, implementation of sustainable tourism principles (Mandić and Petrić 2020), and synergy among stakeholders involved in governance processes to enhance benefit sharing (Heslinga et al. 2017). Currently, the protected area governance model is far from being a central state-based responsibility. It has become a polycentric regime under which powers are distributed among different level governments, private and community-based stakeholders (Ly and Xiao 2016), resulting in various forms of collaboration. In line with that, a paradigm shift is taking place in PA suggesting a transition from a top-down to participatory bottom-up approaches to planning, management, and governance (Islam et al. 2018a).

Given the above, this chapter aims to analyse the key elements and structures of the governance and management of PA while focusing on both conservation effectiveness and socio-economic (including tourism) sustainability goals. As stated by Gurung (2010), governance 'exerts a major influence on the achievements of management objectives, effectiveness, equity and sustainability of protected areas' (p. 5). With this regard, PA governance is observed as a broader framework within which management is exercised. Effectiveness requires the participation of all stakeholders (the core principle of adaptive management [AM]), as well as collaboration and learning (the fundamental principles of adaptive co-management [ACM]). Within this chapter, special attention is paid to the explanation of protected area governance and management assessment methodologies and their interrelation. Finally, the governance-management assessment method developed by Shields et al. (2016) is applied to the case of Krka NP in Croatia to discuss its fundamental shortcomings with respect to governance-management and recommendations for their improvement.

2.2 Governance and Management of PA—From Theoretical Concepts to Quality Assessments

2.2.1 Understanding the Concept of PA Governance and Its Assessment

The Aichi Target 11 Dashboard[3] suggests that the proportion of the global terrestrial area covered by protected areas is 15%, near achieving the terrestrial target of 17%. Global marine area coverage is 7%, targeting a total of 10%. The proportion of the PA where management effectiveness (PAME) evaluations have been undertaken is pretty low, i.e. 5% out of targeted 17% for terrestrial, and 1% out of the targeted 10% for marine PA. At the same time, the need for protected area management improvement is rapidly growing, especially in the light of the increasing interest in visitation of these areas.

Management must be accountable for governance through clear governance structures and processes (Booker and Franks 2019). Abrams et al. (2003) indicate that managers of PA consider operational level problems being closely related to broader governance issues requiring a critical examination of existing laws, policies, programmes, regulations, organisational cultures, and professional attitudes.

Governance is distinct from management (Booker and Franks 2019). It refers to all processes of governing, based on social practices and activities, and undertaken by either a government, or market, or network, whether over a family, formal or informal organisation or territory, through laws, norms, power, or language (Bevir 2012). It is about how power is exercised, decisions are taken, the resources allocated,

[3]https://www.protectedplanet.net/target-11-dashboard, retrieved on May 9, 2020.

and citizens or other stakeholders participate in these processes (Graham et al. 2003; Worboys et al. 2015; Islam et al. 2018b; Booker and Franks 2019). On the other hand, management is about how to achieve the outlined objectives and includes defining and allocating lower level objectives, responsibilities, and accountabilities (Booker and Franks 2019).

Borrini-Feyerabend et al. (2013) indicate that PA governance may take place at several levels often interacting with each other in different ways, i.e. horizontally (e.g. through collaboration and exchanges or voluntary); vertically (through hierarchy); formally (e.g. by-laws); or informally (e.g. because of relationships and trust). By doing so, they may negotiate in the best possible way outcomes that conserve biodiversity while also providing for sustainable resource use (Dearden et al. 2005; de Koning et al. 2017). With this regard, four broad PA governance types are recognised, depending on who is responsible for a given PA, e.g. governance by the government, by private, shared/multi-stakeholder government, or by indigenous people and local communities (Abrams et al. 2003; Dudley 2008; Borrini-Feyerabend et al. 2013).

There are significant regional differences in approaches to governance, with shared governance ranging from less than 1% in Europe and the Polar regions to 10% in West Asia. Approximately 22% of North American PAs are under private governance, while the figure is less than 1% in Europe and West Asia. On the other hand, 9% of the PAs in Latin America and the Caribbean are reported as under the governance of indigenous peoples and local communities, while the figure is less than 3% for all other regions (UNEP-WCMC, IUCN, NGS 2018). More than 96% of European PAs are under a public governance regime. Despite the differences among the PA governance types, what is common to all of them is the need to have multiple stakeholder groups' support in decision-making, planning, implementation, monitoring, and evaluation, which is the fundamental precondition for any governance system to be 'good' (Islam et al. 2018b; Eagles 2009; Gossling and Hultman 2006; Kaltenborn et al. 2011).

Considering they are dependent on tourism revenues, PAs should be governed to enable local stakeholders to 'foster different valuable forms of commitments, synergies and collaborations between public/private actors and assisting policy-makers to implement sustainable development' (Presenza et al. 2015: 480). However, governance models that can best support both PA governance principles and sustainable tourism development are limited (Islam et al. 2018b), although more recent paradigms for PA management call for integrative approaches to conservation and development. Islam et al. (2018b) suggest that the ACM could support the integration of conservation and sustainable tourism in PA, by facilitating the achievement of crucial governance principles. However, achieving participation and collaboration of the stakeholders, the core principles of ACM, requires an empowered and well-informed community, able to understand complex relationships between regulations and sustainability achievement (Presenza et al. 2015).

Governance quality has been assessed in the governance literature by several assessment frameworks, with most of them being based on the principles of good governance to ensure sustainable management or tailored to the requirements of environmental problems or specific needs (Lordkipanidze et al. 2018). Moreover, no

Table 2.1 The 'good governance' principles

The good governance principles	The UNDP principles on which they are based
Legitimacy and voice	Participation Consensus orientation
Direction	Strategic vision, including human development and historical, cultural and social complexities
Performance	Responsiveness of institutions and processes to stakeholders Effectiveness and efficiency
Accountability	Accountability to the public and to institutional stakeholders Transparency
Fairness	Equity Rule of Law

Source Graham et al. (2003)

universally accepted method identifies the requirements of a good governance regime for the PAs (Lockwood 2009). Also, most of the identified evaluative frameworks do not address governance systems holistically (Potts et al. 2016).

The quality of governance is most often evaluated against 'good governance' principles, developed by the United Nations Development Programme (UNDP 1997). Graham et al. (2003) were among the first to explicitly articulate five principles of good governance concerning PA (Table 2.1). They were followed by several authors, including Abrams et al. (2003), Borrini-Feyerabend et al. (2004, 2013), Heylings and Bravo (2007), Shipley and Kovacs (2008), and Eagles (2009). Lockwood (2010) suggested seven principles of good governance, e.g. legitimacy, transparency, accountability, inclusiveness, fairness, connectivity, and resilience. In their '*Governance Assessment for Protected and Conserved Areas*' (GAPA) manual, Franks and Booker (2018) introduced 11 good governance principles, different from those proposed by the UNDP. With respect to each of the 11 principles of GAPA, they elaborated the significant issues to be assessed and a desired level of achievement (pp. 109–111).

The good governance principles may be further described (elaborated) by many measurable indicators, aimed at helping governance performance evaluation. Abrams et al. (2003) and Borrini-Feyerabend et al. (2013) provided more than 100 such indicators, associated with five of the good governance principles of Graham et al. (2003) (Table 2.1). Shields et al. (2016) have identified a significantly smaller number of broadly useful indicators (20) based on a three-round Delphi survey with the Australian PA managers and experts. Along with that, instead of the nine original UNDP governance principles (Table 2.1), the authors employed 10, by splitting the 'efficiency and effectiveness' principle into 'efficiency' and 'effectiveness' as separate requirements.

2.2.2 The Approaches to PA Management

PA management is about actions taken and means used to deliver social, economic, and environmental outcomes for the benefit of all stakeholders. There are several management planning frameworks as well as tools (Fig. 2.2) that assist protected area management organisations to undertake their operations, through a systematic approach to management across what may be large, diverse and decentralised national and sub-national protected area systems (Sandwith et al. 2016). These frameworks aim to guide those responsible for decision-making. The guidance is delivered through management plans. Their purpose is to ensure that exceptional areas are managed in such a way as to achieve specific objectives: to gain public involvement; to develop a shared understanding of a vision for a PA; and to provide public accountability (Mccool 2016). It is generally accepted that the most effective management plans are concise documents that identify the key features or values of the protected area, clearly establish the management objectives and indicate the actions to be implemented (Bushell and Bricker 2017). It contains a variety of topics, one of which is tourism. As stated by Eagles et al. (2002: 41), a plan can describe, among other things, 'how tourism and associated development will be managed'. However, the PA management plans are usually more oriented to the natural resources' management than to the objectives for tourism and their achievement. Therefore, the issue of tourism in protected areas is most importantly addressed in the policies relating to tourism and recreation within the management plan (Eagles et al. 2002).

The sole existence of the management plan is not enough, considering that PA management planning is continuously challenged by, among other factors, climate change, invasive species, visitor impacts, vandalism, poaching, pollution, development, civil unrest, and disasters (IUCN 2017). The decisions in a time of such uncertainty are usually made according to one of the five main approaches, namely, minimum safe standard, precautionary principle, minimax regret criterion, scenario planning, and AM (Prato 2012), with the last one being the most advocated in the relevant PA management and nature-based tourism-related literature (Mandić 2019). While it would be interesting to discuss all five approaches in detail, due to the scope

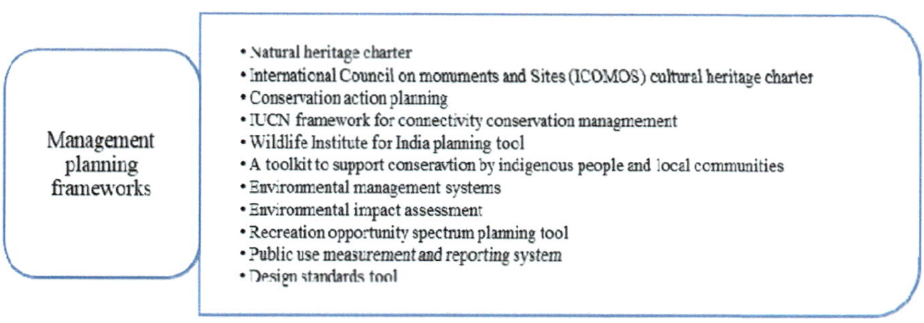

Fig. 2.2 Protected area management and planning frameworks (*Source* Adapted from Sandwith et al. 2016)

Fig. 2.3 The process of adaptive management (*Source* Adapted from Williams et al. 2007)

of this study, we have decided to focus on just AM. For a more detailed examination of other decision-making approaches, see Prato (2012).

AM is a scientific, structured approach for planning and managing natural resource systems based on a perception of ecosystems as being complex, dynamic, and unpredictable (Gunderson 2018; Williams 2011).

The concept of AM in environmental and planning studies was initially proposed by Holling (1978), as an approach to assessing the potential impacts of a specific development project on a natural habitat. AM was initially introduced to fill in the main gaps in existent management approaches, being: to bridge diverging assumptions of resource dynamics; to integrate differing perspectives among scientific disciplines; and to fill the gap between knowledge and action (Gunderson 2018). The AM is a six-step learning cycle (Fig. 2.3) that emphasises the participation of all relevant stakeholders in conflict management, acknowledging that many factors influence the condition of an ecosystem outside the manager's jurisdiction, requiring a broad, systemic, or strategic approach. Despite its simplicity, confusion persists about the methods entailed, the management context to be used, and the extent to which it has been applied successfully (Rist et al. 2013). For example, Prato (2012) suggests that AM is not appropriate for management decisions with very little uncertainty about the efficacy of measure employed. The concept is challenging to implement and appropriate only where plenty of scientific uncertainty exists, sufficient resources allow for experimentation with multiple treatments, completing hypotheses are present, finite and testable and there is leadership that can overcome vested and self-serving stakeholders (Allen and Gunderson 2011). Additionally, a mandate to take action and reliable institutional capacity and commitment to deliver the mission must exist (Williams et al. 2007). Due to its misleading simplicity, AM is too often seen as the only solution for socio-economic challenges. However, in many cases, it is not an ideal response because of limited replication of the experiments, entrenched management, and socio-economic systems (Allen and Gunderson 2011). As indicated by Williams et al. (2007), the process of AM will be successful if the stakeholders are actively involved in and committed to the process; the progress towards management objectives is made; results from monitoring and assessment are used to adjust and improve management decisions, and implementation is consistent with applicable laws. Several pathologies and challenges influence the success of the application of this concept, including stakeholder engagement, difficulties with experiments, surprises, following of prescriptions, learning and discussion, use of knowledge to modify policy and management, avoiding hard truth, lack of leadership and focus on planning, not action (Allen et al. 2011). Rist et al. (2013) argue that many of these barriers are related to the challenges stemming from the management implementation generally.

AM is founded on the notion of the complexity of socio-ecological systems, i.e. they are too uncertain, and unpredictable to be controlled by top-down management regulations. Thus, it prescribes adapting rather than trying to manipulate socio-ecological systems. In other hands, it seeks to manage change rather than resist it, and by that sees many possible equilibrium points, not just one, which is of the essence of resilience approach (Bown et al. 2013). The diversity and resilience are the AM cornerstones, signifying the capacity of a complex ecosystem to withstand fluctuation and still maintain its integrity. Prato (2012) distinguishes active and passive AM with the former being able to produce statistically reliable results about ecosystem responses to management actions, which can be generally applied to other protected natural areas. Recently, Bown et al. (2013) revisit the concept of adaptive ACM, which emerged in the late 1990s. In its essence, ACM synthesises co-management (CM) and AM, to remedy the deficiency within CM, particularly, the lack of the adaptive capacity in the decision-making system. However, the authors emphasise that the concept has proven difficult to put in practice, mainly due to the inability to balance the AM focus on ecological resilience and CM focus on human empowerment. In light of a recent debate on the resilience of social-ecological systems, Farhad et al. (2017) demonstrated that ACM could provide the opportunity to navigate the trade-offs between specified and general resilience by shaping new multi-level governance system and sharing of power between stakeholders involved. Furthermore, the study performed by Plummer et al. (2017) in four UNESCO Biosphere Reserves in Canada and Sweden, provided the first empirical evidence that ACM works; however, much work remains to be done to demonstrate consistently and systematically its ability to deliver outcomes.

Very little has been written on AM and ACM in tourism studies. Although Islam et al. (2018a) and Larson and Poudyal (2012) have carried out valuable work on ACM as a novel approach in nature-based destination management, several questions remain unanswered. The study of Islam et al. (2018a) challenges the assumption that ACM provides the opportunity for enhancing tourism governance practices in protected area contexts, but at the same time fails to integrate either management models for the delivery of tourism services (see Spenceley et al. 2019) or management frameworks to address tourism impacts and visitation (see Mccool et al. 2007) into the conceptual framework. Additionally, the study fails to recognise that some of the ACM principles (e.g. learning, collaboration) are the cornerstone of the recreation and tourism planning frameworks (e.g. Recreation opportunity Spectrum [ROS] or Limits of acceptable change [LAC]). Islam et al. (2018b) also analysed the extent to which the ACM approach could support the achievement of critical governance principles in PAs. The authors concluded that the method was successful in facilitating the participation of stakeholders, a collaboration between them, accountability and transparency, the rule of law, power, and social learning. However, methodological deficiencies limit the generalisation of the results. Larson and Poudyal (2012) delivered an AM approach based on the United Nations World Tourism Organisation's (UNWTO's) sustainable tourism framework. This rather integrative approach accounted for diverse perspectives and embraced the various stakeholders. The authors concluded that an adaptive resource management approach might help planners guide tourism growth. Despite their contribution, both studies, focused on the integration of AM into tourism development and destination management framework, to some extent, violate the principle of AM. As explained by Rist et al. (2013), AM could be used in some challenging management contexts only if less complicated aspects of uncertainty are targeted, and a significant amount of management resources are available, which is hard to combine within integrated sustainable tourism and destination management planning. For example, Lai et al. (2016) discussed the potential of ACM to forge collaborative natural resources management via a social learning model in mountain destinations. The study demonstrated the utility of social representation theory to unpack the cultural, political, and economic aspects of complex interactions between resource stakeholders as participants of ACM. Another attempt to integrate AM into nature-based tourism management was made by Schianetz and Kavanagh (2008). The authors conclude that tourism destinations need to be viewed and studied as complex adaptive systems and that the sustainability indicator system (SIS) needs to be applied in the context of an adaptive management approach, indicating that SIS becomes a learning tool to improve the understanding of the behaviour and threshold levels of the social-ecological system. Recently, several researchers have discussed the AM and ACM in the context of tourism resilience (e.g. Bangwayo-Skeete and Skeete 2020; Dai et al. 2019) and climate change adaptation (e.g. Csete and Szécsi 2015; Landauer et al. 2018; Rutty et al. 2015; Scott and Becken 2010).

2.2.3 How to Evaluate PA Management Effectiveness?

Management effectiveness is a relatively new concept, mentioned for the first time in 1972 by a Conservation Foundation Task Force on the National Parks System, which recommended the adoption of an annual system of park environmental reports to assure continued monitoring of each park's internal and external environmental factors (Hockings et al. 2004). According to Coad et al. (2015: 2), the primary purpose of such evaluations are (a) to improve PA management, (b) to allocate resources more effectively (c) to provide accountability and reporting at local, national, or international levels; and (d) to increase community awareness of PA management and issues. PAME assessments are generally conducted by PA managers or government agency employees and donor institutions, including NGOs (Geldmann et al. 2015). Since the mid-1990s, various methodologies have been developed for assessing PAME, including Management Effectiveness Tracking Tool (METT), Rapid Assessment and Prioritisation of Protected Area Management (RAPPAM), Spatial Monitoring and Reporting Tool (SMART), Integrated Management Effectiveness Tool (IMET). Out of 75 methodologies listed on the Global Database for Protected Areas Management Effectiveness (GD-PAME),[4] the most used is METT. In terms of area, the METT has been used in over a fifth of the world's terrestrial protected areas (Stolton and Dudley 2016). Most of the listed methodologies are based on the IUCN World Commission on Protected Areas (WCPA) framework (Hockings et al. 2006), which means that they have a common underlying approach and mostly standard criteria, although the indicators and assessment methods will vary. Dudley (2008) also stresses that the assessment systems could be divided into two main types, i.e. those that rely on expert knowledge and those that use data monitoring, stakeholder surveys, and other quantitative or qualitative data sources (or both combined).

Hockings et al. (2006) suggest that the WCPA framework for PAME evaluation includes six critical elements shown by management cycle (Fig. 2.4):

- Context: elaboration of values, threats and external influences on management;
- Planning: the creation of vision, goals, objectives, and strategies to conserve values and reduce threats;
- Inputs (resources): availability of resources (although PA management plan rarely provides specific commitments of funds and staff, it establishes the basis for short-term or annual operational planning in which decisions about the allocation of resources are made);
- Processes (policies and procedures): implementation of the actions and strategies indicated by planning documents and through reactive or opportunistic management actions;
- Outputs: the achievement of work programmes, products, and services, usually outlined in management plans and annual work programming;
- Outcomes: the achievement of objectives, goals, and changes in values.

The six elements of the management cycle reflect three broad areas of management, i.e. design (context and planning), appropriateness/adequacy (inputs and processes), and delivery (outputs and outcomes) (Dudley 2008).

Despite the fact that PAME assessment would ideally incorporate components that cover each of the elements of evaluation, Hockings et al. (2004) suggest that managers usually use only information related to outputs and outcomes. Input and process information is used mainly for reporting to the advisory committees. Based on more than 6300 assessments of PAME, Leverington et al. (2010) have defined headline indicators, representing the major themes and elements used in various assessment

[4] https://pame.protectedplanet.net/.

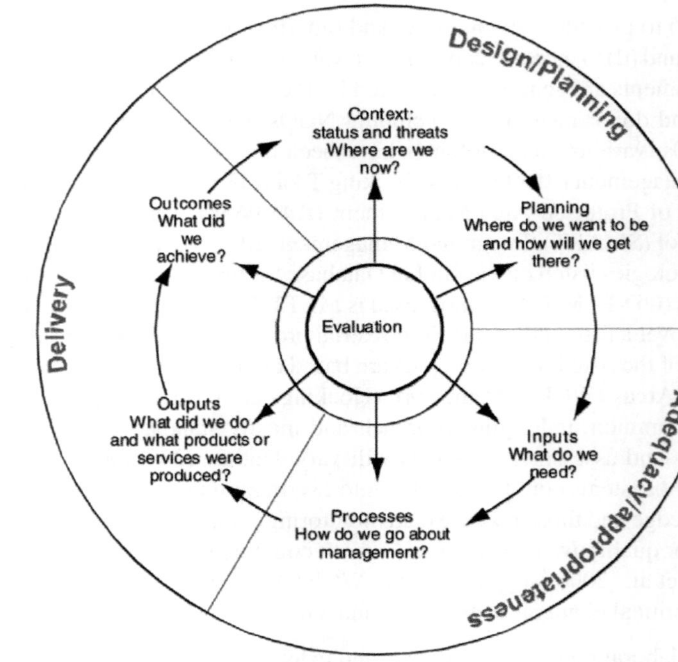

Fig. 2.4 The IUCN WCPA framework for assessing management effectiveness of protected areas (*Source* Retrieved from Hockings et al. 2006)

approaches. Most of the PAME methodologies use a hierarchical approach with between two and five levels of the organisation. Hockings et al. (2015) refer to these levels as topics and indicators and state that methodologies based on the IUCN WCPA framework usually organise indicators based on the cycle of management topics (elements).

2.2.4 Relationship Between the PA's Governance and Management Effectiveness Evaluations

The relationship between management and governance effectiveness is increasingly being explored. Vansteelant and Burgess (nd) emphasise that many PAME assessment methodologies include a few questions about governance, some integrate governance and equity considerations, and some even develop governance-related elements. However, PAME assessment methodologies do not address governance to the degree that enables sufficient understanding or action (Vansteelant and Burgess, nd). In Leverington et al. (2010), governance appears as only one of 30 headline indicators servicing PAME evaluations at a system level, as the process indicator named 'Effectiveness of governance and leadership'. By using the DPSIR framework,[5] Eklund and Cabezza (2017) tried to explain how governance quality affects specific PA management outcomes. Based on a three-round Delphi survey involving 33 middle- to senior-level staff from Western Australian PA agencies, Shields et al. (2016) identified 20 good governance indicators in the form of questions (aligned with ten governance principles). The authors connected them with the accompanying output measures, e.g. management plans, annual reports, stakeholder engagement, and audits. Lockwood (2010) suggests placing good governance principles 'above' the PAME evaluation elements of context, planning, inputs, process, outputs, and outcomes, demonstrating the need to consider governance in all six elements. Despite the common opinion that governance effectiveness evaluation is usually seen as complementary to the PAME, Lockwood (2010) indicates that it may be useful to deliver it separately.

Although PAs have to face continuously increasing visitation, the number of studies discussing the degree to which 'good governance' criteria may influence management and planning for nature-based tourism is limited (Hübner et al. 2014). For example, Su et al. (2007) proposed an analytical framework for the analysis of nature-based tourism governance in the current social, economic, and political contexts of China. Eagles (2008) reviewed different management approaches used for conservation and provision of tourism services, by consideration of generally accepted criteria for governance. Buteau-Duitschaever et al. (2010) compared visitor perspectives of the governance of two of Canada's largest park systems, i.e. the parastatal model of Ontario Provincial Parks and the public and for-profit combination model of British Columbia Provincial Parks. Suntikul et al. (2010) explored the roles of public and private actors in the development and management of national parks in Vietnam, focusing specifically on the conflicts caused by political, social, cultural, and economic factors. Eagles (2013) referred to park tourism governance as one of 10 critical areas that need to be more investigated. Islam et al. (2018a, b) identified opportunities for enhancing tourism governance practices in PAs through an ACM approach, to integrate conservation and sustainable tourism development goals.

2.3 Governance-Management Assessment in Krka National Park

Below is the governance-management assessment of Krka NP in Croatia. The assessment is performed using the Shields et al. (2016) approach, which evaluates the quality of governance through management output performance measures.

---

[5]DPSIR (drivers, pressures, state, impact, and response) is the systems-thinking framework that assumes cause-effect relationships between interacting components of social, economic, and environmental systems (Bradley and Yee 2015).

2.3.1 Study Site

Krka NP is one of the eight Croatian national parks. It is located in the vibrant Central Dalmatia region, covering an area of 109 km^2 (Fig. 2.5). The area, consisting of exquisite natural, and cultural heritage assets, is earmarked primarily for scientific, cultural, educational, recreational, and tourism activities.

Krka NP is located in the region, which accounts for 47% of the total number of overnights (Croatian Bureau of Statistics 2020). Considering its small distance from the central Dalmatian coastal destinations, the park witnesses an enormous pressure of daily visitors. The fact that the infrastructure intended for visitors occupies only 5% of the area of Krka NP makes these visitor numbers even more threatening. In 1988 Krka NP had 385,837 visitors (Radeljak and Pejnović 2008). In 2006, a few years after the independence war in Croatia, the number grew to more than 660,000. Finally, in 2018 this PA recorded 1,354,802 visitors (Ministry of Tourism

Fig. 2.5 The map of Krka National Park, Croatia (*Source* https://www.croatiatraveller.com/National%20Parks/Krka.htm)

2019), which is a 105% increase from 2006. More than 90% of visitors are recorded between May and September.

2.3.2 *The Study Approach*

The pressure of seasonal visitation poses a challenge to the PA authority. This chapter builds on the approach of Shields et al. (2016) to discuss the efficiency of coping with them. Shields et al. (2016) provided 20 governance indicators and output measures, suggesting the quality of governance as a part of the PA process of management evaluation. For the benefit of the analysis, we have advanced the original list of 20 governance indicators, to include those related to financial plans and forecasts (Table 2.2). We considered that due to the growth of self-financing through tourism revenues, it is worth including it in the good governance evaluation process, to show its dedication to the enhancement of this objective.

Within Table 2.2, under 10 governance principles (categories), 21 questions outline good governance indicators described as specific activities undertaken by the Krka NP management authority. The output measures indicate whether the activity was fulfilled. Pluses and minuses associated with each of the outputs indicate whether a specific activity/process was performed, in total or partially.

Hockings et al. (2006: 24–25) stress that outputs are the penultimate part of the assessment—determining if PA managers and other stakeholders achieved what they set out to do. Information on outputs are generally found in annual reports, carried out by PA management authorities. Reviews of work programme achievement and expenditure are standard internal management tools, while broad-scale reviews of implementation of planning commitments are often used as a significant element in external audits or programme reviews because they are necessary for establishing accountability. Hockings et al. (2006) also indicate that output assessment does not address the issue of the plans' appropriateness (adequacy), but it merely reveals if they are implemented. Additionally, the adequacy of planning systems and the plans themselves are better assessed by process and outcome approaches to evaluation. Worth noting is that output measures associated with each of the activities/processes related to a particular governance indicator are not explicitly associated with tourism. However, tourism-related goals and policies are an intrinsic part of all output measures, especially the management plan.

2.4 Discussion

Results of the Shields et al. (2016) research revealed that existence of the PA Management Plan (MP) and the local stakeholders' engagement in its creation are the core indicators of *Public participation* as a governance principle. Integrating stakeholders' opinions is necessary, not only because they may provide PA managers with useful

Table 2.2 Indicators for the good governance of protected areas with output measures

UNDP Good governance principles/characteristics [Graham et al.'s (2003) principles]	Good governance principles' indicators (described as activities and processes)	Output measure					
		Management Plan	Annual reports		Stakeholder engagement	Audits	Other
			Park specific	State/regional			
Public participation *(Legitimacy and voice)*	Are there opportunities for the public to be involved in decision-making, including management plans?	+			+		
	Is there an advisory committee for the park consisting of key stakeholder groups (including local government, landholders, tourism operators, researchers, conservation, 'friends of' groups, etc.)?				–		
Consensus orientation *(Legitimacy and voice)*	Has a framework been developed for decision-making that incorporates stakeholder engagement and/or comment, and do they have the right of appeal?	+/–			–		
	Have stakeholder groups been identified for key engagement requirements (e.g. management plans) and are they advised of any decisions/outcomes?	+/–			+/–		

(continued)

Table 2.2 (continued)

UNDP Good governance principles/characteristics [Graham et al.'s (2003) principles]	Good governance principles' indicators (described as activities and processes)	Output measure					
		Management Plan	Annual reports		Stakeholder engagement	Audits	Other
			Park specific	State/regional			
Strategic vision *(Direction)*	Is there a publicly available plan/strategic direction in place for the protected area based on current 'best practice' protected area management guidelines (including stakeholder engagement)? Does this plan outline/cover any legal and/or other requirements?	+					
	Is adaptive management part of the process of this strategic direction/plan (i.e. measure, review, evaluate, respond), including publishing the results (e.g. an annual report)?	+/−	+/−				
	Is there budgetary planning and forecasting?						+

(continued)

Table 2.2 (continued)

UNDP Good governance principles/characteristics [Graham et al.'s (2003) principles]	Good governance principles' indicators (described as activities and processes)	Output measure					
		Management Plan	Annual reports		Stakeholder engagement	Audits	Other
			Park specific	State/regional			
Responsiveness *(Performance)*	Does the protected area management/strategic plan follow the adaptive management process (i.e. measure, review, re-evaluate, report)?	+					
	Is there a report on the process/progress of management/strategic plan (e.g. annual report)?		+/−	−/+			
	Is there an asset management system to assist with infrastructure/capital works planning, insurance, etc.?						+/−
Effectiveness *(Performance)*	Is there an annual report that highlights the level of achievement of proposed strategic targets including biodiversity conservation, visitor experiences/expectations, etc.?		+/−	−			
	Are there internal and external auditing processes in place to reveal the degree and success of the implementation of strategic/management plans?					−	
Efficiency *(Performance)*	Is the protected area managed under a single authority or agency?						+

(continued)

Table 2.2 (continued)

UNDP Good governance principles/characteristics [Graham et al.'s (2003) principles]	Good governance principles' indicators (described as activities and processes)	Output measure					
		Management Plan	Annual reports		Stakeholder engagement	Audits	Other
			Park specific	State/regional			
	Does the protected area have internal and external auditing processes in place to identify areas where efficiencies can be made?					+/−	
	Are there opportunities for work to be conducted using partnerships with stakeholders (e.g. traditional owners, volunteers' groups, schools, etc.)?				+/−		
Accountability *(Accountability)*	Is there an annual report published that reports on managerial activities and accountability (including financial management, strategic goals/targets, external audit results, etc.)?		+/−	+/−			
	Does the protected area operate within a well-developed framework that is available to the public e.g. management plan that identifies policy, review, systems, etc.?	+					

(continued)

Table 2.2 (continued)

UNDP Good governance principles/characteristics [Graham et al.'s (2003) principles]	Good governance principles' indicators (described as activities and processes)	Output measure					
		Management Plan	Annual reports		Stakeholder engagement	Audits	Other
			Park specific	State/regional			
	Are there opportunities for stakeholders and/or the public to participate in protected area management and/or provide feedback?				+/–		
Transparency *(Accountability)*	Does the protected area publish an annual report including finances, staff numbers, visitor numbers, management plan, stakeholder consultation/engagement, etc.?		+/–	+/–			
Equity *(Fairness)*	Does the protected area employ and develop the park in accordance with local legal requirements concerning equity (including employment within the protected area, access to interpretation, etc.)?						–
Rule of law *(Fairness)*	Does the protected area outline the local/state/federal/international legislation it is governed by and include in its annual report its compliance with these (including any fines/charges within the park)?	+	–	–			

Source Own elaboration based on Shields et al. (2016)

information, especially during the process of MP drafting, but also because they may help PA management authorities become more effective, and increase the acceptance and support of outcomes (Getzner et al. 2012). Concerning the two indicators related to the principle of public participation, Krka NP performs quite well (Table 2.2); the Croatian Nature Protection Law (Official Gazette 80/13, 15/18, 14/19, 127/19), the primary document related to nature protection, indicates (article 138) that PA management is performed based on the MP. It also stresses (article 199), that during the preparation of an MP, public participation must be ensured. Stakeholders were involved in the delivery of the MP for Krka NP (the first one for the period 2006–2010 and the second one from 2011 to 2021). Since the Nature Protection Law does not strictly require the managing institutions to have advisory committees permanently, the Krka NP has only a management board consisting of four members, a director and an expert leader in charge of nature protection activities. Thus, it might be concluded that although regarding the public participation, the legal requirements are fulfilled, there is space for improvement. The Nature Protection Law should introduce the advisory committee as a legally binding body acting permanently during the MP life cycle. This is also a fundamental precondition for the AM to be employed.

Performance with respect to *Consensus orientation* should be evaluated according to the existence of a decision-making framework that incorporates stakeholder engagement and by the level of the stakeholders' engagement in advising on the decisions and discussing the outcomes (Shields et al. 2016). The Nature Protection Law requires no strict framework for decision-making that incorporates stakeholder engagement after the MP is delivered. Therefore, it might be concluded that the governance principle on consensus orientation is inefficiently exercised.

A PA's governance and management cannot be efficient unless all stakeholders have a shared *Strategic vision*, based upon an understanding of the historical, cultural, and social complexities in which that perspective is grounded (Shields et al. 2016). Thus, the management authority should provide policy directions built on national and international laws and regulations. The Krka NP management authorities satisfy this requirement. Due to the legal deficiencies, Krka NP fails to apply AM in the strategic planning (i.e. measure, review, evaluate, respond), as communication with the stakeholders stops as soon as the MP has been delivered. Concerning the need to publish the reports, Krka NP performs well as it regularly publishes its *Annual programme for protection, maintenance, preservation, promotion, and use of Krka National Park (Annual programme)* as well as *Annual financial plan* and *Financial Plan for the three years* (currently 2020–2022). However, none of these documents delineates the mechanism of the permanent stakeholders' engagement, which is why the principle of strategic vision cannot be seen as successfully implemented.

The principle of *Responsiveness* says that institutions have to serve stakeholders. Its performance is considered efficient if the AM is applied, if there is a report on the MP progress and if there is an asset management system applied. An analysis of the available documents for the NP of interest indicates that AM is not fully employed. Namely, the report *Annual programme* only partially delineates the progress of the MP, while no asset management system exists as such. To be precise, in the *Annual Programme*, there is just a little information regarding the usage and needs for technical resources. Hence, the principle of responsiveness does not seem to be satisfactorily addressed.

Efficiency and *Effectiveness* principles require the PA authorities to be cost-effective and efficient, to have the capacity to deliver tasks, to be effectively coordinated, able to demonstrate progress, and to provide information to allow assessments of their performance (Abrams et al. 2003). When it comes to the case study area, neither principle is adequately fulfilled, effectiveness being worse achieved. Within the *Annual Programme*, the statement on progress is poorly elaborated. It is also not explained how the described activities enable the achievement of a particular strategic goal outlined in the MP. Also, to the authors' best knowledge, the results of two critical nature-based tourism development-related documents, i.e. *Action plan for visitor management in the Krka NP 2019–2031* and the *Analysis of the visitors' survey*, were not used to feed the annual reports regarding MP achievements (specifically those referring to nature-based tourism-related goals).

The principle of *Efficiency* seems to have been implemented better. This might be because Krka NP is managed by a Public Institution, i.e. a state governed authority. Currently, there are no available documents as internal audits, suggesting that the performance assessment has not been delivered. Regarding cooperation with stakeholders, Krka NP authorities regularly organise meetings, education and workshops for traditional agricultural producers and local businesses. They also support the community in the preparation of projects to foster the development of the local economy.

Accountability is related to allocation and acceptance of responsibility for decisions and actions (Lockwood 2010). It depicts the interrelation between the managing institution and 'higher level' government authorities. Despite the existence of the MP and of several other documents, none of them sufficiently elaborates relevant issues concerning managerial activities and accountability. Also, there are no opportunities for stakeholders and community to participate in PA management after MP is established and to get the feedback on its performance regularly.

Transparency is a requirement, grounded in ethics, and stakeholders' right to be able to access all the decisions about a PA (Lockwood 2010). Governance authorities are expected to report on their progress regularly. Once stakeholders are involved, their opinion has to be accounted for transparently, and evaluation results have to be made available. Publication of evaluation results has to be done correctly to avoid misinterpretations (Getzner et al. 2012). Park authorities in Krka NP regularly publish previously mentioned *Annual Programme* that gives an overview of the ongoing activities and those expected to start in the current year. However, some previously mentioned issues, particularly in the area of stakeholders' engagement and consultations, are either non-existent or insufficiently elaborated.

Abrams et al. (2003) stress that all people should have equal opportunities to improve or maintain their well-being based on their work. The available documents supporting governance outputs in Krka NP do not provide evidence that PA management authority prioritises or follows local legal requirements concerning *Equity*. It does not necessarily mean that the authority is substantially deficient in this respect, but the information concerning this topic is not available.

The *Rule of law* principle requires laws to be fair and enforced impartially, while governing mechanisms should distribute costs and benefits equitably without any discrimination. The Krka NP Management Plan outlines different level legislation as well as other higher level strategic documents; however, in its Annual reports, there was no evidence of its compliance with the higher level documents, which is why we take this principle as being put partially into practice.

2.5 Conclusion

Shields et al. (2016) suggest that for over half of the UNDP governance principles, an MP was the identified output measure, except for effectiveness and efficiency where it was an audit document. The analysis presented in Table 2.2 shows that Krka NP managing institution has delivered two MPs, thus satisfying most of the good governance principles, including public participation, consensus, strategic vision, responsiveness, accountability and the rule of law. Internal and external audits revealing the success of MP implementation were not delivered, thus showing poor performance regarding efficiency and effectiveness. This is particularly important to emphasise as the *learning by doing* principle is the foundation of AM. At the same time, devotion to AM is crucial to be able to face the recent challenges the PA faces, including those related to excessive visitation. The exceptions were the principle of transparency, where output measures were annual reports, and the principle of equity, where the measure was the compliance with local equity requirements. In Krka NP, governance transparency is not utterly satisfying. The NP possesses an MP and an annual report, but these documents do not provide evidence of compliance with local legal requirements concerning equity and employment within the PA. Regarding budget planning and forecasting, the analysis of the essential documentation of the Krka NP proved the existence of the *Annual Financial Plan* and *Financial Plan for the three years (2020–2022)*; however, due to low transparency, these principles are only partially satisfied.

The overview of the fundamental governance indicators and their measures presented in Table 2.2 demonstrate insufficient devotion of the Krka NP management authority to compliance with UNDP good governance principles. The recent decade has seen significant advances in both governance and management effectiveness. However, there is always room for further improvements, especially in terms of implementing the ACM approach, particularly regarding stakeholder engagement. In the context of a Croatian PA, this will require changes of laws and regulations, particularly regarding the public involvement during the MP lifecycle. The legal requirement to collaborate with stakeholders would ensure their commitment, knowledge sharing, learning enhancement and foster the adaptation of ACM. The ever-growing visitation and engagement of the local community in tourism business within and adjoining the Krka NP, call for a more intense formal and informal communication between community stakeholders and the managing institution to prevent potential hostility. The Krka NP managing institution should anticipate these power relations and work to raise awareness and motivate the community to become involved in relevant local co-management organisations. This can be achieved by implementing different communication strategies from formal meetings and workshops to informal discussions and events. The participatory approach advocated by ACM could foster the creation of the new institutional arrangements through various co-management initiatives enabling better interaction between different stakeholder groups, including community and the PA, as well as the tourism industry and the PA. Diminishment of the communication gaps and information asymmetry will foster accountability and transparency. Currently, the managing authority of Krka NP has four members, none of them an expert for communication with the stakeholders.

Increasing visitation relates to impacts which are not always to the advantage of the community or the protected site. In many cases, they reflect the deterioration of local ecosystems. Thus, the voices of the community must be heard. Krka NP should engage in activities fostering the empowerment of local stakeholders. According to Islam et al. (2018b), the challenges with accountability, transparency, and power affect the governance principle *'Rule of law'*. The introduction of the ACM will inspire positive outcomes in terms of law enforcement. The organisation of campaigns, meetings, and workshops might foster in the stakeholders a better understanding of their role within the PA system.

Finally, although learning is implicitly and explicitly evident in each of the governance principles previously outlined, it is the adaptive and iterative aspects of ACM, which may facilitate incremental improvements in each of the governance principles. Since 2014 Krka NP authorities have organised meetings of traditional agricultural producers, working to educate and support them through interactive workshops and lectures on different topics, among others: eco-agriculture, beekeeping, cheese making, thus contributing to their empowerment. However, there are no particular activities concerning the education of or cooperation with local stakeholders operating tourism businesses, for example, accommodation, catering, guiding, adventure tours. Thus, we conclude that there is still a whole range of activities to be undertaken by the Krka NP management institution to satisfy the ACM requirements fully. Although the legacy of the top-down approach is hard to overcome (Islam et al. 2018b), by applying the ACM approach more intensely, various dimensions of governance, particularly participation, accountability and transparency, the rule of law, power, and vision and consequently efficiency and effectiveness will be facilitated in Krka National Park.

References

Abrams P, Borrini-Feyerabend G, Gardner J, Heylings P (2003) Evaluating governance—a handbook to accompany a participatory process for a protected area, Ottawa: Parks Canada and TILCEPA—Theme on Indigenous and Local Communities, Equity and Protected Areas of IUCN CEESP/WCPA. Available at: https://portals.iucn.org/library/node/12430. Retrieved 10 Oct 2020

Allen CR, Fontaine JJ, Pope KL, Garmestani AS (2011) Adaptive management for a turbulent future. J Environ Manage 92(5):1339–1345. https://doi.org/10.1016/j.jenvman.2010.11.019

Allen CR, Gunderson LH (2011) Pathology and failure in the design and implementation of adaptive management. J Environ Manage 92(5):1379–1384. https://doi.org/10.1016/j.jenvman.2010.10.063
Balmford A, Green JMH, Anderson M, Beresford J, Huang C, Naidoo R, Walpole M, Manica A (2015) Walk on the wild side: estimating the global magnitude of visits to protected areas. PLoS Biol 13(2):e1002074. https://doi.org/10.1371/journal.pbio.1002074
Bangwayo-Skeete PF, Skeete RW (2020) Modelling tourism resilience in small island states a tale of two countries. Tour Geogr 0(0):1–22. https://doi.org/10.1080/14616688.2020.1750684
Bevir M (2012) Governance: a very short introduction. Oxford University Press, Oxford, UK
Booker F, Franks P (2019) Governance assessment for protected and conserved areas (GAPA). Methodology manual for GAPA facilitators. IIED, London. Available at: https://pubs.iied.org/17655IIED/. Retrieved 10 Oct 2020
Borrini-Feyerabend G, Pimbert M, Farvar MT, Kothari A, Renard Y (2004) Sharing power: learning by doing in co-management of natural resources throughout the world. IIED and IUCN CEESP/CMWG, Cenesta, Tehran. Available at: https://pubs.iied.org/pdfs/G01089.pdf. Retrieved 25 Sept 2020
Borrini-Feyerabend G, Dudley N, Jaeger T, Lassen B, Pathak Broome N, Phillips A, Sandwith T (2013) Governance of protected areas: from understanding to action. Best Practice Protected Area Guidelines Series No. 20. IUCN, Gland, Switzerland, xvi + 124 pp. Available at: https://www.iucn.org/sites/dev/files/content/documents/governance_of_protected_areas_from_understanding_to_action.pdf. Retrieved 2 Nov 2020
Bown NK, Gray TS, Stead SM (2013) Co-management and adaptive co-management: two modes of governance in a Honduran marine protected area. Mar Policy 39:128–134. https://doi.org/10.1016/j.marpol.2012.09.005
Bradley P, Yee S (2015) Using the DPSIR framework to develop a conceptual model: technical support document. US Environmental Protection Agency, Office of Research and Development, Atlantic Ecology Division, Narragansett, RI. EPA/600/R-15/154. http://www.epa.gov/nscep/index.html
Buteau-Duitschaever WC, McCutcheon B, Eagles PFJ, Havitz ME, Glover TD (2010) Park visitors' perceptions of governance: a comparison between Ontario and British Columbia provincial parks management models. Tour Rev 65(4):31–50. https://doi.org/10.1108/16605371011093854
Bushell R, Bricker K (2017) Tourism in protected areas: developing meaningful standards. Tour Hosp Res 17(1):106–120. https://doi.org/10.1177/1467358416636173
Centre for the Promotion of Imports from developing countries (CBI) (2020) The European market potential for nature and ecotourism. Available at: https://www.cbi.eu/market-information/tourism/nature-tourism/nature-eco-tourism/market-potential. Retrieved 25 Oct 2020
Centre for Responsible Travel (2018) The case for responsible travel: trends & statistics 2018. Available at: https://www.responsibletravel.org/docs/The_Case_for_Responsible_Travel_2018_FINAL_FOR_WEB.pdf. Retrieved 10 Oct 2020
Coad L, Leverington F, Knights K, Geldmann J, Eassom A, Kapos V, Kingston N, de Lima M, Zamora C, Cuardros I, Nolte C, Burgess ND, Hockings M (2015) Measuring impact of protected area management interventions: current and future use of the Global Database of Protected Area Management Effectiveness. Phil. Trans. R. Soc. B 370. https://doi.org/10.1098/rstb.2014.0281
Croatian Bureau of Statistics (2020) The first release, tourist arrivals and nights in 2019 No: 4.3.2., God LVI. Available at: https://www.dzs.hr/Hrv/publication/FirstRelease/results.asp?pString=Turizam&pSearchString=%Turizam%. Retrieved 10 Oct 2020
Csete M, Szécsi N (2015) The role of tourism management in adaptation to climate change—a study of a European inland area with a diversified tourism supply. J Sustain Tour 23(3):477–496. https://doi.org/10.1080/09669582.2014.969735
Dai S, Xu H, Chen F (2019) A hierarchical measurement model of perceived resilience of urban tourism destination. Soc Indic Res 145(2):777–804. https://doi.org/10.1007/s11205-019-02117-9
de Koning M, Nguyen T, Lockwood M, Sengchanthavong S, Phommasane S (2017) Collaborative governance of protected areas: success factors and prospects for Hin Nam no national protected

area, Central Laos. Conservat Soc. 15:87–99. Available at: http://www.conservationandsociety.org/text.asp?2017/15/1/87/201396. Retrieved 10 Nov 2020

Dearden P, Bennett M, Johnston J (2005) Trends in global protected area governance, 1992–2002. Environ Manage 36(1):89–100. https://doi.org/10.1007/s00267-004-0131-9

Dudley N (2008) Guidelines for applying protected area management categories. IUCN, Gland, Switzerland, x + 86 pp. Available at: https://portals.iucn.org/library/sites/library/files/documents/PAG-021.pdf. Retrieved 25 Oct 2020

Eagles PFJ (2009) Governance of recreation and tourism partnerships in parks and protected areas. J Sustain Tour 17(2):231–248. https://doi.org/10.1080/09669580802495725

Eagles PFJ (2008) Governance models for parks, recreation and tourism. In: Hanna KS, Clark DA, Slocombe DS (eds) Transforming parks and protected areas: management and governance in a changing world. Routledge, London, pp 39–61

Eagles PFJ, McCool SF, Haynes CDA (2002) Sustainable tourism in protected areas: guidelines for planning and management. IUCN, Gland, Switzerland and Cambridge, UK, xv + 183 pp. Available at: https://portals.iucn.org/library/sites/library/files/documents/PAG-008.pdf. Retrieved 25 Oct 2020

Eagles PFJ (2013) Fiscal implications of moving to tourism finance for parks: Ontario Provincial Parks. Manag Leis 19(1):1–17. https://doi.org/10.1080/13606719.2013.849503

Eagles PFJ (2014) Research priorities in park tourism. J Sustain Tour 22(4):528–549. https://doi.org/10.1080/09669582.2013.785554

Eklund J, Cabeza MA (2017) Quality of governance and effectiveness of protected areas: crucial concepts for conservation planning. NY Acad. Sci. 1399:27–41. https://doi.org/10.1111/nyas.13284

European Commission (2016) Flash Eurobarometer 432 study on Preferences of Europeans towards tourism. Available at: https://data.europa.eu/euodp/data/dataset/S2065_432_ENG. Retrieved 2 Nov 2020

European Environment Agency (EEA) (2020) Common Database on Designated Area (CDDA). Available from: https://www.protectedplanet.net/3373. Retrieved 10 Oct 2020

Farhad S, Gual MA, Ruiz-Ballesteros E (2017) Land use policy how does adaptive co-management relate to specified and general resilience. An approach from Isla Mayor, Andalusia, Spain. Land Use Policy 67(May):268–276. https://doi.org/10.1016/j.landusepol.2017.05.038

Franks P, Booker F (2018) Governance assessment for protected and conserved areas (GAPA): early experience of a multi-stakeholder methodology for enhancing equity and effectiveness. IIED Working Paper, IIED, London. Available at: http://pubs.iied.org/17632IIED. Retrieved 10 Oct 2020

Geldmann J, Coad L, Barnes M, Craigie ID, Hockings M, Knights K, Burgess ND (2015) Changes in protected area management effectiveness over time: a global analysis. Biol Cons 191:692–699. https://doi.org/10.1016/j.biocon.2015.08.029

Getzner M, Jungmeier M, Pfleger B (2012) Evaluating management effectiveness of national parks as a contribution to good governance and social learning. Prot Area Manag. https://doi.org/10.5772/50092

Gossling S, Hultman J (2006) Ecotourism in Scandinavia: Lessons in theory and practice. CABI International

Gunderson L (2018) Adaptive management and integrative assessments. Encycl Ecol, 234–238. https://doi.org/10.1016/B978-0-444-63768-0.00635-1

Graham J, Amos B, Plumptre TW (2003) Governance principles for protected areas in the 21st century. Institute on Governance, Ottawa, Ontario. Available at: https://www.files.ethz.ch/isn/122197/pa_governance2.pdf. Retrieved 19 Nov 2020

Gurung HB (2010) Trends in protected areas. Gold CRC for Sustainable Tourism, Coast, Australia. Available at: https://sustain.pata.org/wp-content/uploads/2015/01/STCRC001_Gurung_Trends-In-Protected-Areas_WEB.pdf Retrieved 2 Nov 2020

Heylings P, Bravo M (2007) Evaluating governance: a process for understanding how co-management is functioning, and why, in the Galapagos Marine Reserve. Ocean Coast Manag 50(3–4):174–208. https://doi.org/10.1016/j.ocecoaman.2006.09.003
Heslinga J, Groote P, Vanclay F (2017) Strengthening governance processes to improve benefit sharing from tourism in protected areas by using stakeholder analysis. J Sustain Tour 9582 (January 2018):1–15. https://doi.org/10.1080/09669582.2017.1408635
Hockings M, Leverington F, Cook C (2015) Protected area management effectiveness. In: Worboys GL, Lockwood M, Kothari A, Feary S, Pulsford I (eds) Protected area governance and management. ANU Press, Canberra, pp 889–928. Available to: https://www.researchgate.net/publication/272819371_Protected_area_management_effectiveness. Retrieved 17 Nov 2020
Hockings M, Stolton S, Dudley N, James R (2009) Data credibility: what are the "right" data for evaluating management effectiveness of protected areas? In: Birnbaum M, Mickwitz P (eds) Environmental program and policy evaluation: addressing methodological challenges. New Directions for Evaluation. Wiley, pp 53–64
Hockings M, Stolton S, Dudley N (2004) Management effectiveness: assessing the management of protected areas? J Environ Planning Policy Manage 6(2):157–174. https://doi.org/10.1080/1523908042000320731
Hockings M, Stolton S, Leverington F, Dudley N, Courrau J (2006) Evaluating effectiveness: a framework for assessing management effectiveness of protected areas, 2nd edn. IUCN, Gland, Switzerland and Cambridge, UK, xiv + 105 pp. Available at: https://portals.iucn.org/library/efiles/documents/PAG-014.pdf. Retrieved 28 Oct 2020
Holling CS (1978) Adaptive environmental assessment and management: an overview. In: Proc. Gulf of Mexico Coastal Ecosystem Workshop, Sept 1979, Port Aransas Texas, pp 185–189
Hübner A, Phong LT, Châu TSH (2014) Good governance and tourism development in protected areas: the case of Phong Nha-Ke Bang National Park, central Vietnam. Koedoe 56(2):Art. #1146, 10 pages. http://dx.doi.org/10.4102/koedoe.v56i2.1146
Islam MW, Ruhanen L, Ritchie BW (2018a) Adaptive co-management: a novel approach to tourism destination governance? J Hosp Tour Manag 37:97–106. https://doi.org/10.1016/j.jhtm.2017.10.009
Islam MW, Ruhanen L, Ritchie BW (2018b) Tourism governance in protected areas: investigating the application of the adaptive co-management approach. J Sustain Tour 26(11):1890–1908. https://doi.org/10.1080/09669582.2018.1526291
IUCN (2017) IUCN World Heritage Outlook 2 (Issue November). https://portals.iucn.org/library/sites/library/files/documents/2017-053.pdf
Kaltenborn MP, Qvenild M, Nellemann C (2011) Local governance of national parks: the perception of tourism operators in Dovre-Sunndalsfjella National Park, Norway. Nor Geogr Tidsskrift—Norwegian J Geogr, 65(2):83–92. https://doi.org/10.1080/00291951.2011.574320
Lai PH, Hsu YC, Wearing S (2016) A social representation approach to facilitating adaptive co-management in mountain destinations managed for conservation and recreation. J Sustain Tour 24(2):227–244. https://doi.org/10.1080/09669582.2015.1062018
Landauer M, Goodsite ME, Juhola S (2018) Nordic national climate adaptation and tourism strategies—(how) are they interlinked? Scand J Hosp Tour 18(sup1):S75–S86. https://doi.org/10.1080/15022250.2017.1340540
Law on Nature Protection (Official Gazette 80/13, 15/18, 14/19, 127/19). Available at: https://www.zakon.hr/z/403/Zakon-o-za%C5%A1titi-prirode. Retrieved 17 Nov 2020
Larson LR, Poudyal NC (2012) Developing sustainable tourism through adaptive resource management: A case study of Machu Picchu, Peru. J Sustain Tour 20(7):917–938. https://doi.org/10.1080/09669582.2012.667217
Leverington F, Lemos Costa K, Courrau J, Pavese H, Nolte C, Marr M, Coad L, Burgess N, Bomhard B, Hockings M (2010) Management effectiveness evaluation in protected areas: a global study. The University of Queensland, Brisbane Australia. Available at: https://www.researchgate.net/publication/46378678_Management_Effectiveness_Evaluation_in_Protected_Areas_A_Global_Study. Retrieved 17 Nov 2020

Ly TP, Xiao H (2016) The choice of a park management model: A case study of Phong Nha-Ke Bang National Park in Vietnam. Tour Manag Perspect 17:1–15. https://doi.org/10.1016/j.tmp.2015.10.004

Lockwood M (2010) Good governance for terrestrial protected areas: a framework, principles and performance outcomes. J Environ Manage 91(3):754–766. https://doi.org/10.1016/j.jenvman.2009.10.005

Lockwood M (2009) Governance assessment of terrestrial protected areas: a framework and three case studies. University of Tasmania, Hobart. Available at: http://www.geog.utas.edu.au/geography/nrmgovernance/. Retrieved 17 Nov 2020

Lockwood M, Davidson J, Curtis A, Stratford E, Griffith R (2010) Governance principles for natural resource management. Soc Nat Resour 23(10):986–1001. https://doi.org/10.1080/08941920802178214

Lopez Gutierrez B, Almeyda Zambrano AM, Mulder G, Ols C, Dirzo R, Almeyda Zambrano SL, Broadbent EN (2019) Ecotourism: the "human shield" for wildlife conservation in the Osa Peninsula, Costa Rica. J Ecotourism, 1–20. https://doi.org/10.1080/14724049.2019.1686006

Lordkipanidze M, Bressers H, Lulofs K (2018) Governance assessment of a protected area: the case of the Alde Feanen National Park. J Environ Planning Manage. https://doi.org/10.1080/09640568.2018.1441014

Mandić A (2019) Nature-based solutions for sustainable tourism development in protected natural areas: a review. Environ Syst Decis. https://doi.org/10.1007/s10669-019-09718-2

Mandić A (2020) Structuring challenges of sustainable tourism development in protected natural areas with driving force—pressure—state—impact—response (DPSIR) framework. Environ Syst Decis. https://doi.org/10.1007/s10669-020-09759-y

Mandić A, Petrić L (2020) The impacts of location and attributes of protected natural areas on hotel prices: implications for sustainable tourism development. In: Environment, development and sustainability (Issue 0123456789). Springer, The Netherlands. https://doi.org/10.1007/s10668-020-00611-6

Mccool SF (2016) Reframing sustainable tourism, vol 2. https://doi.org/10.1007/978-94-017-7209-9

Mccool SF, Clark RN, Stankey GH (2007) An assessment of frameworks useful for public land recreation planning. Recreation, 136

Ministry of Tourism (2019) Tourism in numbers. Available at: https://www.htz.hr/sites/default/files/2019-09/HTZ%20TUB%20ENG_2018.pdf. Retrieved 17 Nov 2020

More T (2005) From public to private: five concepts of park management and their consequences. Georg Wright Forum 22(2):12–20. Available at: http://www.georgewright.org/222more.pdf. Retrieved 10 Oct 2020

Newsome D (2020). The collapse of tourism and its impact on wildlife tourism destinations. J Tour Futur (Vol. ahead-of-print, No. ahead-of-print). https://doi.org/10.1108/JTF-04-2020-0053

Plummer R, Baird J, Dzyundzyak A, Armitage D, Bodin Ö, Schultz L (2017) Is adaptive co-management delivering? Examining relationships between collaboration, learning and outcomes in UNESCO biosphere reserves. Ecol Econ 140:79–88. https://doi.org/10.1016/j.ecolecon.2017.04.028

Potts R, Vella K, Dale AP, Sipe N (2016) Evaluating governance arrangements and decision-making for natural resource management planning: an empirical application of the governance systems analysis framework. Soc Nat Resour 29(11):1325–1341. https://doi.org/10.1080/08941920.2016.1185557

Prato T (2012) Increasing resilience of natural protected areas to future climate change: a fuzzy adaptive management approach. Ecol Model 242:46–53. https://doi.org/10.1016/j.ecolmodel.2012.05.014

Presenza A, Abbate T, Micera R (2015) The Cittaslow movement: opportunities and challenges for the governance of tourism destinations. Tour Plan & Dev 12(4):479–488. Available at: https://doi.org/10.1080/21568316.2015.1037929. Retrieved 10 Oct 2020

Radeljak P, Pejnović D (2008) Utjecaj turizma na održivi razvoj funkcionalne regije Nacionalnog parka Krka. Godišnjak Titius (Split). 1(1):329–361. Avaialble at: https://hrcak.srce.hr/111984. Retrieved 10 Oct 2020
Rist L, Felton A, Samuelsson L, Sandström C, Rosvall O (2013) A new paradigm for adaptive management. Ecol Soc 18(4). https://doi.org/10.5751/ES-06183-180463
Rutty M, Hall CM, Grossling S, Scott D (2015) The global effects and impacts of tourism. In: The Routledge handbook of tourism and sustainability. https://doi.org/10.4324/9780203072332.ch3
Sandwith T, MacKinnon K, Enkerlin Hoeflich E (2016) Protected area governance and management, Graeme LW, Lockwood M, Kothari A, Feary S, Pulsford I (eds). ANU Press. https://doi.org/10.26530/oapen_569111
Schianetz K, Kavanagh L (2008) Sustainability indicators for tourism destinations: a complex adaptive systems approach using systemic indicator systems. J Sustain Tour 16(6):601–628. https://doi.org/10.1080/09669580802159651
Schägner JP, Brander L, Maes J, Paracchini ML, Hartje V (2016) Mapping recreational visits and values of European National Parks by combining statistical modelling and unit value transfer. J Nat Conserv 31:71–84. https://doi.org/10.1016/j.jnc.2016.03.001
Scott D, Becken S (2010) Adapting to climate change and climate policy: progress, problems and potentials. J Sustain Tour 18(3):283–295. https://doi.org/10.1080/09669581003668540
Shields BP, Moore SA, Eagles PFJ (2016) Indicators for assessing good governance of protected areas: insights from park managers in Western Australia Parks, vol 22, pp 37–50. Available at: https://parksjournal.com/wp-content/uploads/2016/03/PARKS-22.1-Shields-et-al-10.2305IUCN.CH_.2016.PARKS-22-1BPS.en_.pdf. Retrieved 10 Oct 2020
Shipley R, Kovacs J (2008) Good governance principles for the cultural heritage sector: lessons from international experience. Corp Governance Int J Bus Soc 8(2):214–228. https://doi.org/10.1108/14720700810863823
Spenceley A, Kohl J, McArthur S, Myles P, Notarianni M, Paleczny D, Pickering C, Worboys GL (2015) Visitor management. In: Worboys GL, Lockwood M, Kothari A, Feary S, Pulsford I (eds) Protected area governance and management. ANU Press, Canberra, pp 715–750
Spenceley A, Snyman S, Eagles PFJ (2019) A decision framework on the choice of management models for park and protected area tourism services. J Outdoor Recreat Tour 26(March):72–80. https://doi.org/10.1016/j.jort.2019.03.004
Stolton S, Dudley N (2016) METT handbook: a guide to using the Management Effectiveness Tracking Tool (METT), WWF-UK, Working. Available at: https://www.biopama.org/sites/default/files/resources-doc/WWF_METT_Handbook_2016_FINAL_0.pdf. Retrieved 10 Oct 2020
Su D, Wall G, Eagles P (2007) Emerging governance approaches for tourism in the protected areas of China'. Environ Manage 39(6):749–759. https://doi.org/10.1007/s00267-006-0185-y
Suntikul W, Butler R, Airey D (2010) Implications of political change on national park operations: Doi Moi and tourism to Vietnam's national parks'. J Ecotourism 9(3):201–218. Available at: http://dx.doi.org/10.1080/14724040903144360. Retrieved 5 Oct 2020
UNDP (1997) Governance for sustainable human development. New York. Available at: http://gis.emro.who.int/HealthSystemObservatory/Workshops/WorkshopDocuments/Reference%20reading%20material/Literature%20on%20Governance/GOVERN~2.PDF. Retrieved 5 Oct 2020
UNEP-WCMC, IUCN, NGS (2018) Protected Planet Report 2018. UNEP-WCMC, IUCN and NGS: Cambridge UK; Gland, Switzerland; and Washington, DC, USA. Available at: https://livereport.protectedplanet.net/pdf/Protected_Planet_Report_2018.pdf. Retrieved 10 Oct 2020
UNWTO (2019) International tourism highlights. Available at: https://www.e-unwto.org/doi/pdf/10.18111/9789284421152. Retrieved 10 Oct 2020
Vanclay F (2015) Social impact assessment: guidance for assessing and managing the social impacts of projects. International Association for Impact Assessment (IAIA), USA. Available at: https://www.rug.nl/research/portal/files/17534793/IAIA_2015_Social_Impact_Assessment_guidance_document.pdf. Retrieved 25 Oct 2020

Vansteelant NG, Burgess ND (nd) Assessment tools for protected areas recommended management effectiveness, social, governance and equity tools for WWF in the Miombo Ecoregion. UNEP-WCMC, Luc Hoffman Institute. Available at: http://luchoffmanninstitute.org/wp-content/uploads/2015/03/Assessment-Tools-for-Protected-Areas.pdf. Retrieved 10 Nov 2020

Weaver DB, Lawton LJ (2017) A new visitation paradigm for protected areas. Tour Manag 60:140–146. https://doi.org/10.1016/j.tourman.2016.11.018

Williams BK (2011) Adaptive management of natural resources-framework and issues. J Environ Manage 92(5):1346–1353. https://doi.org/10.1016/j.jenvman.2010.10.041

Williams BK, Szaro RC, Shapiro CD (2007) Adaptive management: the US department of the interior technical guide. Adaptive Management Working Group, US Department of the Interior, Washington, DC. Available at: https://www.doi.gov/sites/doi.gov/files/migrated/ppa/upload/TechGuide.pdf Retrieved 15 Oct 2020

Worboys GL, Lockwood M, Kothari A, Feary S, Pulsford I (eds) (2015) Protected area governance and management. ANU Press, Canberra. http://dx.doi.org/10.22459/PAGM.04.2015

Chapter 3
Managing Overtourism in Nature-Based Destinations

Ante Mandić and Izidora Marković Vukadin

Abstract Tourism has grown and evolved significantly in past decades, and some of the destination hassles (e.g. crowding, rubbish,facility accessibility and conflicts) become more pronounced. Along with that, PAs face biological, social and economic fragmentation, suggesting that to be effective, the existing approaches to nature-based tourism management require improvement. This chapter aims to critically discuss the usability of tourism and visitor use management and planning frameworks to address the challenges associated with overtourism. The analysis suggests there is no one for all solution in terms of visitor use framework, but constituents of existing frameworks can be crucial for mitigating influences related to extensive visitation. The established frameworks should be advanced by consideration of new theoretical and practical advances, employing the system approach in which PAs are seen in the interrelation with other ecosystems. PA managers require tools and resources, which are necessary to prevail the pressure before they even happen. Greater involvement of stakeholders, goal-orientation and monitoring is needed.

Keywords Overtourism · Nature-based destinations · Visitor management · Carrying capacity · Protected areas

3.1 Introduction

Sustainability in tourism context relates to planning to operate within carrying capacity (CC) limits of the destination and its resilience capabilities to avoid a state of overtourism (Doods and Butler 2019). Although this concept is mostly associated with urban areas, e.g. Venice, Barcelona, Prague or Dubrovnik, its adverse impacts are often pronounced in nature-based destinations. Tourists elect to visit exceptional

---

A. Mandić (✉)
Faculty of Economics, Business and Tourism, University of Split, Split, Croatia
e-mail: ante.mandic@efst.hr

I. Marković Vukadin
Institute for Tourism, Zagreb, Croatia
e-mail: izidora.markovic@iztzg.hr

A. Mandić and L. Petrić (eds.), *Mediterranean Protected Areas in the Era of Overtourism*, https://doi.org/10.1007/978-3-030-69193-6_3

places, out of which protected natural areas (PNA) are high on the list of priorities. According to the latest Eurobarometer (European Commission 2016), the natural features of a destination are the main reasons for wanting to return to the same place for a holiday for more than half of EU travellers. Although large PNA tends to permit the wide dispersal of visitors, to make money and ensure visitor flows, it is necessary to concentrate tourists at access points and provide them with information and services in a cost-effective way (Wall 2019). Thus, tourism growth, its economic impacts, management implications and park tourism competencies and new management structures are only some challenges that need to be addressed (Eagles 2002). The accessibility of PNAs, as well as their increasing dependence on tourism revenues (Mandić 2020), challenges the capability of many park management institutions (Eagles 2009). Thus, in many highly attractive PNAs visitors' numbers, density and potential overcrowding are becoming a serious concern. In order to successfully protect an area, influence over the activities taking place within is necessary. Therefore, the management of PNAs depends on the management of economic (and other) activities taking place therein, whereby activities should either be restricted or encouraged, depending on the needs of the space (Koderman et al. 2020).

As a destination experiences higher intensity of tourism development, the inherent conflict between maintaining a healthy natural environment and economic development also increases (Mandić 2019). The high sensitivity on the underlying phenomenon underlines the necessity of overtourism management in PNAs to be a matter of great urgency. However, this should not be considered an easy job. Balancing visitation and conservation through proper planning, is a complicated task for PNA managers (Bushell and Bricker 2017), considering they deal with different elements of integrated tourism product (Mandić 2019), while concentrations of visitors tend to have adverse impacts on protected ecosystems (Olive and Marion 2009). Thus, the mitigation of overtourism in PNAs requires the understanding of the complexity of ecosystems, and acknowledgement that relationships between causes and consequences are not purely linear. In light of such challenges, the management process involving the assessment and evaluation of the condition of the area, the definition of the management objectives and activities that need to be carried out, and following the implementation of policies, with constant monitoring and evaluation of the effectiveness and adjustment, is required (Dudley 2008). Islam et al. (2018) stressed out a paradigm shift in PNAs with the transition from a traditional top-down to participatory bottom-up approaches to planning, management and governance, which reflects the changing expectations of governance towards systems that can empower and benefit local communities (Eagles 2014). In light of such change, adaptive co-management (ACM) is suggested as an approach that might improve the current governance of tourism in PNAs (Plummer and Fennell 2009). However, ACM should not be taken as a panacea, especially in the absence of its fundamental principles (Islam et al. 2018). Thus, the resilience of the PNAs in a case of overtourism will require close relation and clear communication (Stanford and Guiver 2016; Wilson et al. 2009) between all stakeholders involved, including tour-operators, and accommodation and relevant service providers.

The critical challenges associated with tourism in PNAs are the provision of visitor use with zero threat to protected features, and the provision of recreation and tourism opportunities with maximum benefits to all stakeholders (Leung et al. 2018). Due to the continuous growth in the number of visitors, PNAs require spatial–temporal management of visitor flows. Much of the discussion on how to achieve this, for decades, has considered the concept of CC. Although the CC does not meet the criterion to be considered a PNA planning framework (McCool and Bosak 2016), it is often discussed as a prerequisite to achieving the balance (Ly and Nguyen 2017). Most recently, due to the emergence of overtourism, the debate on planning frameworks has been revisited. Leung et al. (2018) within IUCNs *Tourism and visitor management in protected areas guidelines* advocate the use of Limits of Acceptable Change (LAC) and Recreation Opportunity Spectrum (ROS) approaches and establishment of visitor use limits, rather than using the concept of CC as a basis. In the academic literature, as well as in practice, the overtourism related ‘solutions’ are often interpreted in accordance with a different type of destination (Peeters et al. 2018b; UNWTO 2018), the status of protection, utilisation and management objectives.

To maximise the potential of PNAs managers and policymakers will seek information about the strengths and weaknesses of their management (Hockings 2003). In light of such debate, and the need for information and guidance, this chapter aims to critically discuss the usability of tourism and visitor use management and planning frameworks to address the challenges associated with overtourism.

3.2 Overtourism in Protected Natural Areas

Over the last decade, while many scholars have maintained their interest in the classical debate concerning the impacts of tourism, some (Goodwin 2019; Milano et al. 2019a; Peeters et al. 2018a; Seraphin et al. 2018) have converged with the narrative of social movements challenging the tourism growth premise, with the subsequent coining the terms ‘overtourism’ and ‘tourismphobia’ (Milano et al. 2019b). In their comprehensive reflection, which is one of first reports of that kind, Peeters et al. (2018b) define overtourism as ‘*a situation in which the impact of tourism, at certain times and in certain locations, exceeds physical, ecological, social, economic, psychological and/or political capacity thresholds*’. Within its definition, UNWTO (2018) points out the negative influence of excessive tourism on two critical groups, i.e. residents and visitors: ‘*… overtourism can be defined as the impact of tourism on a destination, or parts thereof, that excessively influences perceived quality of life of citizens and/or quality of visitors experiences in a negative way*’. This definition is in line with the one proposed by Goodwin (2019): ‘*overtourism describes destinations where hosts or guests, locals or visitors, feel that there are too many visitors and that the quality of life in the area or the quality of the experience has deteriorated unacceptably*’. Although the phenomenon of overtourism is not novel, the term is used to describe the consequences of tourism is some destinations, and thus can be considered a new issue for future studies, particularly in relation to new models of tourism development (Capocchi et al. 2019).

Growth, social unrests and *overcrowding* are three main concepts associated with overtourism in relevant academic literature and sector reports. *Growth* is recognised as the primary enabler of overtourism (Doods and Butler 2019; Higgins-Desbiolles et al. 2019; Milano et al. 2019a; Peeters et al. 2018a). Tourism growth is, among others, spurred with accessibility and affordability of travel, consumers prioritising travel and leisure experiences, social media, the traditional focus of tourism sector on volume, urbanisation, bucket-list tourism, peer-to-peer accommodation and extensive group travel (Jordan et al. 2018). Within the PNAs, tourism growth is the consequence of increasing demand for tourism and recreation, as well as reliance on nature-based tourism as an essential tool for economic development (Eagles 2002), and the perception that tourism could help fund ever-growing number of PNAs (Whitelaw et al. 2014). The last one is associated with a reduction of government funding, leading to increased operational reliance on visitor-based revenue despite the potential of increased visitation to undermine further the vital ecological functions of protected areas (Weaver and Lawton 2017). Recently, the study on recreational visits and values of European National Parks in 26 countries has demonstrated that the total annual visits amount to more than 2 billion annually, or in a consumer surplus of approximately 14.5 billion euros annually (Schägner et al. 2016). Critique of overtourism calls into question the growth paradigm itself, and the extent to which tourism as we know it can remain sustainable in the face of a mounting range of negative impacts (Fletcher et al. 2019). This critique, i.e. contemporary debate on the effects of over-visitation and overcrowding shifted towards overtourism has resulted in a revisit of degrowth discourse in tourism studies (Hall 2009; Higgins-Desbiolles et al. 2019; Milano et al. 2019a). From a tourism standpoint, degrowth is based on the ideology of opposition to conventional mass tourism and the prevention of the exploitation of the local community (Andriotis 2018), advocating community-based and responsible tourism.

Social unrests, protests and resistance against tourism, tourismphobia and touristphobia in numerous destinations across the globe, have become a trademark of overtourism reflecting the consequences of mass tourism and the answer of local communities. Recently, Seraphin et al. (2019) emphasise that the changes of attitudes of locals towards visitors could be characterised by four archetypes, namely, locals who are helpless victims, peaceful activists, vandals and resilient locals. Overall, the authors suggest that locals are volatile groups, whose resilience could be developed through an ambidextrous management approach which reflects the balance between exploitation and exploration. Within the park tourism and nature-based tourism literature, the attention has been given to the capacity building for local and indigenous communities (e.g. Bello et al. 2016, 2017; Peng et al. 2016; Spenceley and Goodwin 2007; Stone and Nyaupane 2018; Zeppel 2010). Locals are characterised by a very close relationship with their territories and natural resources. Generally, they advocate for collective rather than individual rights to their land, water and natural resources, and such a collective approach tends to maintain the integrity of territory, avoid ecological fragmentation and foster long-term objectives—all key requirements for biodiversity conservation (Sandwith et al. 2016). The studies focused on local communities within PNAs have recognised that PNAs managers often overlook the social and indigenous values of parks and cultural landscapes (Zeppel 2010), as well as that voices of indigenous communities, need to be recognised in the management of parks (Hannam 2005). Furthermore, community attitudes towards tourism development are correlated with impacts (Bello et al. 2016; Peng et al. 2016), while studies (Akyeampong 2011; Serenari et al. 2017; Spenceley and Goodwin 2007; Zapata et al. 2011) have demonstrated the isolated efforts from individual tourism companies have a little tangible impact on the majority of people, but the impact is substantial for the few people who directly benefit. The potential sociological impacts of the overtourism in the context of local and indigenous communities, especially in the context of European PAs in vibrant destinations are still not addressed.

Overcrowding is a phenomenon primarily associated with negative experiences emerging from the presence of too many tourists at certain places and times (Peeters et al. 2018b). Naturally, overcrowding perception varies with individuals and is dependent on the type of tourism development in the areas (Santana-Jiménez and Hernández 2011). Peeters et al. (2018b) suggest that (over)crowding refers to a psychological response to density, i.e. to the feeling of having a lack of privacy, or unwanted interactions, which is in a coastal and island destinations social concern. Crowding of recreation sites causes perceptual impacts, decreasing the quality of a visitors experience, resulting in the adoption of coping mechanisms such as activity substitution or spatial and temporal displacement (Moyle and Croy 2007). Although the influence of crowding on visitor satisfaction has been established, it is essential to acknowledge the contradictions of reports. For example, a recent study by Luque-Gil et al. (2018) suggest that problems of overcrowding do not produce dissatisfaction among the majority of park users. Their conclusions seem to be supported with several other relevant studies, suggesting different levels of tolerance by different types of park users, as well as the 'Expectancy Theory' (Schreyer and Roggenbuck 1978) suggesting that individuals expectations regarding crowding are related to socio-economic or environmental variables. Within nature-based destinations, especially vibrant PNAs, overcrowding is additionally associated with environmental issues, considering that the increasing visitor use can and often does cause increasing impacts in the form of damage to fragile soils and vegetation and conflicting uses (Buultjens et al. 2005). The issue of crowding in outdoor recreation has been the subject of 'Normative approach',[1] which was at some point expanded to include potential indicators of quality (Manning and Lawson 2002). The managerial perspective of crowding reflects the history of development and application of the concept of CC as a management tool. Although it is a useful measure of dealing with issues and impacts that are directly related to use levels, i.e. crowding, Cole and Carlson (2010)

[1]Norms are defined as standards that individuals and groups use for evaluating behaviour and social and environmental conditions. If visitors have normative standards concerning relevant aspects of recreation experiences, then such norms can be measured and used as a basis for formulating standards of quality. Norms are important aspect within research focused on pro-environmental behaviour (Steg and Vlek 2009).

suggest that setting a numerical visitor capacity should never be the first visitor use management approach to consider.

3.2.1 Impacts, Indicators and Management Implications

In PNAs, visitor numbers are a concern, considering that growing numbers pose an environmental threat, and declining numbers induce fear of insufficient funding to operate and maintain the park (Wall 2019). Thus, the management of overtourism in such destinations requires the understanding of the local system and the complexity of interactions within. Overtourism, in general, relates to increasing pollution, increasing demand for usage of infrastructure, visual intrusion, tourist concentrations and congestion, damage of sites, overcrowding, inflation, social change due to economic dependence, touristification of residential areas, marginalisation of residents, hostility and criminality, modification of the recreational areas and loss of cultural identity (Peeters et al. 2018b). (Excessive) tourism development in PNAs relates to the various adverse impacts emerging from diverse tourism activities (e.g. transportation, littering, vandalism, development, use of resources, hunting fishing, pollution, etc.), inducing the deterioration of, among others, air, water, soil, habitats, wildlife, tradition and host communities in general, through its impacts on cultures, psychology, crime, roles, employments and diversification (Leung et al. 2018).

The importance of implementation of indicators in order to strengthen socioecological systems under the intensive influence of tourism (Lacitignola et al. 2007; Krajinović 2015) is widely recognised. In the initial stages of critical discussions on sustainability indicators, Gössling et al. (2006) support the diversification of, as they call it, measurements of socio-economic development because using tourist arrivals numbers as the only indicator omits the complexity of the tourism generated income in PNAs. The challenges of adjusting tourism sustainability indicators on destination level have been extensively addressed in a significant number of papers (Blancas et al. 2010; Torres-Delgado and Saarinen 2014; Agyeiwaah et al. 2017).

The main challenge for the sustainable development of tourism in PNAs is to balance the flow and behaviour of visitors with the protection goals set up for the area at different political levels (FEDERATION 2012). The foundation for reducing tourism impacts on local communities and visitors is to develop partnerships for conservation to transform attitudes, daily behaviours and business practices (Bushell and Bricker 2017). The balancing is in principle related to CC of the PA, which is a matter of visitor flows, and not a question of merely establishing visitor numbers. The management of visitors focuses on manipulating number, spatial distribution and behaviour of visitors (Wall 2019). To prevent, address and reduce the impact of tourism and recreation, numerous planning and management frameworks have been developed, including, CC, ROS, LAC, Visitor impact management (VIM), visitor activity management process (VAMP) and Tourism Optimisation Management Model (TOMM). Contemporary outdoor recreation management frameworks are built on a procedural foundation of formulating indicators and standards, monitoring indicator variables, and applying management practices to ensure that standards are maintained (Manning 2011). Below, the study delivers a critical overview of tourism and visitor use management and planning frameworks and discuss their usability to address the challenges related to overtourism.

3.3 Visitor Use Planning Frameworks—Critical Overview

A variety of impact factors can influence the level and extent of ecological change that occurs in PNA due to the level of visitor use, therefore understanding these factors and how they are interrelated can help managers prevent undesirable impacts (D'Antonio and Monz 2016). To broaden their understanding of causes and consequences, PNA managers rely on visitor use planning frameworks (Barros et al. 2020). Initially vital, the CC concept is today just a constituent of visitor use frameworks. Tourism, as one of the most progressive human activities of the late twentieth century and early twenty-first century, has also led to the commitment to set boundaries (Doods and Butler 2019). Since, in economic terms, tourist attractions are non-reproducible resources, they are treated as common-pool resources, where the market mechanism does not show its usual allocative properties (UNEP/MAP/PAP 1997). Therefore, the relationship between the number of tourists and the reduced quality of the tourist experience is significant. Natural resources are fundamental for tourism development, and their conservation is essential (Marković Vukadin 2017). Long-term conservation of natural resources in vibrant PNAs requires setting up limitations and indicators related to the CC (Carić and Marković 2011; Butler 2019).

CC has attracted intensive focus as a research topic and as well as a management concept in parks and outdoor recreation (nature-based tourism) (Manning 2007). Despite that, its application has often had limited success. The principal difficulty lies in determining how much impact or changes the nature recourse, society, management and visitor experience should endure. Emphasise of most of the models is on different types of CC. For example, Cornejo-Ortega et al. (2011) refer to Physical Carrying Capacity (CCF), Real Carrying Capacity (CCR) and Effective Carrying Capacity (CCE). Physical Carrying Capacity (CCF), reflects the maximum limit of visits, that physically could be done in a day. For the calculation of Real Carrying Capacity (CCR), the CCF was modified by a series of corrections factors such as social involvement, erodibility, accessibility, precipitation of flooding (FCane), biological and vegetation. Finally, there is CCE, which represented the maximum number of visitors allowed at the sites of the area for public use, and relates the CCR with the management capacity (CM; defined as the best condition that the administration should have to practice the activities and meet the goals satisfactorily). In parallel to the more established biological or ecological CC based on CCR, the term 'social CC' emerged to describe the threshold above which the comfort and satisfaction of visiting and/or local people in a given space declines due to perceived crowding (Graefe et al. 1994; Llausàs et al. 2019).

KNOWLEDGE FROM THE PRACTICE

The Medes Islands case study suggests that, paradoxically, its attractiveness might derive from a combination of two elements. First, regardless of previous successes and failures reported in the literature, the CC construct is perceived as a scientifically sound approach, gaining the support of stakeholders ranging from biological scientists to natural park rangers and NGOs. Second, the very liquidity of the concept and uncertainty over its operationalisation make it vulnerable to being emptied of meaning and, ultimately, detached from scientific rigor, a discourse that serves the interests of the most powerful and well-connected stakeholders. Political ecology approach was adopted to the case study presented in this research paper to determine the factors behind the adoption of a CC-based strategy for managing scuba diving tourism pressures in a biodiversity-rich but degraded MPA and its concretion into a visitor cap set at 74,876 annual visitors. The research has revealed an acute dissonance between formal motivations and the stated goal of setting up access restrictions based on monitored environmental conditions (Llausàs et al. 2019).

Additional to establishing a final number of CC, six visitor-use planning frameworks emerged as the universal park management strategies, namely, ROS (Clark and Stankey 1979), LAC (Stankey et al. 1985), VIM (Graefe et al. 1990), VERP (NPS 1997), VAMP (Environment Canada and Park Service 1991), TOMM (UTOK 2000) and Priority Actions Program Regional Activity Center framework (UNEP/MAP/PAP 1997). A short overview of each can be found in the sequel.

3.3.1 Recreational Opportunity Spectrum (ROS)

ROS is a conceptual guideline that has been used widely by land management agencies (Brown et al. 1978; Wall 2020), and which was established to take into account the different settings used by recreationists, based on physical, social and managerial components. ROS starting assumption is that recreation experiences are influenced by the settings in which recreation activities occur. In this context, settings are defined by environmental, social and managerial conditions which can be used to create a diversity of recreation opportunities (Brown et al. 1978). Brown et al. (1978) also defined six setting classes to envelop the range of recreation setting opportunities. The different classes include (1) Primitive, (2) Semi-primitive non-motorised, (3) Semi-primitive motorised, (4) Rustic, (5) Concentrated and (6) Modern urbanised. The six ROS classes have different physical, social and managerial setting component characteristics. The five criteria for ROS classes (remoteness, size, evidence of humans, social setting, and managerial setting) are essential aspects of ROS factors. The ROS classes consist of five major factors of the ROS scale: (1) naturalness of the area, (2) access to the recreation site, (3) contact with other people, (4) amount of management and regulation and (5) amount and type of facilities. Brown and Ross (1982) indicate that desires for specific experiences were associated with preferences for settings as defined by the ROS and that homogenous groups of recreationists have a more consistent experience than all of the recreationists together. This suggests that specific recreation groups, such as climbers, may have more similar experience values than a group of hikers and kayakers, due to differences in desired experiences between different recreation activity groups (Tanakanjana 2007). Based on the ROS knowledge Ecotourism Opportunity Spectrum (ECOS) was developed, with three major delineation factors that have been identified, including remoteness, naturalness and experience (Boyd and Butler 1996).

Academia and practice have acknowledged that refining the ROS and combining it with LAC merges economic and carrying capacity theory with resource stewardship and provides a tool necessary for accurate management (Lichtkoppler and Clonts 2019. The ROS can help to mitigate conflict emerging through interactions of tourism and nature. The framework addresses the diversity of recreation settings and user characteristics by providing relevant management guidelines which consider both land capabilities and recreator needs. However, an incorrect and idiosyncratic ROS interpretation can produce some invalid guidelines. ROS management may also reduce goal incompatibility by establishing or reinforcing expectations about settings and the type and quality of recreation experiences available in particular settings (Daniels and Krannich 2019).

3.3.2 *Limit of Acceptable Change (LAC)*

LAC was developed by the US Forest Service and focused on the impact of human activities on the environment (Mexa and Coccossis 2004). The LAC approach seeks to define those conditions that are considered desirable in the tourism area and set out strategic management to achieve the objectives. This approach does not use quantitative results as boundaries but is based on the concept of the use of space where a set of desired sociological, environmental, physical and economic impacts is maintained. Standards are set, and indicators are used to identify unacceptable conditions, and actions are designed in response to situations that require modification. Indicators can refer to different components of tourism, e.g. traffic congestion, overcrowding of the beach area.

The LAC process consists of nine phases (Stankey et al. 1985) (Fig. 3.1) which can be cyclic.

At the end of the cycle, it is recommended to repeat the process after a particular time spent in monitoring the conditions, for verification, as well as possible further improvement of the conditions. It is important to stress out that as formulated initially (Stankey et al. 1985), the LAC process is driven by issues more than goals. Therefore, Cole and McCool (2020) propose to simply add a new first step to the LAC process—a step that involves defining goals and desired conditions, making it more adequate for operational management. Still, LAC gives its best performance if the issue or

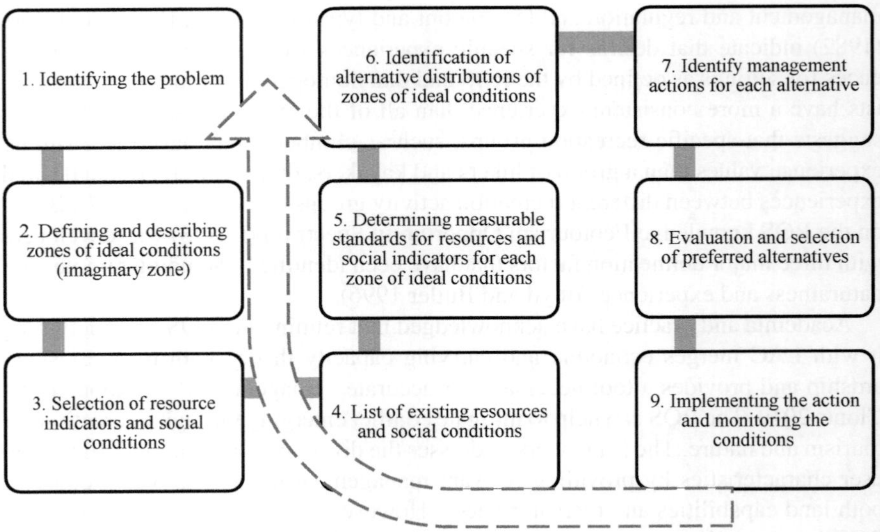

Fig. 3.1 Elements of LAC process (*Source* Based on Stankey et al. 1985)

conflicts between management goals is existing. Therefore it can be stated that its most significant flaw is that it is not suited for management of parks with minor conflicts and possibilities for management compromises (Eagles and McCool 2002). The framework can be quite useful in the management of tourism in PA but can also fail to be as successful in managing moderate changes which could occur in the first stages of overburdening of space or ecosystem. According to IUCN, this approach has been evaluated as useful in facing challenges of overtourism in PAs and is therefore recommended to be used in further management of conflicts. When applying LAC, management objectives are formulated as statements about the desired conditions of PAs and outdoor recreation, including the level of protection of resources and the type and quality of the recreation experience so that conservation is always a priority (Leung et al. 2018).

KNOWLEDGE FROM THE PRACTICE

Since 2006 Pangandaran (Java) has become a model of sustainable tourism development. A programme aimed to empower local communities and to prepare work plans and activities to enrich the various potential development of sustainable tourism. This destination has applied LAC in two cycles with aims to identify and measure the acceptable changes of tourism development in Pangandaran to remain adaptable as a tourist attraction. It is interesting to point out that it was learned that LAC framework is highly dependent on the government and the involvement of local communities in tourism development planning. Some important considerations establishment of indicators can be (Komsary et al 2018): (1) Zoning of tourism business conducted by the local community; (2) Standard rule for community involvement in tourism activities such as performing arts and culture.; (3) The age limit for workers involved in tourism; (4) Development of dedicated recreation area for local people who are not directly involved in tourism activities. Often the development of tourism infrastructure in an area of conflict with resident due to the comfort of residents decreased with increasing activities in the area, while the public can not enjoy such facilities; (5) Continuous and consistent guidance of local communities about the consequences of the presence of tourism in an area.

3.3.3 Visitor Impact Management—VIM

VIM is an evolved form of LAC and differs in that it is more space-oriented. This approach determines the unacceptable impact of visitors on the area, identifies the possible conditions of these impacts and defines several measures that will combat these intolerable impacts. As the problem-solving strategy, VIM provides a response to industry dynamics, from a long-term perspective (Graefe et al. 1990). It also does not search for quantitative values but identifies a set of standards that serve to compare with current conditions. The basis of this management method is an adaptation process that describes the desired conditions and assesses current activities as a basis for determining management objectives. It does not consider tourism as a separate economic activity, but integrates it into other related socio-economic activities, leading to the creation of comprehensive development plans (Mexa and Coccossis 2004). Despite more comprehensive than LAC framework, it is not that widely used and it is often just pointed out as a principle of action, but not as a viable framework due to its complexity (Dowling and Newsome 2017). Some attempts to simplify the VIM framework occurred (Farrell and Marion 2002) intending to make it more operational for PNAs managers.

3.3.4 The Visitor Experience and Resource Protection—VERP

VERP was developed by the US National Park Service to address the challenges related to the CC, impact of the visitors on the environment, as well as on the visitors' experience (Hof and Lime 1997; Fefer et al. 2018). This framework can be applied as part of the general management of the park (in spatial planning documents), or to address specific challenges within the PA system. Quality standards

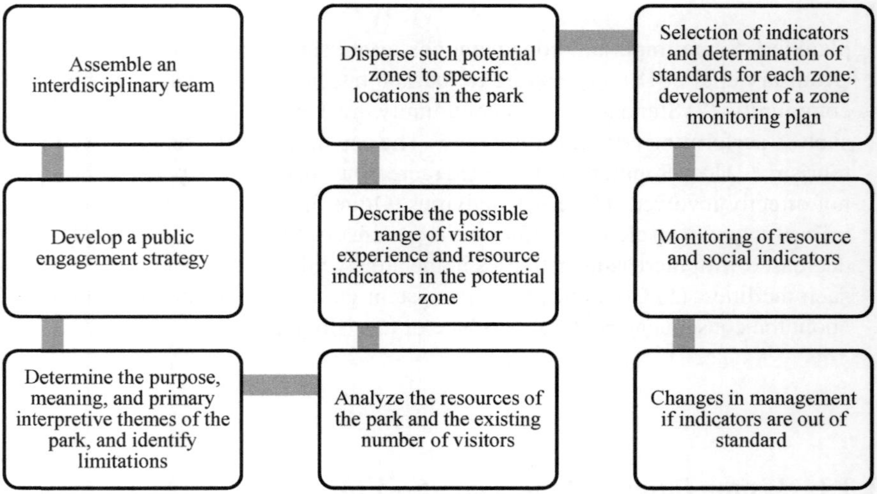

Fig. 3.2 The steps in the VERP process (*Source* Based on Maninng, 2007)

define the minimum acceptable indicator of the state of different variables. Once the final indicators and standards are formed, the indicators of the variables are monitored, and further management actions are employed to maintain the desired level of standards (Manning et al. 1995; Manning 2007). This method of management is most appropriate for areas where landscape views are essential, and crowding issues are emphasised. The constant monitoring from the first to the last phase of VERP, ensure that more realistic indicators and standards are determined (Fig. 3.2).

KNOWLEDGE FROM THE PRACTICE

One of the most important examples of VERP implementation is that of Arches National Park in Utah, involving the development of two-phase research programme to help implement VERP. The first phase focuses on identifying potential indicators of the quality of the visitor experience. At this stage, interviews were conducted with visitors, National Park employees and the local community to identify indicators. In the second phase, quality standards were set for these indicators. At this stage, interviews and questionnaires were also used, as well as visual methods to determine the range of conditions for indicators. Based on the obtained results, quality standards were formed on the basis of which changes were made in the management of the park to meet the needs of visitors, and the environment (Manning 2007). The complex application was also done in USA in Yosemite NP on area-by area and plan by plan basis (Bacon et al. 2006; Fefer et al. 2018). In its implementation it has achieved park planning activities, determined solid indicator variables and standards of quality, achieved public involvement and outreach and informed management action.

It is highly important to stress out that this concepts and frameworks have been widely incorporated in EU NP management plans (during last 10–15 years) and some very recent examples such as NP Krka (Croatia) can be stressed out (Carić et al. 2019). It should be stressed out that in this specific application of VERP a great number of indicators is monitored throughout the making of the plan, but this kind of complex planning could also represent a further obstacle because of the difficulties in further implementation cycles.

3.3.5 Tourism Optimisation Management Model—TOMM

TOMM was developed for assessment, monitoring and management with the aim of long-term destination protection. In this approach, the local community participates in developing scenarios and determining the desirable economic, environmental, infrastructural conditions, and marketing and management actions that should be employed (Miller and Twining-Ward 2005; Higgins-Desbiolles 2011). The process also identifies what needs to be monitored (and acceptable ranges of these indicators and performance) to determine whether these optimal conditions have been achieved (Smallwood et al. 2011). Unlike governance frameworks such as LAC and VIM, where the focus is on setting boundaries for impact management, TOMM emphasises the optimisation and sustainability of tourism and the community and sets acceptable ranges within which this should happen (Higgins-Desbiolles 2018). TOMM focuses on an integrated approach to tourism management and alleviates concerns about growth constraints in the tourism sector by:

- Avoiding the use of the terms 'prohibitions' and 'borders', which are interpreted negatively in the tourism industry and discourage further growth
- Focusing on the entire tourism system, and not only on the market and its ecological components;
- Envisioning the involvement of all stakeholders, through a partnership approach and the incorporation of the system through community processes
- Satisfying all stakeholders, operating at the regional level, ranging from protected areas to property owners in tourism (Twyford 2001).

An even more sophisticated model derived from TOMM is the Integrated monitoring and adaptive management system, highlighting (Guo and Chung 2017) (1) sustainability of all three tiers; (2) optimal conditions (no limits); and (3) simple reporting (through structuring an acceptable range of changes within the model). The model is widely used in Australia and to some extent in Canada. The primary constraint for implementation of TOMM is the complexity, as it requires a relatively long process of education, training and resources (Miller and Twining-Ward 2005).

3.3.6 Priority Actions Program Regional Activity Center Framework—PAP/RAC

PAP/RAC of the Mediterranean Action Plan of UNEP (United Nations Environment Program) has been implementing the priority action '*Development of Mediterranean tourism in accordance with the environment*' within 14 Mediterranean countries.[2] There are four basic tasks of PAP/RAC (Trumbić 2004):

1. Integral planning of Mediterranean basin development and management.
2. Programme for the observation and research of pollution in the Mediterranean basin.
3. Development of legislative regulations.
4. Institutional and financial organisation.

In the PAP/RAC methodology, the following parameters are taken as a framework for determining the boundary (PAP/RAC 1996):

1. Physical—ecological parameters (among others, functional physical capacity, ecological capacity, natural heritage capacity, coastline length, climatic elements, geological characteristics). Most of them are easily measurable, numerical values, with the parameters related to infrastructure, that can be easily changed by policies and investments, such as roads, water supply network, landfills, shops, banks.
2. Socio-demographic parameters (population, working-age population, the share of highly educated and several sociocultural parameters, including tourist experience, the identity of the local community, attitude towards tourists). Policy responses could modify sociocultural parameters, but significantly less than infrastructural parameters.
3. Political and economic parameters—refer to the investments and economic measures related to the development of tourism. They often represent a corrective parameter for the first two parameters, and sometimes play a crucial role.

These parameters show that the methodology is adaptable, i.e. it depends on political and economic preferences, which is not always right (Caric and Markovic 2011), as under the pressure of such influences, it is possible to expect that the non-economic elements of the CC will be pushed aside.

[2]More info at: http://paprac.org/focal-points.

3.4 Meeting Modern Management Needs

Since all of the frameworks were developed decades ago, when management tools and practice along with existing pressures and impact were somewhat different from today's needs, it is vital to evaluate how they meet modern management needs. However, it should be acknowledged that some frameworks did partially evolve not only to address the needs of PA and destinations but also the complexity of change in ecosystems. The visitor use frameworks have been much discussed in recent literature; however, it is our conclusion that there has been no significant change to existing methodologies. Moreover, the initial methodologies adopted by researchers and practitioners, have, in essence, remained the same as 30 or 40 years ago. This is, especially pronounced in Europe (Wall 2020), where the new age emerging of overtourism in PAs has made this topic highly relevant. Therefore, it is necessary to consider some of the elements that are comparable at the level of all analysed frameworks.

The CC has dominated nature-based tourism management for decades; however, as it does not meet the criteria to be considered a planning framework (McCool 2016), the decision was made to exclude it from further analysis. Furthermore, the definitions, as well as the methodologies of CC in tourism, differ depending on the author's perception of the limiting factor, or combination of factors, which is most important in determining the carrying capacity (Coccossis 2004). The methodology most often depends on the factors that receive the most attention. Likewise, not all methodologies apply to all areas; the carrying capacity and visitor use of the strict reserve (IUCN Ia category) and the significant landscape (IUCN IV category) will differ (Growcock and Pickering 2011; Steven et al. 2011; Doods and Butler 2019). For this reason, the following section gives an overview of the general concepts of visitor use frameworks and the assessment of the disadvantages and advantages of the concepts.

3.4.1 *Methodology of Evaluation*

The evaluation focused on the applicability and appropriateness of the implementation of selected frameworks proposed in previous research (Graefe et al. 1990; Cole and McCool 1997; Farrel and Marion 2002; Butler 2019) to address the challenges related to overtourism in PAs. The analysis included three stages. The first stage was the selection of adequate criteria which could be used to assess the frameworks. The criteria were drawn from participatory planning models (Bello et al. 2016; Tebet et al. 2018), IUCN-WCPA framework for assessing management effectiveness of protected areas and protected areas systems (Hockings et al. 2006) (tab 3) and WWF's Management Effectiveness Tracking Tools (METT) (WWF 2007). In total, nineteen criteria were drawn related to six topics, *context*, *planning*, *inputs*, *process*, *outputs* and *outcomes* (Table 3.1). The participants of the expert group were

Table 3.1 IUCN-WCPA framework for assessing management effectiveness of protected areas and protected area systems

	Design	Appropriateness/adequacy			Delivery	
Elements fo management cycle	Context	Planning	Inputs	Process	Outputs	Outcomes
Focus of evaluation	Assessment of importance, threats and policy environment	Assessment of PA design and planning	Assessment of resources needed to carry out management	Assessment of how management is conducted	Assessment of the implementation of management programmes and actions; delivery of products and services	Assessment of the outcomes and the extent to which they achieved objectives
Criteria that are assessed	Significance/values Threats Vulnerability Stakeholders National context	PA legislation and policy PA system design PA design Management planning	Resources available to the agency Resources available to the PA	Suitability of management process and the extent to which established or accepted processes are being implemented	Results of management actions Services and products	Impacts: effects of management concerning objectives

Source Adapted from Hockings et al. (2006)

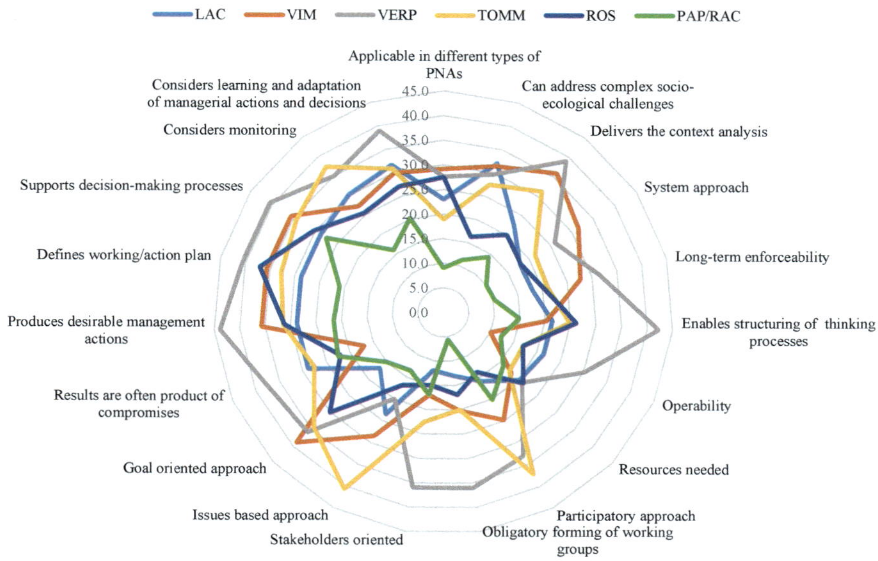

Fig. 3.3 Overall assessment of visitor frameworks based on established criteria (*Source* Authors)

invited to rate each of the proposed frameworks on a 7-point Likert scale considering the (1) focus of the framework on the proposed criteria, and the (2) importance of the criteria in sustainable nature-based tourism planning. These rates were used to calculate scores and critically discuss the frameworks.

The four experts included in the analysis are nature-based tourism researchers and practitioners with extensive experience in the implementation of analysed frameworks. The final scores of the evaluation are based on all four evaluations and their average values (Fig. 3.3 and Table 3.2).

3.4.2 Evaluation and Critical Overview of Frameworks

Management and planning of visitor use within PAs is a complex process. Thus, keeping the process of implementation and adjustment of framework simple as possible should be a priority. Frameworks should enable monitoring of the change of ecosystem throughout time, while applied in different types of PAs, both, depending on the protected features (for example, geoparks, marine areas, mountain parks), as well as the level of protection (for example, national parks, nature parks, reserves). The analysis suggests that with specific modifications, the existing frameworks could be applied in different types of PAs. The PAP/RAC approach is the only one exclusively developed for coastal areas (Tab. 2). The high complexity of the process, as well as resources required (financial, human and time), are the primary constraints

Table 3.2 The assessment of the visitor use frameworks

Category	Criteria	LAC	VIM	VERP	TOMM	ROS	PAP/RAC
		Score	Score	Score	Score	Score	Score
Context	Applicable in different types of PNAs	23,0	29,3	27,6	19,0	27,5	9,0
	Can address complex socioecological challenges	32,1	31,5	29,7	27,5	16,3	11,3
	Delivers the context analysis	22,3	35,9	39,1	31,3	20,0	14,3
	System approach	17,8	31,5	26,0	21,4	18,0	10,0
	Context total score	23,6	32,1	30,7	24,6	20,3	11,1
Planning	Long-term enforceability	18,0	27,6	31,5	21,0	20,2	10,3
	Enables structuring of thinking processes	21,4	20,3	42,3	24,8	26,1	14,9
	Operability	21,4	10,0	30,2	17,0	17,0	12,3
	Planning total score	20,3	18,8	34,5	21,0	21,0	12,4
Inputs	Resources needed	20,3	18,0	21,0	17,5	21,3	15,9
	Inputs total score	20,3	18,0	21,0	17,5	21,3	15,9
Process	Participatory approach	15,9	24,8	33,0	37,1	13,8	20,2
	Obligatory forming of working groups	13,1	20,3	36,0	20,0	17,0	5,6
	Stakeholders oriented	12,0	17,0	35,9	22,5	15,0	17,0
	Issues-based approach	23,6	28,5	20,0	40,6	16,9	13,5
	Goal-oriented approach	16,9	39,0	36,0	34,4	30,2	15,0
	Results are often product of compromises	28,9	17,0	37,5	27,6	22,0	22,5
	Process total score	18,0	24,0	32,8	30,0	19,0	15,2
Outputs	Produces desirable management actions	28,8	35,8	43,9	31,5	31,3	21,6
	Defines working/action plan	28,8	35,8	40,6	32,8	37,1	21,0
	Supports decision-making processes	28,8	35,8	40,6	34,4	30,3	27,5
	Outputs total score	28,8	35,8	41,7	32,9	32,9	23,5
Outcomes	Considers monitoring	30,3	27,3	35,0	37,5	25,5	15,9
	Considers learning and adaptation of managerial actions and decisions	31,6	30,0	39,0	30,9	27,0	20,3
	Outcomes total score	31,0	28,7	37,1	34,3	26,3	18,1

Maximum score = 49

●—High score; ●—Low score

Source Authors

of the implementation of existing frameworks. These limitations may also lead to the rejection of the implementation of policy measures, which are dependent on the implementation of the concrete framework.

Factors related to path dependency, neoliberal governance frameworks, uneven distribution of power among stakeholders and regulatory weaknesses were found to be the most influential in facilitating increased visitor pressure (Guo and Chung 2017; Llausàs 2019). Although frameworks discussed were developed to manage PAs in an environmentally sustainable way (Butler 2019), they commonly fail to equally address the interests of all stakeholders (Guo and Chung 2017), which limits their potential to address specific phenomenon as overtourism. Setting up a tourist CC, i.e. limiting the number of visitors in sensitive locations where a range of interests converge could cause social tensions and conflict (Few 2000; McCarthy 2002; Robbins 2012). Although all frameworks instrumentally support the CC approach, this analysis suggests that traditional scientific ecological knowledge plays only a specious role in decision-making.

Most of the frameworks discussed rely on management-by-goals approach and are iterative, creating the foundation for adaptivity (Cole and McCool 1997). LAC, PAP/RAC and VIM are initially problem-driven and include identifying management priorities in the first phases. The processes are triggered by issues, such as crowding and trampling. The problem-driven frameworks are less able to achieve the resilience of the PA system, as they wait for the problem to happen. The VERP, on the other hand, integrates proactivity, as decisions regarding the visitor use, are made based on desired future condition on PA (NPS 1997; Fefer et al. 2018). TOMM offers a holistic approach in the pursuit of sustainable tourism and could be valuable to mitigate the pressures related to extensive visitation in PAs. However, the complexity of implementation, i.e. operability, is a significant challenge, often impossible to overcome within PA managing agencies lacking financial and human resources.

The analysis suggests that VERP, TOMM and ROS show higher usability to mitigate the pressures related to excessive visitation (Fig. 3.1). The resources needed are a primary constraint for implementation of all framework discussed. PAP/RAC methodology has shown to be more simple to implement; however, the complexity of socioecological challenges it can tackle, the inclusion of stakeholders in the process and long-term enforceability limits its applicability.

LAC (highest score), VIM, VERP and TOMM, have demonstrated the ability to address the complex socioecological challenges as overtourism in PNAs. Their implementation could contribute to reconciling conservation and recreation goals; however, specific improvements, particularly those related to process simplification, are needed. VIM and VERP enable system approach, which is strongly advocated by researchers, practitioners, as well as Europarc Federation and US Forest service. When it comes to planning, a solid framework should enable structuring of the thinking process as a prerequisite for effective management and stakeholder involvement, and ultimately in mitigating crisis and conflicts. VERP, and partially TOMM and ROS fulfil these criteria. All frameworks, except PAP/RAC, have a solid score considering outputs and outcomes, which suggested their appropriateness to address specific pressure related to visitor concentrations.

Out of six categories, VERP has the best rating in four (context, planning, inputs and outcomes), while PAP/RAC methodology has the lowest rating in three categories (planning, inputs and outcomes). All framework record the lowest scores in criteria, which are the part of the process and inputs elements of the management cycle, which confirms previous research emphasising resources needed (both, financial and human) (Miller and Twining-Ward 2005), and the insufficient involvement of stakeholders in the decision-making process (Llausàs et al. 2019) as the main drawbacks of existing frameworks.

3.5 Conclusion

The idea of using CC as a management framework builds on the assumption that natural ecosystems are stable over time and their condition remains unaffected by either change in environmental conditions or by human action unless their intrinsic CC is exceeded, in which case they would rapidly degrade (Llausàs et al. 2019). Abundant scientific literature has challenged the validity of these assumptions, demonstrating the complexity of population dynamics in natural ecosystems and suggesting the need for a new paradigm or a 'new ecology' that abandons the alleged notion of a natural balance and its associated CC and tipping points (Zimmerer 2000; Worboys et al. 2015). The existing visitor use frameworks have tried to answer the 'call' and did it successfully for several decades. However, recent changes require new solutions. It is our humble opinion that established frameworks should be advanced by consideration of new theoretical and practical advances employing the system approach in which PAs are seen in the interrelation with other ecosystems. PA managers require tools and resources, which are necessary to prevail the pressures before it even happens. The analysis presented in this chapter suggests there is no one for all solution in terms of visitor use framework, but some constituents of established frameworks could be useful to mitigate challenges related to overtourism. Tourism has grown and evolved significantly in past decades, and some of the existing and known phenomena, like overcrowding, become more pronounced. Along with that, PAs are facing biological, social and economic fragmentation. All of this supports the conclusion that to be effective; the existing approaches require change, related mainly to the management process, where greater involvement of stakeholders, a goal-orientation arising from management outcomes, and monitoring of change through time is necessary. Finally, the reactive monitoring of policy measures employed is required as a foundation for adaptive management, and as a mean of creating new values for the visitors.

References

Agyeiwaah E, McKercher B, Suntikul W (2017) Identifying core indicators of sustainable tourism: a path forward? Tour Manag Perspect 24:26–33. https://doi.org/10.1016/j.tmp.2017.07.005

Akyeampong OA (2011) Pro-poor tourism: residents' expectations, experiences and perceptions in the Kakum National Park area of Ghana. J Sustain Tour 19(2):197–213. https://doi.org/10.1080/09669582.2010.509508

Andriotis K (2018) Degrowth in tourism: conceptual, theoretical and philosophical issues. CABI

Bacon J, Roche J, Elliot C, Nicholas N (2006) VERP: putting principles into practice in Yosemite National Park. Visitor Impact Monitoring 23(2)

Barros A, Aschero V, Mazzolari A, Cavieres L, Pickering C (2020) Going off trails: how dispersed visitor use affects alpine vegetation. J Environ Manage 267:110546. https://doi.org/10.1016/j.jenvman.2020.110546

Bello FG, Carr N, Lovelock B (2016) Community participation framework for protected area-based tourism planning. Tour Plan Dev 13(4):469–485. https://doi.org/10.1080/21568316.2015.1136838

Bello FG, Lovelock B, Carr N (2017) Constraints of community participation in protected area-based tourism planning: the case of Malawi. J Ecotourism 16(2):131–151. https://doi.org/10.1080/14724049.2016.1251444

Blancas FJ, Caballero R, Gonzalez M, Lozano-Oyola M, Perez F (2010) Goal programming synthetic indicators: an application for sustainable tourism in Andalusian coastal counties. Ecol Econ 69(11):2158–2172

Boyd WS, Butler RW (1996) Managing ecotourism: an opportunity spectrum approach. Tour Manag 17(8):557–566

Brown P, Driver B, McConnell C (1978) The opportunity spectrum concept in outdoor recreation supply inventories: background and application. Proceedings of the Integrated Renewable Resource Inventories Workshop. USDA Forest Service General Technical Report RM-55, pp 73–84

Brown P, Ross D (1982) Using desired recreation experiences to predict setting preferences. In: Forest and river recreation: research update. Agricultural Experiment Station Miscellaneous Report 18, University of Minnesota, St. Paul, MN, pp 105–110

Bushell R, Bricker K (2017) Tourism in protected areas: developing meaningful standards. Tour Hosp Res 17(1):106–120. https://doi.org/10.1177/1467358416636173

Butler R (2019) Tourism carrying capacity research: a perspective article. Tour Rev 75(1):207–211. https://doi.org/10.1108/tr-05-2019-0194

Buultjens J, Ratnayake I, Gnanapala A, Aslam M (2005) Tourism and its implications for management in Ruhuna National Park (Yala), Sri Lanka. Tour Manag 26(5):733–742. https://doi.org/10.1016/j.tourman.2004.03.014

Capocchi A, Vallone C, Amaduzzi A, Pierotti M (2019) Is 'overtourism' a new issue in tourism development or just a new term for an already known phenomenon? Curr Issues Tour 0(0):1–5. https://doi.org/10.1080/13683500.2019.1638353

Carić H, Jakelić M, Jerković I, Klanjšček T, Tomljenović R, Vidović H (2019) Studija upravljanja posjetiteljima NP "Krka". Javna ustanova "Nacionalni park Krka", Hrvatska

Carić H, Marković I (2011) Integralno planiranje u turizmu kao temelj održivog razvoja, u: Izazovi upravljanja turizmom. Institut za turizam, Zagreb

Clark RN, Stankey GH (1979) The recreation opportunity spectrum: a framework for planning, management, and research. USDA Forest Service, General Technical Report (PNW-98)

Coccossis H (2004) Sustainable tourism and carrying capacity: a new context. In: Coccossis H, Mexa A (eds) The challenge of tourism carrying capacity assessment: theory and practice. Ashgate, Aldershot, pp. 3–15. https://doi.org/10.4324/9781315240817

Cole D, Carlson T (2010) Numerical visitor capacity: a guide to its use in wilderness. USDA Forest Service - General Technical Report RMRS-GTR, 247, 1–21. https://doi.org/10.2737/rmrs-gtr-247

Cole, D. & McCool, S. (1997). Limits of acceptable change and related planning processes: a workshop. United States Department of Agriculture, Forest Service, pp. 1–2. Available at: http://www.leopold.wilderness.net/pubs/313.pdf.

Cornejo-Ortega J, Chávez Dagostino R, Cupul A (2011) Estimating carrying capacity in a natural protected area as a conservation strategy. https://doi.org/10.13140/2.1.5037.0886

Daniels S, Krannich R (2019) The recreational opportunity spectrum as a conflict management tool. In: Vining J (ed) Social science and natural resource recreation management. Routledge, New York. https://doi.org/10.4324/9780429306365-11

D'Antonio A, Monz C (2016) The influence of visitor use levels on visitor spatial behavior in off-trail areas of dispersed recreation use. J Environ Manage 170:79–87. https://doi.org/10.1016/j.jenvman.2016.01.011

Doods R, Butler RW (2019). The enablers of overtourism. In: Doods R, Butler RW (eds) Overtourism: issues, realities and solutions. De Gruyter, Boston, pp 6–27

Dowling RK, Newsome D (2017) Geotourism destinations—visitor impacts and site management considerations. Czech J Tour 6(2):111–129. https://doi.org/10.1515/cjot-2017-0006

Dudley N (ed) (2008) Guidelines for applying protected area management categories. Gland, Switzerland: IUCN

Eagles PFJ (2009) Governance of recreation and tourism partnerships in parks and protected areas. J Sustain Tour 17(2):231–248. https://doi.org/10.1080/09669580802495725

Eagles PFJ (2002) Trends in park tourism: economics, finance and management. J Sustain Tour 10(2):132–153. https://doi.org/10.1080/09669580208667158

Eagles PFJ (2014) Research priorities in park tourism. J Sustain Tour 22(4):528–549. https://doi.org/10.1080/09669582.2013.785554

Eagles PFJ, McCool SF (2002) Tourism in national parks and protected areas: planning and management. CABI Publishing, New York

Environment Canada and Park Service (1991) Selected readings on the visitor activity management process. Environment Canada, Ottawa, ON

European Commission (2016) Preference of Europeans towards tourism (Issue February). http://ec.europa.eu/COMMFrontOffice/PublicOpinion/index.cfm/Survey/getSurveyDetail/yearFrom/1973/yearTo/2016/search/tourism/surveyKy/2065

Farrell T, Marion J (2002) The protected area visitor impact management (PAVIM) framework: a simplified process for making management decisions. J Sustain Tour 10(1):31–51. https://doi.org/10.1080/09669580208667151

FEDERATION E (2012) Guide to sustainable tourism in protected areas. 3–148

Fefer J, De Urioste-Stone S, Daigle J, Silka L (2018) "Understanding the perceived effectiveness of applying the visitor experience and resource protection (VERP) framework for recreation planning: a multi-case study in US National Parks". Qualitative Report 23(7):1561–1582. Retrieved from https://nsuworks.nova.edu/tqr/vol23/iss7/6

Few R (2000) Conservation, participation, and power: protected-area planning in the coastal zone of belize. J Plan Educ Res 19(4), 401–408. https://doi.org/10.1177/0739456X0001900409.

Fletcher R, Murray Mas I, Blanco-Romero A, Blázquez-Salom M (2019) Tourism and degrowth: an emerging agenda for research and praxis. J Sustain Tour 27(12):1745–1763. https://doi.org/10.1080/09669582.2019.1679822

Goodwin H (2019) Overtourism: causes, symptoms and treatment. April, pp 110–114

Gössling S, Bredberg M, Randow A, Sandström E, Svensson P (2006) Tourist perceptions of climate change: a study of international tourists in Zanzibar. Curr Issues Tour 9:419–435. https://doi.org/10.2167/cit265.0

Graefe AR, Confer Jr JC, Drogin E, Titre J (1994) Re-examining the crowding model: a comparative analysis. Paper presented the 5th International Symposium on Society and Resource Management. Colorado State University, Fort Collins, CO

Graefe AR, Kuss FR, Vaske JJ (1990) Visitor impact management: the planning framework. National Parks and Conservation Association, Washington, DC

Growcock A, Pickering CM (2011) Impacts of small group short term experimental camping on alpine and subalpine vegetation in the Australian Alps. J Ecotourism 120:86–100. https://doi.org/10.1016/j.jenvman.2009.09.022
Guo W, Chung S (2017) Remaking tourism carrying capacity frameworks for geoparks. DEStech Trans Soc Sci, Educ Hum Sci, 197–205. https://doi.org/10.12783/dtssehs/asshm2016/8357
Hall CM (2009) Degrowing tourism: décroissance, sustainable consumption and steady-state tourism. Anatolia 20(1):46–61. https://doi.org/10.1080/13032917.2009.10518894
Hannam K (2005) Tourism management issues in India's national parks: An analysis of the Rajiv Gandhi (Nagarahole) National Park. Curr Issues Tour 8(2–3):165–180. https://doi.org/10.1080/13683500508668212
Higgins-Desbiolles F (2011) Death by a thousand cuts: governance and environmental trade-offs in ecotourism development at Kangaroo Island, South Australia. J Sustain Tour 19(4–5):553–570. https://doi.org/10.1080/09669582.2011.560942
Higgins-Desbiolles F (2018) Event tourism and event imposition: a critical case study from Kangaroo Island, South Australia. Tour Manag 64:73–86. https://doi.org/10.1016/j.tourman.2017.08.002
Higgins-Desbiolles F, Carnicelli S, Krolikowski C, Wijesinghe G, Boluk K (2019) Degrowing tourism: rethinking tourism. J Sustain Tour 27(12):1926–1944. https://doi.org/10.1080/09669582.2019.1601732
Hockings M (2003) Systems for assessing the effectiveness of management in protected areas. BioScience 53(9):823–832. https://doi.org/10.1641/0006-3568(2003)053[0823:SFATEO]2.0.CO;2
Hockings M, Stolton S, Leverington F, Dudley N, Courrau J (2006) Evaluating effectiveness: a framework for assessing management effectiveness of protected areas, 2nd edn. IUCN, Gland, Switzerland and Cambridge, UK
Hof M, Lime D (1997) Visitor experience and resource protection framework in the national park system: rationale, current status, and future direction. In: McCool S, Cole D (eds) Proceedings—limits of acceptable change and related planning processes: progress and future directions, comps. General Technical Report INT-GTR-371. US Department of Agriculture–Forest Service, Rocky Mountain Research Station, Ogden, Utah
Islam MW, Ruhanen L, Ritchie BW (2018) Adaptive co-management: a novel approach to tourism destination governance? J Hosp Tour Manag 37:97–106. https://doi.org/10.1016/j.jhtm.2017.10.009
Jordan P, Pastras P, Psarros M (2018) Managing tourism growth in Europe the ECM toolbox. 30. https://fr.calameo.com/read/0006740147d7bd41b5afc
Koderman M, Opačić VT, Marković Vukadin I (2020) Tourism development and green horizons in protected areas of Croatia and Slovenia: synthesis and perspectives
Komsary K, Tarigan W, Wiyana T (2018) Limits of acceptable change as tool for tourism development sustainability in Pangandaran West Java. IOP Conference Series: Earth and Environmental Science 126(012129). https://doi.org/10.1088/1755-1315/126/1/012129
Krajinović V (2015) Challenging the key issues in measuring sustainable tourism development. Acta Turistica 27(1):63–91
Lacitignola D, Petrosillo I, Cataldi M, Zurlini G (2007) Modelling socio-ecological tourism-based systems for sustainability. Ecol Model 206(1):191–204
Leung Y, Spenceley A, Hvenegaard G, Buckley R, Editors V, Groves C, Editor S (2018) Tourism and visitor management in protected areas: guidelines for sustainability. In: Tourism and visitor management in protected areas: guidelines for sustainability (Issue 27). https://doi.org/10.2305/iucn.ch.2018.pag.27.en
Lichtkoppler R, Clonts H (2019) Recreation opportunity spectrum reevaluated: its application to the Eastern US. In: Vining J (ed) Social science and natural resource recreation management. Routledge, New York. https://doi.org/10.4324/9780429306365-8

Llausàs PA, Subirós J, Fraguell Sansbelló RM, Pueyo-Ros J (2019) Carrying capacity as a tourism management strategy in a marine protected area: a political ecology analysis. Conserv Soc 17(4):266–376. https://doi.org/10.4103/cs.cs_18_154

Luque-Gil AM, Gómez-Moreno ML, Peláez-Fernández MA (2018) Starting to enjoy nature in Mediterranean mountains: crowding perception and satisfaction. Tour Manag Perspect 25(November 2017):93–103. https://doi.org/10.1016/j.tmp.2017.11.006

Ly TP, Nguyen THH (2017) Application of carrying capacity management in Vietnamese national parks. Asia Pac J Tour Res 22(10):1005–1020. https://doi.org/10.1080/10941665.2017.1359194

Mandić A (2019) Nature-based solutions for sustainable tourism development in protected natural areas: a review. Environ Syst Decis. https://doi.org/10.1007/s10669-019-09718-2

Mandić A (2020) Structuring challenges of sustainable tourism development in protected natural areas with driving force—pressure—state—impact—response (DPSIR) framework. Environ Syst Decis. https://doi.org/10.1007/s10669-020-09759-y

Manning RE (2007) Parks and carrying capacity: commons without tragedy. Island Press, Washington, Covelo, London

Manning RE (2011) Indicators and standards in parks and outdoor recreation. In Phillips R, Budruk M (eds) Quality-of-life community indicators for parks, recreation and tourism management. Springer, The Netherlands, pp 11–22

Manning RE, Lawson SR (2002) Carrying capacity as "informed judgment": the values of science and the science of values. Environ Manage 30(2):157–168. https://doi.org/10.1007/s00267-002-2772-x

Manning R, Lime D, Hof M, Freimund W (1995) The visitor experience and resource protection (VERP) process: the application of carrying capacity to Arches National Park. The George Wright Forum 12(3):41–55. Retrieved February 17, 2021, from http://www.jstor.org/stable/43597433

Marković Vukadin I (2017) Sustainability issues in management of tourism in protected areas: case study of Plitvice Lakes National Park. In: Dwyer L, Tomljenović R, Čorak S (eds) Evolution of destination planning and strategy. Palgrave Macmillan. https://doi.org/10.1007/978-3-319-42246-6_10

McCarthy L (2002) The brownfield dual land-use policy challenge: reducing barriers to private redevelopment while connecting reuse to broader community goals. Land Use Policy 19(4):287–296. https://doi.org/10.1016/s0264-8377(02)00023-6

McCool SF (2016) Tourism in protected areas: frameworks for working through the challenges in an era of change, complexity and uncertainty. In: McCool S, Bosak (eds) Reframing sustainable tourism. Environmental challenges and solutions. Springer, Dordrecht, vol 2. https://doi.org/10.1007/978-94-017-7209-9_7

McCool SF, Bosak K (eds) (2016) Reframing sustainable tourism. Springer, Netherlands. https://doi.org/10.1007/978-94-017-7209-9

Mexa A, Coccossis H (2004) Tourism carrying capacity: a theoretical overview. In: Cocosis H, Mexa A (eds) The challenge of tourism carrying capacity assessment: theory and practice. Ashgate, Aldershot, pp 37–55. https://doi.org/10.4324/9781315240817

Milano C, Novelli M, Cheer JM (2019a) Overtourism and degrowth: a social movements perspective. J Sustain Tour 27(12):1857–1875. https://doi.org/10.1080/09669582.2019.1650054

Milano C, Novelli M, Cheer JM (2019b) Overtourism and tourismphobia: a journey through four decades of tourism development, planning and local concerns. Tour Plan Dev 16(4):353–357. https://doi.org/10.1080/21568316.2019.1599604

Miller G, Twining-Ward L (2005) Monitoring for a sustainable tourism transition: the challenge of developing and using indicators. CAB International, Oxon. https://doi.org/10.1079/9780851990514.0000

Moyle B, Croy G (2007) Crowding and visitor satisfaction during the off-season: Port Campbell National Park. Ann Leis Res 10(3–4):518–531. https://doi.org/10.1080/11745398.2007.9686779

National Park service (1997) "VERP the visitor experience and resource protection (VERP) framework a handbook for planners and managers". National Park Service, U.S. Department of Interior. Denver Service Center

Olive ND, Marion JL (2009) The influence of use-related, environmental, and managerial factors on soil loss from recreational trails. J Environ Manage 90(3):1483–1493. https://doi.org/10.1016/j.jenvman.2008.10.004
Peeters P, Gössling S, Klijs J, Milano C, Novelli M, Dijkmans C, Eijgelaar E, Hartman S, Heslinga J, Isaac R, Mitas O, Moretti S, Nawijn J, Papp B, Postma A (2018) Research for TRAN Committee—overtourism: impact and possible policy responses. October. http://www.europarl.europa.eu/RegData/etudes/STUD/2018/629184/IPOL_STU(2018)629184_EN.pdf
Peeters P, Klijs J, Novelli M (2018) Requested by the TRAN committee Research for TRAN Committee—overtourism: impact and possible policy responses. February 2019
Peng J, Chen X, Wang J (2016) Applying relative deprivation theory to study the attitudes of host community residents towards tourism: the case study of the Zhangjiang National Park, China. Curr Issues Tour 19(7):734–754. https://doi.org/10.1080/13683500.2013.877876
Plummer R, Fennell DA (2009) Managing protected areas for sustainable tourism: prospects for adaptive co-management. J Sustain Tour 17(2):149–168. https://doi.org/10.1080/09669580802359301
PAP/RAC (1996) Approaches for zoning of coastal areas with reference to Mediterranean Aquaculture, Split, pp 35
Robbins P (2012) Political ecology: a critical introduction. Blackwell, Malden
Sandwith T, MacKinnon K, Enkerlin Hoeflich E (2016) Protected area governance and management, Worboys GL, Lockwood M, Kothari A, Feary S, Pulsford I (eds). ANU Press. https://doi.org/10.26530/oapen_569111
Santana-Jiménez Y, Hernández JM (2011) Estimating the effect of overcrowding on tourist attraction: the case of Canary Islands. Tour Manag 32(2):415–425. https://doi.org/10.1016/j.tourman.2010.03.013
Schägner JP, Brander L, Maes J, Paracchini ML, Hartje V (2016) Mapping recreational visits and values of European National Parks by combining statistical modelling and unit value transfer. J Nat Conserv 31:71–84. https://doi.org/10.1016/j.jnc.2016.03.001
Schreyer R, Roggenbuck JW (1978) The influence of experience expectations on crowding perceptions and social-psychological carrying capacities. Leis Sci 1(4):373–394. https://doi.org/10.1080/01490407809512896
Seraphin H, Ivanov S, Dosquet F, Bourliataux-Lajoinie S (2019) Archetypes of locals in destinations victim of overtourism. J Hosp Tour Manag (December):0–1. https://doi.org/10.1016/j.jhtm.2019.12.001
Seraphin H, Sheeran P, Pilato M (2018) Over-tourism and the fall of Venice as a destination. J Destin Mark Manag 9(September 2017):374–376. https://doi.org/10.1016/j.jdmm.2018.01.011
Serenari C, Peterson MN, Wallace T, Stowhas P (2017) Private protected areas, ecotourism development and impacts on local people's well-being: a review from case studies in Southern Chile. J Sustain Tour 25(12):1792–1810. https://doi.org/10.1080/09669582.2016.1178755
Smallwood CB, Beckley LE, Moore SA, Kobryn HT (2011) Assessing patterns of recreational use in large marine parks: a case study from Ningaloo Marine Park, Australia. Ocean & Coast Manag 54(4):330–340. https://doi.org/10.1016/j.ocecoaman.2010.11.007
Spenceley A, Goodwin H (2007) Nature-based tourism and poverty alleviation: Impacts of private sector and parastatal enterprises in and around Kruger National Park, South Africa. Curr Issues Tour 10(2–3):255–277. https://doi.org/10.2167/cit305.0
Stanford D, Guiver J (2016) Driving pro-environmental change in tourist destinations: encouraging sustainable travel in National Parks via partnership project creation and implementation. J Sustain Tour 24(3):484–505. https://doi.org/10.1080/09669582.2015.1122018
Stankey GH, Cole DN, Lucas RC, Petersen ME, Frissell SS (1985) The limits of acceptable change (LAC) system for wilderness planning. General Technical Report, Intermountain Forest and Range Experiment Station, USDA Forest Service (INT-176)
Steg L, Vlek C (2009) Encouraging pro-environmental behaviour: an integrative review and research agenda. J Environ Psychol 29(3):309–317. https://doi.org/10.1016/j.jenvp.2008.10.004

Steven R, Pickering C, Castley G (2011) A review of impacts of nature based recreation on birds. J Environ Manag 92:2287–2294. https://doi.org/10.1016/j.jenvman.2011.05.005
Stone MT, Nyaupane GP (2018) Protected areas, wildlife-based community tourism and community livelihoods dynamics: spiraling up and down of community capitals. J Sustain Tour 26(2):307–324. https://doi.org/10.1080/09669582.2017.1349774
Tanakanjana N (2007) Recreation opportunity classification and challenges in maintaining recreation diversity in Thailand's National Parks. In: Proceedings of the 2007 George Wright Society Conference, pp 151–156
Tebet G, Trimble M, Medeiros RP (2018) Using Ostrom's principles to assess institutional dynamics of conservation: Lessons from a marine protected area in Brazil. Mar Policy 88:174–181
Torres-Delgado A, Saarinen J (2014) Using indicators to assess sustainable tourism development: a review. Tour Geogr. https://doi.org/10.1080/14616688.2013.867530
Trumbić I (2004) Tourism carrying capacity in mediterranean. In: Coccossis H, Mexa A (eds) The challenge of tourism carrying capacity assessment: theory and practice. Ashgate, Aldershot, pp 181–201. https://doi.org/10.4324/9781315240817
Twyford K (2001) Protected area management: principles and practices. In: Worboys G, Lockwood M, De Lacy T (eds) Protected area management: principles and practices. CD Compendium. Oxford University Press, South Melbourne.
UNEP/MAP/PAP (1997) Guidelines for carrying capacity assessment for tourism in mediterranean coastal areas. Priority Action Programme, Regional Activity Centre, Split
UNWTO (2018) 'Overtourism'?—Understanding and managing urban tourism growth beyond perceptions, executive summary. World Tourism Organization (UNWTO); Centre of Expertise Leisure, Tourism & Hospitality; NHTV Breda University of Applied Sciences; and NHL Stenden University of Applied Sciences. https://doi.org/10.18111/9789284420070
UTOK (2000) Annual Report Tourism Optimisation Management Model TOMM Kangaroo Island South Australia 2000. Retrieved February 27, 2020, from http://www.utok.cz/
Wall G (2019) Perspectives on the environment and overtourism. In: Dodds WR, Butler R (ed) Overtourism, issues, realities and solutions. De Gruyter Studies in Tourism, pp 3–27
Wall G (2020) From carrying capacity to overtourism: a perspective article. Tour Rev 75(1):212–215. https://doi.org/10.1108/TR-08-2019-0356
Weaver DB, Lawton LJ (2017) A new visitation paradigm for protected areas. Tour Manag 60:140–146. https://doi.org/10.1016/j.tourman.2016.11.018
Whitelaw PA, King BEM, Tolkach D (2014) Protected areas, conservation and tourism—financing the sustainable dream. J Sustain Tour 22(4):584–603. https://doi.org/10.1080/09669582.2013.873445
Wilson E, Nielsen N, Buultjens J (2009) From lessees to partners: exploring tourism public-private partnerships within the New South Wales national parks and wildlife service. J Sustain Tour 17(2):269–285. https://doi.org/10.1080/09669580802495774
Worboys G, Lockwood M, Kothari A, Feary S, Pulsford I (2015) Protected area governance and management. ANU Press, https://doi.org/10.26530/oapen_569111
WWF (2007) Management effectiveness tracking tool. WWF, Gland
Zapata MJ, Hall CM, Lindo P, Vanderschaeghe M (2011) Can community-based tourism contribute to development and poverty alleviation? Lessons from nicaragua. Curr Issues Tour 14(8):725–749. https://doi.org/10.1080/13683500.2011.559200
Zeppel H (2010) Managing cultural values in sustainable tourism: conflicts in protected areas. Tour Hosp Res 10(2):93–115. https://doi.org/10.1057/thr.2009.28
Zimmerer KS (2000) The reworking of conservation geographies: nonequilibrium landscapes and nature-society hybrids. Ann Assoc Am Geogr 90(2):356–369. https://doi.org/10.1111/0004-5608.001

Chapter 4
Mitigating the Pressures: The Role of Participatory Planning in Protected Area Management

Smiljana Pivčević, Josip Mikulić, and Damir Krešić

Abstract This chapter aims to contribute to the knowledge on participatory planning (PP) in the protected area (PA) management focusing on areas whose sustainability is significantly affected by excessive tourism activity. The existing literature has shown PP to be both crucial for successful PA management as well as one of the weaker links in current PA management processes. However, the analysis of PPs key features and their implementation in the PAs are not adequately covered in the literature. Thus, this chapter analyses the role and critical elements of successful PP process in PA using the case studies of two Croatian national parks (NP) that have been under significant visitor pressure over the past years (pre-COVID-19). Based on several criteria devised through a critical review of PP literature, an assessment of PP models in two NP is performed, and critical points requiring improvements identified.

Keywords Participatory planning · Protected areas · National parks: management · Croatia

4.1 Introduction

The recent decades have brought a shift in the perception of PAs, not any longer as unspoiled patches of nature, rather as sustainable human living spaces (Nastran 2015). Their role has been transformed from that of preserving the environment (Sekhar 2003; Yergeau 2020) to that of supporting local development (Bello et al.

S. Pivčević (✉)
Faculty of Economics, Business and Tourism, University of Split, Split, Croatia
e-mail: smipi@efst.hr

J. Mikulić
Faculty of Economics and Business, University of Zagreb, Zagreb, Croatia
e-mail: jmikulic@net.efzg.hr

J. Mikulić · D. Krešić
Institute for Tourism Zagreb, Zagreb, Croatia
e-mail: damir.kresic@iztzg.hr

A. Mandić and L. Petrić (eds.), *Mediterranean Protected Areas in the Era of Overtourism*, https://doi.org/10.1007/978-3-030-69193-6_4

2016). Tourism is one of the few permitted as well as one of the most widespread uses of PAs worldwide (Buckley 2000; Chung et al. 2018). This shift is a result of a combination of two major trends (i) the neoliberalist call for PAs to be turned from "sacred" into "cash" cows (Müller 2014) due to increasing demands for their existence to be economically justified (Balmford et al. 2009); and (ii) increasing interest in experiencing natural environments (Buckley 2003; Tverijonaite et al. 2018). Visitor growth in PAs has brought economic and welfare benefits (Job et al. 2017; Richardson et al. 2012; Yergeau 2020), but also an overuse of the natural environment (Balmford et al. 2015; Stemberk et al. 2018), a decline in the quality of local community life and visitor experience (Monteiro 2017) and ultimately, the risk of overtourism (Peeters et al. 2018). The final balance of the impacts of tourism on PAs depends on the industry's compatibility with its conservation objectives. This is determined by management planning (Balmford et al. 2009) which needs to assure Pas' long-term sustainability (Eagles et al. 2002; Job et al. 2017). Stakeholder involvement in decision processes related to environmental issues is seen as a mechanism for collective responsibility (Van Den Hove 2000), as a democratic right (Rasheed and Abdulla 2020; Reed 2008) and as an avenue for improving decision-making outcomes and sustainability (Jones et al. 2013; Luyet et al. 2012; Wondirad and Ewnetu 2019). As a result, PP-based models of governance have been the subject of research in a variety of PAs (Badola et al. 2018; Belkayali and Kesimoğlu 2015; Wondirad and Ewnetu 2019), including marine and coastal areas (Jones et al. 2013; Russi 2020; Djosetro and Behagel 2020), as well as World Heritage sites (Adie and Amore 2020; Rasoolimanesh et al. 2017; Rasoolimanesh and Jaafar 2017). These studies elicit PP as one of the critical prerequisites of effective PA management in the desired models of governance such as co-management, collaborative governance and more recently advocated adaptive co-management (Alipour and Arefipour 2019; Benedetto et al. 2016; Djosetro and Behagel 2020; Islam et al. 2018a, b). Although highlighting its relevance in effective governance, the existing studies fail to analyse the key features of successful PP and its implementation in PA practice in more detail.

This chapter aims to fill this literature gap by focusing exclusively on PP to analyse its key features and elements necessary for effective PA management. Thus, a comprehensive literature review is conducted to elicit the participatory planning (PP) features, typologies and critically reflect the key elements required for successful PP in the context of tourism-related challenges in PAs. The case study method is used to analyse the PP state of the art in the two most visited PAs in Croatia. Based on the analyses and results obtained, key possible drawbacks in the PP process are highlighted as well as recommendations for its improvement.

4.2 Participatory Planning in Protected Areas

4.2.1 The Importance of Participatory Planning in Protected Area Management

In ecosystems that tend to be fragile, management intervention is required to address or prevent undesirable changes (Addison et al. 2015). Establishing protected areas (PAs) is commonly considered a key strategy for natural resource, biodiversity and landscape conservation, typically in the form of a park or reserve (e.g. Mukul et al. 2017; Molina-Murillo et al. 2016). Likewise, PA-based tourism is considered a viable means to foster local development that goes in hand with environmental conservation goals (Bello et al. 2016).

In contrast to top-down governance, which circumvents local communities in decision-making (Nita et al. 2018), a bottom-up participatory PA management approach involves a range of relevant stakeholders covering, for instance, local communities, scientists and environmental interest groups. More holistically, potential stakeholders in a power-sharing network for PA management purposes include (Gil et al. 2011:1327): (i) public regional administration entities; (ii) public local administration; (iii) research centres; (iv) land-users; (v) landowners within or adjacent to the PA and (vi) environmental and rural development non-governmental organizations (NGOs). The latter has proven to be particularly influential and effective in promoting environmental policies at all levels of governance, ranging from local to global (Nita et al. 2018). If present in and around a PA site, particular attention needs to be devoted to indigenous communities (Kamal and Lim 2019; Major et al. 2018).

To establish a supportive attitude towards PAs among the range of stakeholders, it is vital to communicate effectively and achieve a common understanding of PA values and benefits, hence laying a robust ground for management and policy development (Ivanić et al. 2017). In this regard, the importance of the perception of people living in protected areas has been particularly highlighted (Pelegrina-López et al. 2018; Hirschnitz-Garbers and Stoll-Kleemann 2011). To gain local community support for PA establishment, locals should be granted possibilities to engage in ecologically friendly activities in areas that either surround the strictly protected areas or that are established as special (e.g. Ayivor and Ntiamoa-Baidu 2015) or buffer zones (e.g. Palomo et al. 2013).

Community participation is almost unanimously regarded as a prerequisite for PA management that is both equitable and effective (Luyet et al. 2012; Stringer et al. 2006; Reed 2008). However, effective community involvement remains a key challenge in PA-based tourism, one that is yet to be achieved at many sites (Bello et al. 2016; Trimble et al. 2014; Gerner et al. 2011). Studies have shown that it is not environmental awareness but rather active participation in the resolution of environmental problems (Dimitrakopoulos et al. 2010) that typically tends to be a problem. To improve the involvement of local people in, e.g. joint meetings or workshops, the PA management should assure appropriate convening of meetings, unbiased and transparent participatory processes, and precise definition of objectives, procedures and desired outcomes (Trimble et al. 2014). It is further essential to grant a fair share of power among stakeholders (Kalternborn et al. 2011), in particular, to prevent elite groups drowning out local community voices, especially in less developed areas (Rashid et al. 2013). Others argue that PP is a form of incentive required for successful and effective governance (Jones et al. 2013; Russi 2020). Within this context, participatory scenario planning has been suggested as a means to achieve consensual management strategies directed towards a common desirable future, initially based on potentially different visions (Palomo et al. 2013).

However, PP also entails pitfalls which need to be taken into account. These refer to (i) the time and costs required (Luyet et al. 2012), (ii) potential stakeholder frustration and fatigue (Rasheed and Abdulla 2020; Reed 2008), (iii) the issue of power dynamics and group thinking (Luyet et al. 2012), (iv) communication problems such as technical language and general language barriers (Glicken 2000; Rasheed and Abdulla 2020), (v) potential new conflicts (Kangas and Store 2003), (vi) involvement of non-representative stakeholders (Reed 2008) or additional empowerment of those already important (Luyet et al. 2012; Rashid et al. 2013) and (vii) data generation and retrieval (Costa et al. 2018). In terms of the data issue, a key challenge for effective PA management and, in particular, visitor management, is the availability of spatial data (Hennig 2017). Combined with other factors, these pitfalls undermine the quality of PP processes and diminish the quality of their outcomes. However, as they are rooted in ineffective stakeholder design, planning and application of participatory processes (Santos et al. 2006), they can to a large extent by prevented by effective planning of the PP.

As a response to poor management effectiveness, especially in developing countries, studies highlight the need for organizational capacity building in co-managed protected areas (Appleton et al. 2017; Mukul et al. 2017). To build the capacity of involved local communities, which has shown to be deficient in several settings, a primary necessity is an appropriate environmental education (Zorrilla-Pujana and Rossi 2014). Moreover, the PA management is advised to foster and make use of traditional ecological knowledge, and even consider subsidy systems to support traditional communities, if necessary (Babai and Molnár 2014; Biró et al. 2014). To improve the effectiveness of PA management, stronger emphasis should also be put on park/reserve staff voice and collecting data at all organizational levels (Allen et al. 2019).

4.2.2 Participatory Planning Typologies and Key Features for Successful Tourism Planning in Protected Areas

The literature lists several typologies of participation, which can be categorized by different criteria, the prevailing being *the degree of participation.* Arnstein (1969), the pioneer and most prominent author on the subject, suggested an eight-tier participation ladder based upon power assignment from the administrative perspective and ranging from non-participation (manipulation of residents), degrees of tokenism (consultation by residents) to maximum participation (citizen control). As different process objectives and context require various forms of participation, (Reed 2008), a "wheel of participation" (Davidson 1998) is often suggested as an adequate alternative. Other authors have also proposed different terms for the steps/rungs of the participation ladder (cf. Biggs 1989; Pretty 1995; Farrington 1998; Lawrence 2006). These models have also inspired tourism research, thus, synthesizing the typologies of Arnstein (1969) and Pretty (1995), Tosun (1999, 2006) suggested three levels of community participation in tourism development—coercive, induced and spontaneous participation (Rasoolimanesh et al. 2017; Zhang et al. 2013). Apart from the variety in terminology, in all typologies, the ladder stands for gradation in which empowerment grows moving up the ladder and along the scale of variables as actors' roles, in process and outcomes, methods, resources and scale change (cf. Lawrence 2006). In the highest level of participation, residents have the power to make decisions and to control the process of development. This can build trust, a sense of belonging and social capital in the community (Rasoolimanesh and Jaafar 2017), as opposed to the "lower" rungs of participation, which generate conflicts and are not as effective (Zhang et al. 2013). A high level of stakeholder participation begins in the early planning stages and ensures the active involvement of all stakeholder groups throughout the entire PP process (Rasoolimanesh et al. 2017; Tosun 2006). These typologies have, not gone uncriticized (Cornwall 2008; NORAD 2013), particularly Arnstein's ladder (Collins and Ison 2009; Tritter and McCallum 2006), for, among other things, their normative nature, lack of context observation and linear approach.

Participation can be also be viewed in terms of the *direction of communication flows* (Reed 2008). Thus, Rowe and Frewer (2000) argue participation to be two-way communication between participants and exercise organizers through dialogue or negotiation, while dissemination to passive recipients constitutes communication and gathering information from participants should be termed consultation. A typology based on *theoretical* basis distinguishes between normative participation (focused on the process and based on the premise of the democratic right to participation) and pragmatic participation (views participation as a means to an end, i.e. higher quality decisions) (Beierle 2002; Reed 2008). However, it is often argued that community participation should not only strive to ensure equitable distribution of resources (Arnstein 1969) but should also enable knowledge transfer and induce community transformation in the long run (Okazaki 2008; Wondirad and Ewnetu 2019).

Furthermore, participation can be classified according to the *objectives of participation.* Thus, "planner-centred" participation is focused on process outcomes while "people-centred" participation aims at building capacity and empowering stakeholders throughout the process (Michener 1998). Finally, Dowers and Hussey (2013, as cited in Dovers et al. 2015) differentiate the levels of the *hierarchy of participation,* i.e. participation in governance arrangements, policy settings and management. These levels differ in terms of the aim, opportunities for participation for different

Fig. 4.1 Hierarchy of Governance and Participation (Dovers et al. 2015)

stakeholders and typical rates of change (years to decades; many months to years; weeks to months respectively) (Fig. 4.1).

In more practical terms, four questions guide the framework behind the design of engagement strategy and methods adequate for a specific context (Dovers et al. 2015; Quesada-Silva et al. 2019) and are thus useful for mapping the PP practices. These questions are. (i) *who* should be engaged, (ii) what is the purpose of engagement (*why*); (iii) *how* is the engagement/participation done and (iv) *when* is participation planned (timing and frequency)?

In line with the scope of this book, PP is viewed within the specific context of tourism-development in PAs. In both subareas, researchers have explored and conceptualized the PP process and its relevant factors (Bichler 2019; Davies et al. 2018; Islam et al. 2018a; Quesada-Silva et al. 2019). However, in terms of conceptualizing the intersection of PP, PA and tourism development, the framework of Bello et al. (2016) is a rare example. The framework identifies seven major participatory planning elements that need to be a part of the three phases of tourism planning/strategy. These elements are the timing of involvement, resource accessibility, representativeness, independence, influence and power, transparency and decision-making structure and further elaboration of the *who*, *why*, *how* and *when* questions in the PP model (Fig. 4.2).

In terms of *timing*, it is general accepted that all citizens should be involved in PP as early as reasonably and practically possible during the preparatory or predesign phase (Garrod 2003; Rowe and Frewer 2000). The same goes for tourism planning in PAs, where the planning bodies need to enable participation in all stages of the process (Bello et al. 2016; Garrod 2003). The *resource accessibility* refers to the need for participants in the tourism planning process to have access to financial, information, human and material resources required for effective participation (Rowe and Frewer 2000). The information resources, i.e. continued access to tourism information and experts are especially highlighted as many studies have found that local

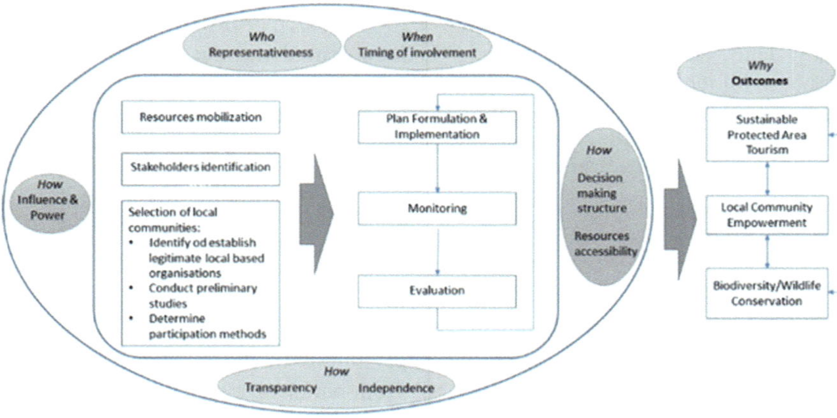

Fig. 4.2 PP Framework for PA-Based Tourism (authors' elaboration based on Bello et al. 2016; Dovers et al. 2015; Quesada-Silva et al. 2019)

people living around PAs, specifically in developing countries, lack the basic knowledge about tourism development (Bello et al. 2016). Thus, education and empowerment through public awareness and education programmes are necessary (Marzuki et al. 2012) especially in the area of tourism–conservation relationship (Garrod 2003) and tourism skills (Bello et al. 2016). The challenge of *representativeness, i.e.* the need for communities to be represented by legitimate individuals serving their interests, is frequently found in PP literature (McCool 2009; Reed 2008). The major challenges in this domain are dilemma between the inclusion of traditional formal versus informal leaders (Stone and Stone 2011); fair representation of all communities (McCool 2009) and groups traditionally less represented such as youth and women (Tosun 2000; Stone and Stone 2011). *Independence* is closely related to this and refers to the lack of any influence from the sponsoring authority, planning body or any other stakeholder (Bello et al. 2016; Reed 2008), which can be achieved through the engagement of an independent and legitimate facilitator (Jamal and Getz 1995 as cited in Bello et al. 2016). *Influence and power* are again inseparable from empowerment and capacity building. They are a prerequisite for citizens to be active partners in the process and also obviate situations in which participation is merely a tool for legitimizing planning authorities' decisions (Rowe and Frewer 2000). *Transparency* during the planning process enables all interested stakeholders to see how the process is developing and how the decisions are being made; it reduces suspicions about the motives of other stakeholders (Bello et al. 2016). PA management regularly needs to communicate to the public all planning procedures and decisions (Rowe and Frewer 2000) regarding tourism development, PA finances and the reasons for individual decisions (Bello et al. 2016), as inadequate information can undermine community participation in tourism planning (Marzuki et al. 2012; Cevat Tosun 2000). Finally, *decision-making structure* denotes an adequate mechanism for the organization of the decision-making process (Rowe and Frewer 2000), which is directly related to transparency. Thus, decision-making should be documented and the information made available to all the citizens (Bello et al. 2016).

In the case studies that follow, the typologies and key elements elaborated are used to post hoc categorize the type of participation that has occurred (Reed 2008).

4.3 Methodology

To contribute to existing knowledge on participatory planning processes in protected areas, this chapter seeks to present two case studies of the most visited NPs in Croatia, i.e. Plitvice Lakes NP and Krka NP. Both parks have been under significant pressure from increasing visitor flows over the past years, and, in particular, in Plitvice Lakes, local communities and even the general public have raised growing concerns regarding the sustainability of tourism activity in and around the park. Given the focus of this chapter, the case studies focus dominantly on the participatory planning models used in both NPs. More specifically, following a brief introduction of the study settings, the analysed participatory models are evaluated in terms of the elements discussed in the theoretical part of the paper.

4.3.1 An Overview of Studied Protected Areas

Plitvice Lakes NP is the oldest, largest, and most visited national park in Croatia. It was proclaimed a natural protected area in 1949, and it was internationally recognized in 1979 when it was added to the UNESCO World Heritage List. According to the Plitvice Lakes NP Management Plan (2019), the Outstanding Universal Value (OUV) of the Park is the "interaction of water, air, geological foundation and organisms which, coupled with specific physiochemical and biological conditions, enabled the formation of the tufa that has created a series of lakes, barriers, cascades and waterfalls by dividing the lakes" (pp. 22). The Park covers an area of 297 km^2 and records slightly less than 2 million visitors annually with an extremely high share of same-day visitors (70%) and high seasonal variations in tourism demand (more than 80% of visits are recorded in the period May–September). The park also plays a pivotal role in the local and regional economy, contributing directly and indirectly to the livelihood of the local population in the wider park area. However, the high share of same-day visitors, emphatic seasonality of tourism demand, increased visitation pressure on tufa barriers and lakes, deforestation and other anthropogenic influences have led to harmful and undesirable impacts on the sensitive and fragile ecosystem of the Park.

Krka NP was proclaimed in 1985 and covers 109 km^2, making it the third-largest PA in Croatia. According to Krka NP Management Plan (Carić et al. 2019), it receives slightly fewer than 1.5 million visitors annually, and it is the second most visited PA in Croatia. Visitation patterns show that individual, same-day visitors mostly visit this PA with the highest number of visitors being recorded during peak tourism season (June to August). High and irregular visitation patterns are primarily the consequence of the PA being located in the southern region of Croatia (Šibenik-Knin County), which has a highly developed tourism industry. The central phenomenon and most important tourist attraction of Krka NP are the travertine barriers forming seven waterfalls on the river Krka with a total drop in altitude of 242 m, with a number of other geomorphological, hydrological and cultural values. Albeit playing an important role in the local and regional economy, this role is not as important as in the case of Plitvice Lakes NP. In the last ten years, the number of visitors to this PA has doubled and tourism activity led to severe impacts on the fragile and delicate ecosystem of the PA, including wastewater disposal and a visible increase of organic matter content in the Krka River, diminishing the biologically and aesthetic value of the PA. Like Plitvice Lakes NP, Krka NP is struggling to strike a delicate balance among tourism activity, well-being of the local community and nature conservation through sustainable management strategies.

4.3.2 Participation Models

Participatory planning (PP) in Croatian natural protected areas (PA) has a long tradition and most Croatian PAs, including Plitvice Lakes and Krka NP, have, at least formally, implemented some kind of PP model. The right of the local community and various local stakeholders to participate in the development and management of protected natural areas is also legally defined, through the Nature Protection Act, Art. 239 (Croatian Parliament 2018), which defines compulsory public participation in the planning and management processes of the PA. PP model is commonly defined within the PA management plan, whose development and implementation are a legal obligation of the park management authority. A significant challenge in the sustainable management of Croatian PAs is harmonizing the conflicting interests of different stakeholders to balance their activities and to achieve social, economic and environmental sustainability. Commonly, Croatian PAs have a number of different stakeholders included in the PP process, namely national, regional and local governments, destination management and marketing organizations (DMMOs), tourism companies, NGOs, academia, media, PA employees, local populations, tourists and visitors. The main goal of the PP process itself is to address the conflicting interests of different groups of stakeholders (mostly economic development vs conservation) and to reach a consensus regarding the current situation and plans and goals, through informed dialogue and continuous partnership.

The analysis of existing PP models in Plitvice Lakes NP and Krka NP adopted in this paper is based on information available in their management plans, previous research (Innes 2004; Kulözü and Tekeli 2014; Kangas et al. 2015) and recommendations of the Food and Agriculture Organization of the United Nations—FAO (Jain and Polman 2003). Since PP is not a final result but a process, it is essential to identify the most important elements of the process and to investigate their role in it. Additionally, it is crucial to recognize that, due to their common goals (achieving a balance between economic development and conservation) PP models are partly similar in all PAs. Still, they are also somewhat site-specific and as unique as the context in which the PP process is conducted (Kulözü and Tekeli 2014).

The analysis shows that PP models in the two analysed PAs are mostly based on the problem-solving approach which utilizes an iterative development process (learning-by-doing) with several mid-term corrections and involves several local stakeholders, knowledgeable in diverse background domains, important for effective management of PA (Lefèvre et al. 2000). The essential elements detected in the PP models of both analysed PAs include (i) identification and prioritization of local stakeholders, (ii) collection of primary data/indicators, (iii) formation of functional working groups, (iv) definition of common objectives/goals, (v) active participation in the process of strategic planning and (vi) implementation and monitoring activities.

Identification and prioritization of local stakeholders is the first step of the PP process. Both analysed PAs use PP models in their management strategies to detect low- and high-influence stakeholders, to prioritize their relevance (opportunities and limitations) for cooperation and possible scope of involvement in park project activities. PP models in both analysed PAs are based on the bottom-up planning approach, which allows active community involvement, decentralized decision-making and generation of new and fresh ideas early in the planning process. Park management authorities (public institutions) are the initiators and coordinators of the PP process, and other involved stakeholders provide a substantial contribution to the preparation, implementation and revision of the PP process. All stakeholders are prioritized and defined as crucial, important, and potentially important, considering their role or potential value-adding role regarding their knowledge and skills contribution, issues-resolving experiences, motivation and creation of dissemination and policies. Finally, different communication strategies are used for different stakeholder groups, depending on the role they play in the PP process, their priority in the process and stages of their involvement in the PP process.

Collection of primary data/indicators is an activity necessary for informed decision-making. Plitvice Lakes and Krka NP collect a wide range of data and indicators to enhance their conservation potential and to implement principles of informed decision-making in their management practices. Basic guidelines regarding the types of data which are to be collected are commonly defined by different official documents, such as PA management plans, various sectoral studies (forestry, hydrology, geology, etc.), spatial and development programmes and workshops, etc. Data and indicators are most commonly collected with the assistance of different stakeholders, and the degree of their commitment to the PP process can have a significant impact on the quality and accuracy of the collected data. Data and indicators collected as part of the PP process could be aggregated in several major groups: (i) conservation of natural values, (ii) conservation of cultural heritage, (iii) visitor management, (iv) data related to support of local community sustainable development, (v) capacity development and management of Public Institution/park management authority and

(vi) PA spatial zoning data. Collection of data itself is also a process that is not focused on a one-time period but is continuous and often cyclic.

Formation of functional working groups is at the centre of the PP process. Both analysed PAs have formed several functional working groups consisting of PA management authority employees, external consultants and government representatives from local, regional and national levels. Multidisciplinarity is one of the most important aspects of a working group since one functional working group usually includes experts from different fields, such as nature conservation, biology, chemistry, geology, spatial planning, economics and tourism.

Definition of common objectives/goals is commonly based on the situational analysis. The general purpose of the PP process is to define a new management framework and to determine goals, activities, and performance indicators that will improve the management infrastructure following the active protection of natural and other values. Most of the defined objectives are long-term objectives that cannot be achieved within a single year, they may differ for different functional zones within PA, and they must be aligned with the central development vision of the PA. Active participation of all relevant stakeholders, as well as a bottom-up approach, is necessary to define goals, activities and indicators that are relevant, achievable and measurable. Both PAs have determined activities, deadlines, human and financial resources required for the implementation of the goals, while their practical performance is unfortunately unknown.

Programmes and activities are defined and organized in such a way as to encourage active stakeholder participation in the process of strategic planning. Individual external stakeholders are invited to participate in the relevant, appropriate phases of the project, which are organized in partnership with the expert team. Local stakeholders are commonly invited through word-of-mouth and written communication (notice and invitation letters with PP summary, advertisements). These interest groups and local population participate through the coordination meetings, internal thematic workshops and advisory meetings in the iterative process after the preliminary reports.

Implementation and monitoring activities are defined through the action plan, which is an integral part of the management plan. The action plan defines the roles of different stakeholders as well as activities, timetables and resources necessary for implementation. During the implementation period, which is ten years for both analysed PAs, intensive and continuous proactive consultations and dialogues with different stakeholder and interest groups are required.

It is important to emphasize that even though PP models are commonly included and integrated into the PA management plans, in reality, there are several problems with regards to PP implementation. Usually, PP models do not sufficiently recognize site-specific features of the PA, and consequently, PP models are generic, meaning that site-specific features are not adequately addressed. The lack of site-specific goals in PP models is demonstrated by Marković Vukadin (2020). Her research finds that out of the 14 main management goals in Plitvice Lakes NP, many are poorly defined and irrelevant, and the majority have been only partly achieved or not achieved at all, indicating highly inefficient park management. Additionally, one of the most important and highly relevant goals in the context of the PP model in Plitvice Lakes NP, named "Cooperation with the local community", was not achieved in the previous management period. Another critical issue is the lack of the quality, transparent and accurate data needed for informed management and objective assessment (Canteiro et al. 2018). What is more, the monitoring process is implemented by the same stakeholders (usually by members of the PA management authority) whose performances are being assessed, so they are evaluating themselves. With all the above in mind, it is reasonable to conclude that PP models in both analysed PAs do not contribute significantly to mitigating tourism-induced pressures, mostly since PP models are implemented on a generic level.

4.4 Discussion and Conclusions

Aiming to fill the literature gap in terms of detailed insights into the critical features of PP in PAs, this section provides conclusions and insights gained through the comparison of the key theoretical determinants/criteria of PP models and the empirical evidence collected from the selected PAs available information and management plans. Thus, in terms of the degree of participation, the PP models studied are characterized by mid-level participation, i.e. "degrees of tokenism" or "consultation by residents" in Arnstein's (1969) ladder. However, it is used only in the preliminary phases of management planning, thus narrowing the full range of options in the "*when*" aspect of PP substantially. Passive participation through periodical reporting is also recognized as one of the PP approaches used (Marzuki et al. 2012). The inclusion of community is mostly spontaneous (Zhang et al. 2013) with broad private initiative and sporadic park management induced actions. In the PA management authority activities, the local community is represented through different interest groups. Still, it lacks opportunities to make decisions in the management process, leading to conflicts with management and making the PP process less effective. The direction of communication flows is based on inclusion in consultation when needed (Rowe and Frewer 2000), i.e. gathering information from participants rather than participation (Reed 2008). Communication and dissemination are characterized by inadequate information-gathering from the local population and consequently lead to less effective and uninformed PA management. In both cases, normative participation focused on process outcomes, such as successful conservation actions is in place and, therefore, is planner-centred (Michener 1998). Although many studies call for PP to be a tool of community and stakeholder empowerment and a leverage for knowledge transfer and community transformation (Lawrence 2006; Okazaki 2008; Wondirad and Ewnetu 2019) and some such activities are found in both PAs, it is not found to be realizing its potential. Finally, in terms of levels of the hierarchy of participation, the PPs in the cases studied are focused exclusively on governance arrangements.

Regarding the four questions framing the practical arrangement of PP (Dovers et al. 2015; Quesada-Silva et al. 2019) in the cases studied: (i) diverse stakeholders take part in the PP process, i.e. *(who)*, (ii) the underlying reason for engagement fulfils the legal requirement to include stakeholders in the planning process *(why)*; (iii) participation is achieved prevailingly through face-to-face modes of group interaction *(how)* and (iv) is undertaken in the process of designing the ten year PA management plans *(when)* as suggested in the literature (Bello et al. 2016; Garrod 2003; Rowe and Frewer 2000). Furthermore, continuous narrower participation exercise is enabled through the working groups, which, however, comprise only selected stakeholders.

In terms of the seven major tourism PP planning elements proposed by Bello et al. (2016) the available data for the majority of them are not sufficient to enable a clear-cut conclusion on their state of the art. This goes for resource accessibility, representativeness, influence and power and, to some extent, for the decision-making structure. This, however, is sufficient to conclude that the transparency of the PP process needs to be improved to reduce the potential suspicions concerning interested stakeholders (Bello et al. 2016) and the resulting reluctance to participate in the tourism planning process (Marzuki et al. 2012; Tosun 2000). This goes especially for Krka NP, whose former and current management plans are not publicly available.

Thus, the overall conclusion is that PP in both PAs studied requires further development and improvement to mitigate the pressures associated with tourism flows. These pressures, as we are witnessing, have changed unexpectedly and dramatically in the recent period. As a result, in the COVID and post-COVID period concerns about overtourism have been replaced by fears of too low visitations, which might seriously hamper the financial stability and the overall viability of many PAs (IUCN 2020). This unprecedented situation reinforces the need for PAs to be sustainable human living spaces (Nastran 2015) and the need for the inclusion of diverse stakeholders in the decision-making process.

Even though PP in PA is a concept widely discussed in the scientific and professional literature as a model providing integrated sustainable management and development of PAs, in practice PP models vary significantly. In both PA analysed, current involvement of some stakeholders, specifically, the local community, is relatively modest (marginal). However, the crucial question, which cannot be answered by the available and analysed data, is whether the local community has the necessary capacity and knowledge to engage meaningfully in the PP process (Bello et al. 2016; Garrod 2003; Marzuki et al. 2012; Zorrilla-Pujana and Rossi 2014). Nevertheless, there are several projects and initiatives contributing to better tourism development and conservation by connecting stakeholders through various working groups. These projects and initiatives enable synergies between stakeholders and lead to better cooperation and faster achievement of stakeholder goals and interests. One such initiative is "Lika Destination", a project including over 100 stakeholders from the Plitvice Lakes region. Although founded by the Lika Local Action Group (LAG), today its primary role is creating new values for the Plitvice Lakes NP. Although this LAG is not formally involved in the PA management, its activities significantly contribute to strengthening the communication and initiative of the local community, consequently leading to a higher degree of local community involvement in the management processes. Regarding PP in Krka NP, it should be pointed out that a new management plan has been recently created based on numerous focus group results and consultations with a variety of PA stakeholders.

In conclusion, the relationship between PP, PA management and tourism development is very dynamic and complex. In line with findings of previous research (Gani et al. 2018), the analysis conducted provides evidence that the practice of PP in the PA management can have positive as well as negative effects on mitigating pressures associated with tourism visitation If the PP process is implemented only to satisfy the formal and legal requirements imposed to the PA management authority, it will probably be ineffective, or can even have negative impacts. Due to a wide array of stakeholders included in the PP process and the growing demand for special interest tourism, PP has the potential to contribute to the complexity of tourism product, hence making PAs more attractive to visitors and increasing the number of visitors and visitation pressures. Tourism driven pressures are especially emphasized in the populated PAs. In such PAs, the livelihood of the local population heavily depends on tourism infrastructure development, which is sometimes in conflict with nature conservation practices. The extent and type of public participation should be defined, having in mind that the PAs are not established primarily for tourism purposes. Namely, tourism is only a means for achieving economic sustainability and an appropriate level of nature protection and conservation. Therefore, sustainable, non-invasive forms of tourism development should be encouraged in the PAs, which require a high level of interpretation, multidisciplinary team and provide added value for fragile and delicate ecosystems of the PAs. An example of such tourism products are workshops and boat tours for school children organized in NP Krka. Such forms of educational and ecotourism imply a high level of PP and coordination in PA since it is necessary to include different stakeholders (such as park management, pedagogues, biologists, geologist, speleologists, tour guides, etc.) in its development and provision. This type of tourism product development through PP leads to self-mobilization (Marzuki et al. 2012; Rasoolimanesh et al. 2017), and creates new values in PA tourism products through a raised level of innovation. However, if the PP is implemented with the intention of empowering local population and different niche stakeholders in the decision-making process, the results of the PP process can be beneficial for the PA as well as for the local community.

References

Addison PFE, De Bie K, Rumpff L (2015) Setting conservation management thresholds using a novel participatory modelling approach. Conserv Biol 29(5):1411–1422

Alipour H, & Arefipour T (2019) Rethinking potentials of Co-management for sustainable common pool resources (CPR) and tourism: The case of a Mediterranean island. Ocean and Coastal Management, 183 (July 2019). https://doi.org/10.1016/j.ocecoaman.2019.104993.

Allen L, Holland KK, Holland H, Nabaala M, Seno S, Nampushi J (2019) Expanding staff voice in protected area management effectiveness assessments within Kenya's Maasai Mara National Reserve. Environ Manage 63(1):46–59

Appleton MR, Toussaint A, Daltry JC (2017) From forestry to protected area and ecosystem management: organisational change in Saint Lucia, West Indies. Parks 32(2):51–62
Arnstein SR (1969) A ladder of citizen participation. J Am Inst Plann 35:37–41
Ayivor JS, Ntiamoa-Baidu Y (2015) Assessing the socio-economic stressors of Ghana's only Strict Nature Reserve-Kogyae. Parks 21(2):85–100
Babai D, Molnár Z (2014) Small-scale traditional management of highly species-rich grasslands in the Carpathians. Agr Ecosyst Environ 182:123–130
Badola R, Hussain SA, Dobriyal P, Manral U, Barthwal S, Rastogi A, Gill AK (2018) Institutional arrangements for managing tourism in the Indian Himalayan protected areas. Tour Manag 66:1–12
Balmford A, Beresford J, Green J, Naidoo R, Walpole M, Manica A (2009) A global perspective on trends in nature-based tourism. PLoS Biol 7(6):e1000144. https://doi.org/10.1371/journal.pbio.1000144
Balmford A, Green JMH, Anderson M, Beresford J, Huang C, Naidoo R, Walpole M, Manica A (2015) Walk on the wild side: estimating the global magnitude of visits to protected areas. PLoS Biol 13(2):e1002074. https://doi.org/10.1371/journal.pbio.1002074
Beierle TC (2002) The quality of stakeholder-based decisions. Risk Anal Inter J 22(4):739–749
Belkayali N, Kesimoğlu MD (2015) The stakeholders' point of view about the impact of recreational and tourism activities on natural protected area: a case study from Kure Mountains National Park, Turkey. Biotechnol Biotechnol Equipment 29(6):1092–1103
Bello FG, Carr N, Lovelock B (2016) Community participation framework for protected area-based tourism planning. Tour Plan Dev 13(4):469–485
Benedetto G, Carboni D, Corinto GL (2016) Governance of Sustainable Tourism in a Vast Area Surrounding a National Park. Procedia Environmental Sciences, 32:38–48. https://doi.org/10.1016/j.proenv.2016.03.010
Bichler BF (2019). Designing tourism governance: the role of local residents. J Destination Manag Mark, 100389. online first. https://doi.org/10.1016/j.jdmm.2019.100389
Biggs S (1989) Resource-poor farmer participation in research: a synthesis of experiences from nine national agricultural research systems. OFCOR Comparative Study Paper, Vol. 3. International Service for National Agricultural Research, The Hague
Biró É, Babai D, Bódis J, Molnár Z (2014) Lack of knowledge or loss of knowledge? Traditional ecological knowledge of population dynamics of threatened plant species in East-Central Europe. J Nat Conserv 22(4):318–325
Buckley R (2000) Tourism in the most fragile environments. Tourism Recreation Res 25(1):31–40
Buckley R (2003) Ecological indicators of tourist impacts in parks. J Ecotourism 2(1):54–66
Canteiro M, Córdova TF, Brazeiro A (2018) Tourism impact assessment: a tool to evaluate the environmental impacts of touristic activities in natural protected areas. Tourism Manag Perspect 28:220–227
Carić H, Jakelić M, Klanjšek T, Ozimec R, Tomljenović R, Vidović H, Jerković I (2019) Studija upravljanja posjetiteljima NP „Krka". Šibenik 2019: Javna ustanova „Nacionalni park Krka"
Chung MG, Dietz T, Liu J (2018) Global relationships between biodiversity and nature-based tourism in protected areas. Ecosyst Serv 34:11–23
Collins K, Ison R (2009) Jumping off Arnstein's ladder: social learning as a new policy paradigm for climate change adaptation. Environ Policy Gov 19(6):358–373
Cornwall A (2008) Unpacking "participation" models, meanings and practices. Commun Dev J 43(3):269–283
Costa DC, Pereira HS, Marchand GAEL, Silva SCP (2018) Challenges of participatory community monitoring of biodiversity in protected areas in Brazilian Amazon. Diversity 10(3):61, 1–11
Davidson S, (1998) Spinning the wheel of empowerment. Planning, 1262, 14–15.
Davies K, Murchie AA, Kerr V, Lundquist C (2018) The evolution of marine protected area planning in Aotearoa New Zealand: Reflections on participation and process. Marine Policy 93:113–127
Dimitrakopoulos PG, Jones N, Iosifides T, Florokapi I, Lasda O, Paliouras F, Evangelinos KI (2010) Local attitudes on protected areas: evidence from three Natura 2000 wetland sites in Greece. J Environ Manag 91(9):1847–1854

Djosetro M, Behagel JH (2020) Building local support for a coastal protected area: collaborative governance in the Bigi Pan Multiple Use Management Area of Suriname. Marine Policy 112(15):103746

Dovers S, Feary S, Martin A, McMillan L, Morgan D, Tollefson M (2015) Engagement and participation in protected area management: who, why, how and when? In: Worboys GL, Lockwood M, Kothari A, Feary S, Pulsford I (eds) Protected Area Governance and Management (pp. 413–440). ANU Press, Canberra

Eagles PF, McCool SF, Haynes CD (2002) Turismo sostenible en áreas protegidas. Directrices de planificación y gestión. Preparado para el Programa de las Naciones Unidas para el Medio Ambiente, la Organización Mundial del Turismo y la UICN–Unión Mundial para la Naturaleza. UNWTO, Madrid

Farrington J (1998) Organisational roles in farmer participatory research and extension: lessons from the last decade. Nat Resour Perspect 27:1–7

Gani AA, Mahdzar M, Razak IRA (2018) Local Community's Opinions on Public Participation Practices at Kuala Gandah Elephant Sanctuary, Pahang. Inter J Acad Res Bus Soc Sci 8(16):356–366

Garrod B (2003) Local participation in the planning and management of ecotourism: a revised model approach. J Ecotourism 2(1):33–53

Gerner J, Heurich M, Günther S, Schraml U (2011) Red deer at a cross-roads—an analysis of communication strategies concerning wildlife management in the 'Bayerischer Wald' National Park, Germany. J Nat Conserv 19(5):319–326

Gil A, Calado H, Costa LT, Bentz J, Fonseca C, Lobo A, et al (2011) A methodological proposal for the development of Natura 2000 sites management plans. J Coast Res SI64:1326–1330

Glicken J (2000) Getting stakeholder participation 'right': a discussion of par-ticipatory processes and possible pitfalls. Environ Sci Policy 3:305–310

Hennig S (2017) OpenStreetMap used in protected area management. The example of the recreational infrastructure in Berchtesgaden National Park. J Protected Mt Areas Res Manag 9:30–41

Hirschnitz-Garbers M, Stoll-Kleemann S (2011) Opportunities and barriers in the implementation of protected area management: a qualitative meta-analysis of case studies from European protected areas. Geogr J 177(4):321–334

Innes JE (2004) Consensus building: clarifications for the critics. Plann Theory 3(1):5–20

Islam MW, Ruhanen L, Ritchie BW (2018a) Adaptive co-management: a novel approach to tourism destination governance? J Hospitality Tourism Manag 37:97–106

Islam MW, Ruhanen L, Ritchie BW (2018b) Tourism governance in protected areas: investigating the application of the adaptive co-management approach. J Sustain Tourism 26(11):1890–1908

IUCN (2020), Conserving Nature in a time of crisis: Protected Areas and COVID-19, https://www.iucn.org/news/world-commission-protectedareas/202005/conserving-nature-a-time-crisis-protected-areas-and-covid-19. Accessed 15th June 2020

Ivanić KZ, Štefan A, Porej D, Stolton S (2017) Using a participatory assessment of ecosystem services in the Dinaric Arc of Europe to support protected area management. Parks 23(1):61–74

Jain SP, Polman W (2003) A Handbook for Trainers on Participatory Local Development. FAO Regional Office for Asia and the Pacific, Bangkok

Job H, Becken S, Lane B (2017) Protected Areas in a neoliberal world and the role of tourism in supporting conservation and sustainable development: an assessment of strategic planning, zoning, impact monitoring, and tourism management at natural World Heritage Sites. J Sustain Tourism 25(12):1697–1718

Jones PJS, Qiu W, De Santo EM (2013) Governing marine protected areas: social-ecological resilience through institutional diversity. Marine Policy 41:5–13

Kaltenborn BP, Qvenild M, Nellemann C (2011) Local governance of national parks: the perception of tourism operators in Dovre-Sunndalsfjella National Park, Norway. Norsk Geografisk Tidsskrift-Norwegian J Geogr 65(2):83–92

Kamal SF, Lim VC (2019) Forest reserve as an inclusive or exclusive space? Engaging orang ASLI as stakeholder in protected area management. J Trop Forest Sci 31(3):278–285
Kangas AS, Kurttila M, Hujala T, Eyvindson K, Kangas J (2015) Participatory planning processes in action. In: Kangas A, Kangas J, Kurttila M, Hujala T, Eyvindson K (eds) Decision support for forest management. Springer Nature, Cham, pp 253–286
Kangas J, Store R (2003) Internet and teledemocracy in participatory planning of natural resources management. Landscape Urban Plann 62(2):89–101
Kulozou N, Tekeli İ (2014) Socio-psychological factors affecting participatory planning processes at interactional level. Megaron 9(1):1–13
Lawrence A (2006) "No personal motive?" Volunteers, biodiversity, and the false dichotomies of participation. Ethics, Place Environ 9(3):279–298
Luyet V, Schlaepfer R, Parlange MB, Buttler A (2012) A framework to implement stakeholder participation in environmental projects. J Environ Manage 111:213–219
Major K, Smith D, Migliano AB (2018) Co-managers or co-residents? Indigenous peoples' participation in the management of protected areas: a case study of the Agta in the Philippines. Hum Ecol 46(4):485–495
Marković Vukadin I (2020) The interrelation between development, management and management issues in Plitvice Lakes National Park. In: Kodrman M, Opačić VT, Challenges of Tourism Development in Protected Areas of Croatia and Slovenia (p in press). University of Primorska Press, Koper
Marzuki A, Hay I, James J (2012) Public participation shortcomings in tourism planning: the case of the Langkawi Islands, Malaysia. J Sustain Tourism 20(4):585–602
McCool SF (2009) Constructing partnerships for protected area tourism planning in an era of change and messiness. J Sustain Tourism 17(2):133–148
Michener VJ (1998) The participatory approach: contradiction and co-option in Burkina Faso. World Dev 26(12):2105–2118
Molina Murillo SA, Fernández Otárola M, Shreeve KN (2016) Understanding the link between benefits from protected areas and their relationship with surrounding communities: an exploration in Costa Rica. Parks 22(1):79–88
Monteiro L (2017) Are protected areas becoming victims of their own popularity? Protected areas in-sight. J Europarc Fed, Special Edition Overtourism, 21–23
Mukul SA, Sohel MSI, Herbohn J, Inostroza L, König H (2017) Integrating ecosystem services supply potential from future land-use scenarios in protected area management: a Bangladesh case study. Ecosyst Serv 26:355–364
Müller M (2014) From sacred cow to cash cow: the shifting political ecologies of protected areas in Russia. Zeitschrift Fur Wirtschaftsgeographie 58(2–3):127–143
Nastran M (2015) Why does nobody ask us? Impacts on local perception of a protected area in designation, Slovenia. Land Use Policy 46:38–49
Nita A, Ciocanea CM, Manolache S, Rozylowicz L (2018) A network approach for understanding opportunities and barriers to effective public participation in the management of protected areas. Soc Netw Anal Mining 8(1):31
NORAD (2013) A Framework for analysing participation in development. Report No. 1/2013. Norwegian Agency for Development Cooperation, Norway
Okazaki E (2008) A community-based tourism model: its conception and use. J Sustain Tourism 16(5):511–529
Palomo I, Martín-López B, Potschin M, Haines-Young R, Montes C (2013) National Parks, buffer zones and surrounding lands: mapping ecosystem service flows. Ecosyst Serv 4:104–116
Peeters P, Gössling S, Klijs J, Milano C, Novelli M, Dijkmans C, Eijgelaar E, Hartman S, Heslinga J, Isaac R, Mitas O, Moretti S, Nawijn J, Papp B, Postma A (2018) Research for TRAN Committee-Overtourism: impact and possible policy responses. Structural and Cohesion Policies, Transport and Tourism, European Parliament, Directorate General for Internal Policies, Policy Department B, Brussels

Pelegrina-López A, Ocaña-Peinado FM, Henares-Civantos I, Rosúa-Campos JL, Serrano-Bernardo FA (2018) Analysing social perception as a key factor in the management of protected areas: the case of the Sierra Nevada Protected Area (S Spain). J Environ Plann Manag 61(1):124–142

Plitvice Lakes National Park Public Institution (2019) Plitvice Lakes National Park Management Plan 2019–2028. Plitvice Lakes National Park, Plitvice Lakes

Pretty JN (1995) The many interpretations of participation. Focus, 4(16):4–5.

Quesada-Silva M, Iglesias-Campos A, Turra A, Suárez-de Vivero JL (2019) Stakeholder participation assessment framework (SPAF): a theory-based strategy to plan and evaluate marine spatial planning participatory processes. Marine Policy 108:103619

Rasheed AR, Abdulla A (2020) Evaluating stakeholder participatory processes in policy development for Marine Protected Areas. Marine Policy 112:103737

Rashid AM, Craig D, Mukul SA, Khan NA (2013) A journey to-wards shared governance: status and prospects for collaborative management in the protected areas of Bangladesh. J For Res 24(3):599–605

Rasoolimanesh SM, Jaafar M (2017) Sustainable tourism development and residents' perceptions in World Heritage Site destinations. Asia Pac J Tourism Res 22(1):34–48

Rasoolimanesh SM, Jaafar M, Ahmad AG, Barghi R (2017) Community participation in World Heritage Site conservation and tourism development. Tour Manag 58:142–153

Reed MS (2008) Stakeholder participation for environmental management: a literature review. Biol Cons 141(10):2417–2431

Richardson RB, Fernandez A, Tschirley D, Tembo G (2012) Wildlife conservation in Zambia: impacts on rural household welfare. World Dev 40(5):1068–1081

Rowe G, Frewer LJ (2000) Public participation methods: a framework for evaluation. Sci Technol Hum Values 25(1):3–29

Russi D (2020) Governance strategies for a successful marine protected area—the case of Torre Guaceto. Marine Policy 115:103849

Santos R, Antunes P, Baptista G, Mateus P, Madruga L (2006) Stakeholder participation in the design of environmental policy mixes. Ecol Econ 60(1):100–110

Sekhar NU (2003) Local people's attitudes towards conservation and wildlife tourism around Sariska Tiger Reserve, India. J Environ Manag 69(4):339–347

Stemberk J, Dolejs J, Maresova P, Kuca K (2018) Factors affecting the number of visitors in national parks in the Czech Republic, Germany and Austria. ISPRS Inter J Geo-Information 7(3):124

Stone LS, Stone TM (2011) Community-based tourism enterprises: challenges and prospects for community participation; Khama Rhino Sanctuary trust, Botswana. J Sustain Tourism 19(1):97–114

Stringer LC, Dougill AJ, Fraser E, Hubacek K, Prell C, Reed MS (2006) Unpacking "participation" in the adaptive management of social–ecological systems: a critical review. Ecol Soc 11(2):39

The Croatian Parliament (2018) Environmental Protection Act. Zagreb: Official Gazette 80/13, 153/13, 78/15, 12/18

Tosun C (1999) Towards a typology of community participation in the tourism development process. Anatolia 10(2):113–134

Tosun C (2000) Limits to community participation in the tourism development process in developing countries. Tour Manag 21(6):613–633

Tosun C (2006) Expected nature of community participation in tourism development. Tour Manag 27(3):493–504

Trimble M, de Araujo LG, Seixas CS (2014) One party does not tango! Fishers' non-participation as a barrier to co-management in Paraty, Brazil. Ocean Coast Manag 92:9–18

Tritter JQ, McCallum A (2006) The snakes and ladders of user involvement: moving beyond Arnstein. Health Policy 76(2):156–168

Tverijonaite E, Ólafsdóttir R, Thorsteinsson T (2018) Accessibility of protected areas and visitor behaviour: a case study from Iceland. J Outdoor Recreation Tourism 24:1–10

Van Den Hove S (2000) Participatory approaches to environmental policy-making: the European Commission Climate Policy Process as a case study. Ecol Econ 33(3):457–472

Wondirad A, Ewnetu B (2019) Community participation in tourism development as a tool to foster sustainable land and resource use practices in a national park milieu. Land Use Policy 88:104155
Yergeau ME (2020) Tourism and local welfare: a multilevel analysis in Nepal's protected areas. World Dev 127:104744
Zhang Y, Cole ST, Chancellor CH (2013) Residents' preferences for involvement in tourism development and influences from individual profiles. Tour Plan Dev 10(3):267–284
Zorrilla-Pujana J Rossi S (2014) Integrating environmental education in marine protected areas management in Colombia. Ocean Coast Man 93:67–75

Part II
Local Community and Well-Being

Chapter 5
Overtourism and the Local Community Well-Being

Ivana Damnjanović

Abstract This chapter aims to provide a general overview of overtourism related-impacts on local communities' well-being in nature-based tourism destinations. It emphasises the interdependencies of multiple stakeholder groups and the need for their collaboration to address the contemporary challenges. To deliver our conclusions, we analyse the Mediterranean area and survey relevant literature addressing overtourism with particular attention paid to the nature-based tourism destinations and local communities' well-being. The recent literature relates the concept of overtourism with detrimental effects on the destination level. While the tourism-induced environmental and economic impacts have been extensively addressed in the relevant literature, the sociocultural impact, especially in the context of European protected areas, remained unattended. Thus, this chapter focuses on the impacts of overtourism on local communities and their quality of life and well-being. Protected areas in vibrant tourism regions, especially those near the coast, need management approaches that will enable usage-protection equilibrium and local communities' well-being. They often represent a setting for communities' rites, rituals, customs, and traditions associated with their spiritual, physical, emotional, and mental health. The overuse by the tourism industry can disrupt the essence of the locale. In these fragile but complex ecosystems, local communities often take on the indispensable responsibility of nature protection stewardship. Therefore, nature-based tourism development should be based on its capacity to annul and prevent the negative impacts of overtourism, especially in the context of local communities. This chapter builds upon scholarly and managerial perspectives to foster the understanding of overtourism-related challenges and discuss potential responses in the Mediterranean protected areas.

Keywords Overtourism · Protected areas · Sustainable tourism · Local community well-being

I. Damnjanović (✉)
Faculty of Health and Business Studies, Singidunum University, Valjevo, Serbia
e-mail: idamnjanovic@singidunum.ac.rs

A. Mandić and L. Petrić (eds.), *Mediterranean Protected Areas in the Era of Overtourism*, https://doi.org/10.1007/978-3-030-69193-6_5

5.1 Introduction

Tourism was on the road of seemingly unstoppable success for decades until 2020. However, success depends on how we define and measure it. Under the lens of sustainability, it is occasionally excessive and fails local communities whose needs dissolve under overtourism. Overtourism is the source of environmental, social, and economic impacts that have detrimental effects on numerous destinations (Damnjanović 2020). A rising concern is related to their local communities' quality of life, including their health, welfare, and general well-being.

Destinations thriving on healthy ecosystems have come into tourists' spotlight. Simultaneously, increasingly desirable authentic tourism experiences depend on destinations' atmosphere based on local culture, lifestyle, and heartbeat. However, sense of place can differ with residents and tourists (Kianicka et al. 2006), so a balance in use by the tourism industry has to be achieved instead of overuse, disrupting the locale's essence. In such traditionally established socioecological systems (Berkes and Folke 1998; Colding and Barthel 2019), local communities take on an indispensable responsibility of nature protection stewardship. Therefore, nature-based destinations worldwide, including terrestrial and particularly marine protected areas, are in dire need of governance and management approaches that enable usage-protection equilibrium (Sarkki 2017; Villasante et al. 2016). Within such an intersection, tourism can contribute towards improving human health and well-being and stimulate appreciation and stewardship of nature (Azara et al. 2018). The sustainable approach to contemporary and future tourism development is founded in its capacity to prevent and annul potential negative impacts of overtourism, especially in preserving local communities' well-being on tourism destinations, protected areas in particular.

So far, extensive research has been conducted on various aspects of overtourism, including its impacts on destination residents (Cheer et al. 2020; Cheung and Li 2019; Goodwin 2017; Milano 2017; Milano et al. 2019a; Muler Gonzalez et al. 2018; Perkumienė and Pranskūnienė 2019; Sari and Nazli 2020), nature-based and protected area tourism (Chung et al. 2018; Hockings et al. 2020; 2018; Leung et al. 2018; Mandić 2019; Spenceley et al. 2017) in overtourism context (Koščak et al. 2020). Also, there is a rich base of research on local communities in tourism (Fiorello and Bo 2012; Lopes et al. 2015) as well as on human and community well-being (Cloutier Cloutier et al. 2019; Sarkki 2017), its connection to nature (Azara et al. 2018; Naidoo et al. 2019; Sandifer et al. 2015) and tourism (Dwyer 2020; Musikanski et al. 2019). This chapter fills the gap in the research of local communities' well-being in nature-based tourism in the era of overtourism.

Therefore, the chapter expands on the topic by determining the interconnection between local communities' well-being and (1) its measurement as a factor of sustainable tourism success, (2) nature-based and protected area tourism, (3) all in the context of overtourism.

5.2 Tourism: Success or Failure?

For several decades, tourism success faltered only due to occasional crises and showed the ability to regenerate regardless of their nature, origin, scope, and consequences (global financial downturn, wars and acts of terrorism, political causes, natural and climate-change disasters, virus outbreaks, etc.). With its ninth consecutive year ahead of average economic growth rate, 10.3% of global GDP and accounting for 1 in 10 jobs around the world (World Travel & Tourism Council [WTTC] 2020), the tourism sector has been looked at as a solution to uneven regional development and an opportunity for local communities' employment. Europe continued its role as the main inbound region with a full decade of sustained tourism growth by hosting every second international tourist arrival in 2019 (World Tourism Organization [UNWTO] 2020b). In these terms, Mediterranean Europe experienced another year as the most successful subregion in which most destinations enjoyed double-digit growth, with Italy, Greece, Portugal, and Croatia at the top of the list (UNWTO 2019a).

In many world regions, tourism is the source of welcome change to the lives of local communities. Being a complex system, it incorporates the needs of people with various roles on destinations (tourists, local communities, tourism, and related businesses, NGOs, government, etc.). Tourism has the power to accordingly respond to them by altering and enhancing the environment, without which many regions would still be underdeveloped: halted extractive and harmful activities, increased accessibility and desirability to (re-)inhabit and visit, rekindled appreciation of traditional lifestyles, increased pride of the locale, reintroduced traditional occupations leading to poverty mitigation and improved social position of vulnerable groups, etc.

However, due to the prevalent framework of measuring tourism's success, which is based solely on economic parameters, positive changes tourism brings, remain under the radar. Simultaneously, in the era of overtourism (Ali 2016; Koens et al. 2018), tourism has at numerous destinations become excessive (Capocchi et al. 2019; Sari and Nazli 2020) and failed destination communities -a manifestation of unsustainable tourism (Mihalic 2020). With the global rise of sustainable tourism sentiment (UNWTO 2020a) as a new perspective, the reality of tourism performance changes significantly. UN sustainable development goals (UN SDGs) redefine success parameters, while Statistical Framework for Measuring Sustainable Tourism (MST) (UNWTO 2017) should extend its success indicators to environmental, social, and cultural dimensions. However, there has been an urge to activate critical thinking concerning UN SDGs and tourism (Boluk et al. 2019). Accordingly, the SDGs are criticised as having a growth mentality without explicit commitment to human prosperity and travelling within the limits of the ecosystems to which we belong (Hall 2019), the result of which is overtourism. Overtourism appearing as a symptom of the economic growth model prevents tourism from contributing to (sustainable) development which has evolved into the concept of societal well-being (Sharpley 2020). It is argued that destination residents' well-being is associated with achieving sustainable tourism development (Dwyer 2020). Also, SDGs do not include subjective indicators of well-being and social connections (Iriarte and Musikanski 2019). The conclusion is that measuring tourism's contribution to the SDGs at various scales essentially depends on the existence of sustainable tourism indicators (Rasoolimanesh et al. 2020) and that they should include various dimensions of well-being (Dwyer 2020).

For example, the Global Sustainable Tourism Council (GSTC) developed destination and industry criteria for measuring sustainability arranged in four pillars: sustainable management, socio-economic, cultural, and environmental impacts (GSTC 2017). Each criterion has its set of clear-cut and easy-to-follow indicators and specific UN SDGs towards which it is directed. Destination criteria contain an entire section dedicated to socio-economic sustainability with two subsections:

- delivering local economic benefits: measuring the economic contribution of tourism; decent work and career opportunities; supporting local entrepreneurs and fair trade;
- social well-being and impacts: community support, preventing exploitation and discrimination, property and user rights, safety and security, access for all.

An example of the GSTC criteria application is Dubrovnik, Croatia, which commissioned the2019 GSTC Destination Assessment. Dubrovnik-Neretva region, a UNESCO-designated natural, cultural, and historical site in which tourism is a key economic pillar, is one of the most prominent tourist destinations in the Mediterranean with almost 1.3 million overnight visitors in 2018 to which the numbers of excursionists and cruise passengers entering the city with a daily maximum of over 9,000 in August 2019 should be added (Pappas 2020). With these numbers and strong seasonality, Dubrovnik has become a globally recognised example of overtourism. The region boasts 40 protected natural areas, some of which are among the top city attractions. This assessment showed both positive and negative aspects of tourism on this destination and mapped the road towards reaching its sustainability.

The application of sustainability principles in tourism both by destinations and businesses allows more efficiency in resource usage, biodiversity conservation, and dealing with climate change-induced challenges, which increases their competitiveness (Calderwood and Soshkin 2019). As many as 101 UNWTO Member States have sustainability as an essential part of their tourism policies (UNWTO and UNEP 2019).

5.3 Local Communities in the Focus of Nature-Based Tourism Products

A corresponding demand has paralleled the rise of sustainability sentiment by tourism organisations and providers. Nature and wildlife tourism are major contributors to economic activity around the world (Hockings et al. 2020, p. 9). To illustrate that, six current global travel trends include (UNWTO 2019a, p. 5):

- travel "to change"—living like a local and looking for authenticity and transformation;
- the pursuit of a healthy life including walking, wellness and sports tourism;
- rising awareness of sustainability, particularly zero plastic and climate change.

These trends require preserved natural surroundings (protected areas in particular) and authentic lifestyles that they harbour. Many such tourism forms (Azara et al. 2018; World Bank and Spenceley 2020) differentiate in the main travel motive, but their scope remains not so clear-cut because they entwine and intersect in several points, including: (1) the need for preserved natural and sociocultural assets; (2) well-being and health purposes of travel; (3) the necessity of sustainability approach to tourism development.

By their definition, many forms of tourism in natural areas are closely related to existing local communities. The roles they take on are often crucial for the quality of delivered tourism experiences. Therefore, local communities' well-being needs to be put into the centre of tourism initiatives and operations. Among such tourism forms, some do not allow overcrowding since it would not fit their tourist profile. Ecotourism, for example, is created for low-scale visitation. It is also desirable as a form of tourism that sustains local people's well-being (The International Ecotourism Society 2015). Responsible tourism focuses entirely on the benefits it provides for local communities in sociocultural, economic, and environmental aspects (Center for Responsible Travel 2019, p. 2).

However, the popularity of certain tourism activities may lead to overtourism on some natural destinations. The experience provided for tourists depends on thriving and healthy local communities, but tourists' behaviour is not necessarily equally responsive. For example, with a 10% annual growth rate, wildlife watching tourism also includes visiting culture, historic buildings, and scenic lookouts (World Tourism Organization [UNWTO] and Guangdong Chimelong Group 2020). Similarly, culturally immersive experience is an inevitable ingredient in adventure tourism products (Adventure Travel Trade Association [ATTA]& The George Washington University International Institute of Tourism Studies[IITS] 2020, p. 8). However, a sustainable approach to tourism increases the competitiveness of such destinations, many of which can be found in the Mediterranean region (ATTA & IITS 2020): France, Spain, Malta, Italy, Cyprus, and Greece on the top 30 list and Slovenia and Israel among the top five developing countries accompanied by Croatia, Montenegro, Turkey, and Albania. Wellness tourism as another globally thriving and prospective form of tourism - traditional in Europe and the Mediterranean region, particularly in France, Italy, Spain, and Turkey (Global Wellness Institute[GWI] 2018)—relies on the well-being of destinations' local communities. An increasing number of destinations realise that a healthy place represents the DNA of authentic wellness offer and prioritise their residents' well-being and environment, an example being Wellness Valley in Romagna, Italy (GWI 2018).

Various nature-based tourism products incorporate direct or less direct contact with local communities and their lifestyles, making them an integral part of the product: community-based and indigenous, rural, ethical, voluntourism, various niche, or special-interest tourism products, etc. It is evident that the link between tourism and the impact on the destination, especially local communities, will become crucial in the time to come.

5.4 Nature Protected Areas Tourism Today

All described types of tourism require natural settings for their production and delivery. Healthy and preserved natural surroundings with thriving local human ecosystems are necessary to respond to initial travel motives and provide genuine, authentic, unique, and memorable experiences that such motives imply in contemporary tourism. Therefore, protected areas are often regarded as desirable tourism destinations that can satisfy a craving for peace, quiet, tranquillity, meaningfulness, balance, transformation in opposition to stressful life and health threats. The ultimate travel goal is improving and maintaining optimal health or preventing disease. In the light of a new global crisis (COVID-19 pandemic), it must be emphasised that the role protected areas will have in life in general and in tourism will be significant. Due to the pandemic, the connection between healthy nature and human health and well-being, especially for its therapeutic effects for our mental health, is highlighted (Hockings et al. 2020, pp. 13–14). This will particularly be important to the increasing number of people living in urban areas where they may feel or actually be deprived of nature's healing power. For them, the proximity to preserved nature can reduce stress, restore brain functions, lower the incidence of more than a dozen diseases, especially in the case of tech-connected and nature-disconnected younger generations (Global Wellness Summit 2019, pp. 57–58). The role of protected areas in creating well-being, preventing public health problems, and promoting an active lifestyle is emphasised by the Europarc Federation's promotion of the "Healthy Parks, Healthy People" 2020 campaign designed even before the COVID-19 pandemic (EUROPARC Federation 2018).

However, there is a possibility that this popularity might backfire and harm the very assets that protected areas try to save. It has been reported that there are approximately 8 billion annual visits to protected areas worldwide and about 3.8 billion in Europe (IUCN 2015). The direct value of wildlife tourism generates 21.8 million jobs (World Travel and Tourism Council [WTTC] 2019). However, the distribution of visits and benefits has not been equal. While many protected areas have remained remote and generally undisturbed, there are numerous examples of those being swarmed by tourists and visitors. In line with the portion of tourism that Europe and the Mediterranean region host, examples of overvisited protected areas are Cinque Terre in Italy or Plitvice National Park in Croatia. This occurrence has raised concerns about overtourism on some heavily visited sites, and meagre visitation numbers on others in which tourism may bring about necessary benefits (Center for Responsible Travel 2019, p. 10). On certain destinations, overtourism is often a result of domestic tourism growth, while international tourism causes it on others (UNWTO and Guangdong Chimelong Group 2020, p. 10).

In the Mediterranean region, 6.01% of which is under protection, 1,062 protected areas are coastal and marine (Claudet et al. 2020). Adding to the concern is the fact that 80% of all tourism occurs within coastal areas (World Wildlife Fund [WWF] 2019), where overuse of water resources, economic and environmental vulnerability is reported (Peeters et al. 2018, p. 34). As such, marine protected areas and islands are particularly vulnerable to overtourism effects, especially in terms of environmental impacts. Ocean health is recognised as critical for human well-being and is threatened by multiple stressors, particularly due to the cultural and recreational services it provides (Claudet et al. 2020).

5.5 Nature Protected Areas Tourism and Local Communities

It has been evident that all benefits emanating from tourism in nature protected areas cannot do without healthy communities living within and in their vicinity. They do not only coexist there but comprise a part of an intricate human–nature ecosystem that has evolved through time (Negev et al. 2019). This is evident from the International Union for Conservation of Nature (IUCN) definition of protected areas (Dudley 2008/2013, p. 8), which implies (Leung et al. 2018, pp. 2–3):

- respecting rights and sociocultural authenticity, cultural heritage and traditional values of local and indigenous communities;
- ensuring viable, long-term socio-economic benefits, including stable employment opportunities and social services for local communities.

However, protected areas differ substantially in their main goals and objectives. This is why ensuring local communities' well-being and their inclusion in tourism value chains through management practices is specific for each IUCN protected area management category. Therefore, it is worth exploring existing differences to emphasise the nature of the relationship between protected areas (and tourism in them) and their local communities (Table 5.1, based on IUCN 2019).

Each protected area category resonates with a certain level of connection with local communities, i.e., preserving their way of life, traditional, cultural, spiritual, and other values through which it caters to their well-being. Although this approach is not driven by tourism, it contributes immensely towards creating a unique and authentic tourism experience that is in demand today through a myriad of travel motives. Therefore, the protection of both natural and sociocultural values against overtourism effects is inherent for protected areas. Simultaneously, many protected areas that are not under threat of overuse can help mitigate overtourism in heavily visited ones by attracting the excessive numbers of tourists.

Higher biodiversity in terrestrial protected areas leads to more nature-based tourism and related socio-economic benefits that contribute to biodiversity conservation (Chung et al. 2018). Similarly, sustainable tourism in marine protected areas

Table 5.1 Focus on Local Communities per IUCN Protected Area Category

Protected Area Category	Relatedness to local communities through objectives/features/roles
Ia: Strict Nature Reserve	Such natural area could be of religious or spiritual significance (such as a sacred natural site), so its objective is to conserve cultural and spiritual values associated with nature
Ib: Wilderness Area	Among its objectives there are: * enabling indigenous communities to maintain their traditional wilderness-based lifestyle and customs, living at low density and using available resources in ways compatible with conservation objectives * protecting relevant cultural and spiritual values and non-material benefits to indigenous populations (e.g., solitude, respect for sacred sites, ancestors, etc.)
II: National Park	The category's objectives include: * taking into account indigenous people and local communities' needs, including subsistence resource use in line with its primary management objective * contributing to local economies through tourism
III: Natural Monument or Feature	An objective of the category is to conserve the traditional spiritual and cultural values of a site/feature: * culturally influenced natural features (e.g., cave dwellings) * natural-cultural sites, e.g., sacred natural sites of importance to faith groups * cultural sites with associated ecology (e.g., archaeological/historical sites inextricably linked to a natural area)
IV: Habitat/Species Management Area	The category may provide flexible management strategies in buffer zones around more strictly protected areas, making them more acceptable to local communities and other stakeholders
V: Protected Landscape/Seascape	The category definition highlights the interaction of people and nature, which has, over time, produced its distinct character with significant ecological, biological, cultural, and scenic value. Therefore, its objectives include: * maintaining a balanced interaction of nature and associated traditional management approaches, societies, cultures, and spiritual values * providing a framework for active community management involvement

(continued)

Table 5.1 (continued)

Protected Area Category	Relatedness to local communities through objectives/features/roles
VI: Protected area with sustainable use of natural resources	The category has cultural values and traditional natural resource management systems built-in. Accordingly, its objectives include: * promoting social and economic benefits of sustainable development to local communities where relevant (especially on the local level); * facilitating intergenerational security for local communities' livelihoods * integration of other cultural approaches, belief systems, and world-views regarding social and economic approaches to nature conservation

Source Based on: *Protected area categories* by IUCN (2019)

can stop coastal degradation by creating synergetic interactions between conservation and ecosystem services (e.g., fisheries) and securing benefits for local communities (Lopes et al. 2015). Conversely, rapid tourism expansion, especially in marine and coastal areas, has often disregarded local communities and their role of ecosystem stewards, thus preventing sustainable development (Lange 2015). To meet its goals, each protected area category needs local communities' active involvement in tourism development and land use planning (Alexander et al. 2018).

5.6 Overtourism and Local Communities

Tourism experience has become the main pivot in decision-making on the demand side and product design and delivery on the supply side. Today it extends to tourists requiring the opportunity to co-create their own experience, especially through digital platforms that contain such amenities that give added value to the devoted travel time (Skift 2019, pp. 10–13; Zhang et al. 2017). As overtourism endangers desired experiences, emerging destinations promise peaceful, immersive, and exciting alternatives with new stories to tell and relationships with places and their communities to build (Skift 2019, p. 34). However, undertourism exists hand in hand with overtourism as another destructive product of mismanaged destinations (Gowreesunkar and Vo Thanh 2020). Undertourism refers to destinations where insufficient tourism opportunities prevent potential benefits, including long-term quality of life (Mihalic 2020). Only a bird' s-eye view of broader destination regions, with a creative management approach and adequate, timely preparation, might allow undertourism to counterpoise overtourism. This is emphasised in the case of protected areas in which providing meaningful and high-quality visitor experience results in an increased sense of nature stewardship (UNWTO & UNEP 2005). Such experiences should not be promised or cannot be delivered through the entire value chain without paying close attention to local communities' well-being.

Local communities' well-being depends on meeting their needs which differ according to their roles, especially in protected areas: tourism and related sectors service providers, decision-makers or stewards, or simply residents of the area. The fact that protected areas are usually rural, remote, or offbeat might imply that their local communities are less prepared or equipped to deal with tourism-related impacts, especially when it is excessive.

Overtourism negative impacts on local communities are perceivable through its definitions:

1. According to European Parliament, "overtourism describes the situation in which the impact of tourism, at certain times and in certain locations, exceeds physical, ecological, social, economic, psychological, and/or political capacity thresholds" of a destination (Peeters et al. 2018).
2. Center for Responsible Travel (2018) defines overtourism as "tourism that has moved beyond the limits of acceptable change in a destination due to quantity of visitors, resulting in degradation of the environment and infrastructure, diminished travel experience, wear and tear on built heritage, and/or negative impacts on residents".
3. UNWTO considers overtourism as "the impact of tourism on a destination, or parts thereof, that excessively influences the perceived quality of life of citizens and/or quality of visitors' experiences in a negative way" (UNWTO 2018).
4. The Responsible Tourism Partnership refers to overtourism as to "destinations where hosts or guests, locals or visitors, feel that there are too many visitors and that the quality of life in the area or the quality of the experience has deteriorated unacceptably" (Goodwin 2017).
5. The International Ecotourism Society (TIES) (2019) defines overtourism as the one "which in its simplest form is tourism that harms communities by overuse or destruction of resources through overcrowding and stems from a lack of concern about destination community health and welfare".

The first definition remains broad regarding the communities' well-being and leaves the negative overtourism impacts implicit. The following one (2) focuses more on the negative physical impacts on the site and the travel experience but recognises those that extend to the residents. Definitions (3) and (4) focus and emphasise the deteriorated quality of life of local communities. The last definition (5) pinpoints local community health and welfare as the principal victims of under-controlled excessive tourism. What can be concluded is that overtourism diminishes the quality of tourist experience, but is predominantly the problem that negatively affects local communities. If the problems underlying decreased well-being and general quality of life of local communities were tackled, tourist experience would be more easily delivered.

Responding to certain overtourism impacts on a destination may only be treating symptoms of deeper underlying issues. Epler Wood et al. (2019) recognise them as "the invisible burden" and define it as "the unaccounted for destination costs [necessary] to provide local infrastructure and the protection of eco and sociocultural systems for tourists and local people" (Epler Wood et al. 2019, p. 7). This burden is placed on local economies, i.e., local communities often unable to cover them. This can further degrade tourism assets precious to local community members' business operations who are thus drawn into this vicious circle. The invisible burden of tourism, particularly regarding the protection of social and natural capital, can influence the health of both local communities and entire destinations. This specifically relates to protected areas, in which a lack of financial resources allocated to vital heritage or environmental assets undermines their market value (Epler Wood et al. 2019, p. 19) and impact their sense of place for both tourists and locals (Guthey et al. 2014, p. 260; Kianicka et al. 2006). On the other hand, the economic impact of protected areas tourism dwarfs current conservation and maintenance investments, which would yield a substantial return (IUCN 2015). Incorporating these invisible costs of tourism into protected areas management, especially in those that suffer from overtourism, may add to the growing value of protected areas to tourists and local communities and enable necessary financial sources to balance tourism and protection.

5.7 Healthy Communities and Tourism in Protected Areas

World Health Organisation's 1984 definition of health focuses on people's physical, mental, and social well-being. Further, well-being is defined as feeling satisfied and happy, developing as a person, being fulfilled, and contributing to the community (Wellness Tourism Worldwide 2011, p. 9). Community well-being results from the combination of social, environmental, economic, political, and/or cultural factors influencing a community's quality of life, thus putting emphasis away from material improvements of a place (Cloutier et al. 2019). Community well-being depends on individual levels of well-being of a community's members and its synergy (Sung and Phillips 2018; VanderWeele 2019). Many definitions of community well-being focus on communities of place, i.e., sharing space and governance for functioning in those places (Sung and Phillips 2018). Such an approach is particularly valid when considering tourism on destinations as places and in interaction with their local communities.

There are a plethora of community well-being indicators (Sirgy 2018), such as those available in the OECD Well-being Framework (2019), most commonly divided into subjective and objective indicators (VanderWeele 2019), or combined, i.e., intersubjective (Choi et al. 2020).

In protected area tourism, healthy communities (people well-being) are as important as healthy nature (environmental well-being), and their interdependence is highlighted (Junot Junot et al. 2018). Sandifer et al. (2015) emphasise strong linkages between human health benefits, resilience and community well-being and nature, biodiversity, and ecosystem services, particularly in coastal communities. Allgood et al.'s (2019) conclusions exemplify that incorporating a variety of aspects of human well-being into community-based wildlife conservation projects contributes to their success.

Finally, Dwyer (2020) recognises how the Beyond GDP approach can measure tourism destination sustainability and resident current and future well-being (economic, human, social, natural capital). Excessive and uncontrolled tourism activities make it more difficult for locals to achieve their own well-being in the place they inhabit. Musikanski et al. (2019) propose a tourism management methodology (Planet Happiness) whose focal point is local communities' well-being and destination sustainability as a key to fighting off overtourism. Table 5.2 provides an overview of various types of local communities' well-being in protected areas and their reference to overtourism.

If it is not excessive, tourism can positively impact local communities in all three sustainability aspects, making it possible for them to maintain the state of general well-being.

From an economic perspective, a community's mere existence in the close vicinity of a protected area provides tourism-induced benefits. Households located within a

Table 5.2 Local Communities in Protected Areas: Well-Being vs Overtourism

Type of well-being	Reference to protected area local communities regarding overtourism
Physical	The protected area should provide its residents with opportunities for maintaining physical health through medical services and healthcare access, availability of medicinal herbs, and other natural, alternative, and traditional health benefits. Tourism impacts and the presence of tourists should not interfere with this aspect, especially not overcrowding
Environmental	Preserved nature, such as clean air, water, and the environment in general, recreational, sports, and other nature-based activities, the ones available to tourists, should be available to local communities The protected area should be careful not to deprive them of these benefits by overusing protected area assets through tourism
Mental	Local communities should be able to practice their traditional and other activities such as meditation, yoga, and related therapies. These should be available separately from or together with tourists, depending on examined residents' needs
Spiritual	Overcrowding in a protected area should not hinder its residents' spiritual connections with nature (prayer, time alone) or with their families. The opportunities offered to tourists should be readily available for locals as well
Emotional	Local communities in a protected area should have conditions for solitude, retreat, stress reduction, etc., as the primary area concern (perhaps, more than equal to tourists). When tourism is excessive, residents will eventually show dissatisfaction
Social	Protected areas should provide unique surroundings for local communities to express social behaviour and habits, such as traditional gatherings, ceremonies, celebrations, and generally, how their society functions It should not be allowed for them to be disrupted through overcrowding

Source Created by Author

10 km radius from a protected area visited by tourists have 17% higher wealth levels and 16% lower likelihood of poverty than similar households living far from protected areas, especially in developing countries (Naidoo et al. 2019, pp. 1–4). Such support for families comes through a share in tourism-derived income (Hockings et al. 2020, p. 9) and their involvement in tourism services provision (e.g., tour guides), particularly in remote regions where employment for local communities is scarce (UNWTO and Guangdong Chimelong Group 2020, p. 11).

On the environmental side, tourism in protected areas has helped to safeguard nature and the provision of related services for local communities, including food and water security, disaster risk reduction, climate mitigation, and adaptation (Dudley et al. 2010) and nurturing their relationship with the land and a sense of pride (UNWTO 2019b, p. 2).

Additionally, the general quality of life is enhanced through, e.g., improvements to infrastructure and telecommunications, education, training (Leung et al. 2018, p. 19), local community empowerment, especially women and youth (UNWTO 2019b, p. 2). Finally, innumerable cultural, spiritual, psychological, emotional, aesthetic, and health values of protected areas for the local communities have been emphasised (Hockings et al. 2020, p. 8; Leung et al. 2018, p. 19; UNWTO 2019b, p. 2).

All stated benefits have recently been recognised through Central Europe Eco-tourism: Tools for nature protection programme (Interreg—CEETO 2018, p. 36) in which several protected areas from the Mediterranean region stand as good practice examples: 3 protected areas from the Italian Emilia-Romagna Region, Nature Park Medvednica in Croatia, and Landscape Park Strunjan in Slovenia (Koščak et al. 2020).

5.8 Overtourism Negative Impacts on Local Communities in Protected Areas

According to Skift (2020, p. 15), “the holy grail of a sustainable yet lucrative tourism industry has, by definition, come to mean tourism that local residents and stakeholders feel good about, too”. However, since numerous destinations, including protected areas, worldwide suffer from overtourism, it is evident that negative impacts on local communities can undermine tourism’s sustainability in all its aspects.

5.8.1 Economic Impacts of Tourism

In light of traditional tourism measuring, we tend to see the economic impacts of tourism as positive through its revenues. However, there are numerous negative ones. They are listed here inspired by and as a sum of several sources with the special attention to protected areas tourism (Interreg—CEETO 2018; Leung et al. 2018; Peeters et al. 2018; WTTC and McKinsey Co. 2017):

- Employment options may exist but are often menial, with low skill requirements and low wages. This additionally applies to rural, marginalised, or vulnerable communities. Crowded destinations attract more people seeking employment from outside the local community. The industry's seasonality causes income-related insecurity and job losses, especially during low seasons and external crises.
- Economic leakage takes away the purpose of tourism and related businesses on a destination since they are often owned by foreign investors instead of local communities, which can add to resident disgruntlement.
- Unequal distribution of economic benefits which leads to inequalities and relative poverty.
- Inflation is caused by increased demand for goods and services, which become overpriced and unaffordable for residents. When the real estate becomes unreasonably priced due to gentrification, local community members might be forced to move out of the area.
- Training and advancement are rarely provided as an opportunity for an unqualified local workforce, as it may be seen as an unnecessary expense.
- The disappearance of traditional local livelihoods results from them being replaced by more promising tourism employment opportunities or being incompatible with the prevalent tourism activities.
- Overdependence on tourism makes local communities economically vulnerable and "unresilient" in the case of a sudden downturn in visitation.

COVID-19 pandemic global crisis demonstrated numerous problems that excessive tourism can cause regarding the economic aspect of local communities' well-being. With tourism on halt, many local communities found their livelihoods threatened, with the problem especially severe in community conservancies and privately protected areas in which staff salaries depend heavily on tourism (Hockings et al. 2020, p. 11). People are forced to turn to alternative subsistence sources, not all of which are legal or non-harmful for local ecosystems. Finally, such problems create a dozen more to which authorities cannot respond accordingly (such as human–wildlife incidents or fires, etc.).

5.8.2 Sociocultural Impacts of Tourism

Negative impacts of excessive tourism also exist in the sociocultural domain, particularly in protected areas. They diminish local communities' quality of life on the psychological and emotional level (internal) and both mundane and comprehensive societal level (external). Once again, the list provided here is a collection inspired by several sources (Center for Responsible Travel 2018; Interreg—CEETO 2018;

Hockings et al. 2020; Leung et al. 2018; Peeters et al. 2018; UNWTO 2019b; WTTC and McKinsey Co. 2017) and is organised in two parts.

Externally caused and manifesting negative impacts include:

- Tourists may overuse infrastructure, consequently decreasing the quality of non-tourism services essential to local communities. Its maintenance effort and cost remain as residents' burden, especially when infrastructure development is distorted to meet tourism needs, far exceeding the inhabitants' needs.
- Priority for resource exploitation given to tourists often leaves residents deprived of basic (energy, water, etc.) or other facilities that influence their life quality more or less directly.
- The destabilisation of communities through crime, gambling, begging, alcohol and drug abuse, sexual exploitation, etc.
- Discrimination or resettlement of local communities for several reasons, e.g., as deemed incompatible with tourism development or incapable of providing sustenance. In some cases, local communities' communal or nomadic lifestyle may represent an obstacle to the way tourism is imagined.
- Change of essence of the place through the influx of people looking for employment in tourism brings new patterns of behaviour, tradition, customs, and habits. In the case of indigenous communities, such displacement is from their ancestral territories.
- Loss of traditional knowledge and cultural values through residents' displacement or loss of traditional employment opportunities.
- Devaluation of tradition through re-enacting religious and other ceremonies and rites for tourists, including their temporal disruptions.
- Deterioration of artistry through volumising production of crafts for tourists.
- Protests and anti-tourism campaigns by local communities due to the complex nature of the harmful effects of overtourism on their lives.
- Destination closures for tourists and non-residents as the last resort decision to allow the environment and communities to recuperate from harmful overtourism effects.

Internally caused and manifesting negative excessive tourism impacts, often under-perceived due to their subtle nature incorporate:

- Changed residents' behaviour by observing tourists and new people inhabiting an area in the hope of achieving what they deem to be higher status. Protected area residents may imitate tourists and become disillusioned.
- Loss of pride and self-esteem of those not actively involved in the tourism industry. It may also originate from working in degraded positions in which they feel as "servants" to tourists, while their education and experience become devalued. Encounters may become superficial, misleading, and their culture misinterpreted.
- The offence is caused to residents when their culture is inappropriately presented or when tourists behave disrespectfully. Even conflicts may arise.
- Loss of spiritual integrity of a destination for local community members through the misbehaviour of some tourists.

- Alienation of residents from their residential neighbourhoods and entire destination because of an unacceptable level of disturbances to their way of life by excessive numbers of tourists, which more often leads to protests, disgruntlement, or even exodus (for example, "the Venice syndrome") (Milano 2017, p. 9).
- Lost sense of belonging and a diminished sense of place (Cheer et al. 2019), especially on vulnerable destinations including protected areas, small islands, and critical cultural heritage locations (Milano et al. 2019b, p. 355). The term solastalgia has recently been connected to overtourism, denoting residents' responses to negative changes in their home environment and their lack of comfort of the current relationship their place previously offered (Lalicic 2020).

It is also worth noting that there is a subjective sense of residents' well-being in the context of tourism, and it determines the perceived quality of life. Not even the initial economic benefits can keep their welcoming sentiment when visitor numbers increase so much as to lead to the plummeting quality of life (Croes et al. 2017). It is also in close relation to residents' willingness to welcome tourists, in addition to the level of community dependence on tourism (Muler Gonzalez et al. 2018).

Many aspects of the relationship between tourists and local communities on destinations overburdened by tourism have been covered by numerous research studies ranging from residents' inclination towards tourism degrowth (Milano et al. 2019a), bolstering social-ecological and community resilience (Cheer et al. 2020) through exploring sustainable degrowth (Claudet et al. 2020; Fletcher et al. 2020).

5.9 Conclusion: Lessons Learned to Pave a Way Forward

Destinations striving to become or remain competitive need to protect their communities and cultural capital and incorporate local stakeholders' concerns into their tourism boards' strategic planning (Skift 2020, p. 15). Overtourism can be overcome by making it possible for local communities to reap direct benefits from tourism and keeping their core values intact (UNWTO 2019b, p. 4) and allowing the balance of equality between the right to travel and residents' rights (Perkumienė and Pranskūnienė 2019). However, many protected area managers worldwide have multiple and sometimes conflicting goals in their hands, making it difficult to ensure that communities are involved and benefit and protected from overtourism (Leung et al. 2018, p. 2). One component of successful revenue-sharing systems in protected areas is community involvement in decision-making (Spenceley et al. 2017). A new transformative relationship with nature has been perceived as a path to recovery after the global tourism crisis, and the European Union, among other regions, has signalled its intention to embrace this opportunity (Hockings et al. 2020, p. 15). Among the main concerns are support, equity, benefit-sharing, and well-being in the local communities living in or near protected areas (Hockings et al. 2020, pp. 16–18).

Diversification may lead to sustainable tourism development (Benur and Bramwell 2015; Weidenfeld 2018). As such, it can be a solution to overtourism in nature-based destinations, particularly in those marine and coastal, such as in the Mediterranean region (Zahra 2017). Simultaneously, promoting physical and psychological health benefits of protected areas has been proposed as a step on the tourism recovery journey (Hockings et al. 2020, pp. 16–17). Accordingly, wellness tourism can carve out a unique niche, reduce seasonality, draw visitors to under-visited regions, and benefit local communities. Thus, as an awarded sustainable destination (Green Destinations 2020; Slovenian Tourist Board 2020), Slovenia recognised its four macro-regions, three of which are wellness-oriented (Wellness Summit, Global Spa and International, SRI 2017, pp. 26–29).

Unawareness of underlying, often invisible problems tourism may cause in protected areas and omitting their integration into management processes may make overtourism a second-tier issue. For a destination unprepared to handle tourism impacts, even one tourist may be one too many. Focusing on local communities in that respect, tourism in protected areas should not be something that happens to them and what they need to suffer from. Tourism needs to cover its own expenses while it nurtures local communities as one of the assets it crucially depends on. Building trust and partnership among tourism stakeholders, especially local communities in protected areas, needs to be based on the recognition of everybody's needs. However, determining that may depend on "speaking everybody's language".

Tourism in protected areas is about people's livelihoods, unique ways of life of both tourists and local communities. This will be prominent in the future, especially in the post-COVID19 and pre-other crisis times. Preserved nature inseparable from the local communities that inhabit it in protected areas will be considered a refuge and retreat from the risks from crowds and high population densities. For this reason, among many others, protected areas should not fall under overtourism. It should also be noted that protected areas are not isolated, especially in the densely populated Mediterranean region, but represent a part of broader destination areas. More often than not, overtourism spills over to protected areas whose purpose makes them even less receptive to such a challenge. Sometimes, however, protected areas can help mitigate overtourism in larger destination areas by attracting an acceptable portion of tourists away from the main locations.

Due to high visitation levels, the Mediterranean region needs a holistic and systemic strategy for the future in which sustainable tourism is its core value. Safe and resilient destinations pay close attention to both tourists' and local communities' well-being. This strategy will also need to contain the instructions for tourism "to lose some weight". However, it can happen for two reasons. One is involuntarily imposed, an example of which is the COVID-19 pandemic, and, as such, may just be temporal. On the other hand, a planned and wayward weight-loss process requires some farsightedness and envisioning. It is based on re-tailoring behaviour, deciding what kind of tourists we target and what services and experiences are provided.

Overtourism is essentially not about the numbers. It is about the balance between what a destination can provide tourists with and focusing not only on avoiding harm but also on extracting positive benefits. On the destination level, a more holistic (spatial and temporal) approach can, after a close listening to all stakeholders, find and maintain the point of equilibrium for all—environment, local communities, and businesses. This problem is sometimes difficult to grasp due to a short-sighted approach and lack of knowledge and awareness. On destinations in general and particularly in protected areas, local biodiversity and preserved local communities are the product they sell. Local communities receive income through tourism and live with not only from nature. In the case of excessive tourism activities, nature needs tourism revenue, local caring inhabitants, and their businesses to keep it safe. Therefore, from a tourism perspective, protecting local communities and the environment is a long-term business investment.

Pandemic as the latest crisis put light on the over-reliance and overdependence of many economies on tourism. It is perceived not only through zero demand but also the impacts, such as no revenue and massive loss of jobs. On the one hand, if there were no tourism, many destinations (protected areas in particular) would not have enough financial, managerial, or political support to exist and thrive together with its communities. On the other, tourism has replaced other more sustainable activities in too many cases—many local communities abandoned traditional ways to make a living. In the "no tourism" crisis, this overdependence on tourism has wreaked havoc for local communities. They changed because and for tourism, and now such a change has lost its purpose.

The crisis-induced halt of tourism might spur a shift of focus: from sheer tourist volume to managing their impacts. Excellence in creating memorable experiences for an increasing number of responsible tourists will become the main selling point. It will allow former investments in marketing to be used for management purposes. The pandemic might solve two problems simultaneously for safety reasons—reduce or eradicate overtourism and bring benefits to "under-touristed" locations. They have the chance to learn from their big brothers' mistakes and manage tourism carefully, fairly, genuinely protecting the environment and local communities through adequately prepping tourists for visiting a sensitive location.

The need for further research regarding sustainability as an approach to the complex relationship between tourism, health, well-being, and protected areas is not novel (Azara et al. 2018). Within such a relationship, research focusing on local communities' roles, impacts, and particularly aspects preventing or contributing to their well-being in the context of overtourism, is necessary. A vast spectrum of potential research topics leaves a door open for further qualitative and quantitative research in this domain.

References

Adventure Travel Trade Association (ATTA), The George Washington University International Institute of Tourism Studies (GW) (2020) Adventure Tourism Development Index (ATDI) 2020. https://www.adventuretravel.biz/research/2020-adventure-tourism-development-index

Alexander K, Ramotadima M, Sanderson CE (2018) The power of consensus: Developing a community voice in land use planning and tourism development in biodiversity hotspots. Ecosyst Serv 30:350–361. https://doi.org/10.1016/j.ecoser.2018.02.008

Ali R (2016) Foreword: the coming perils of overtourism. In: Ali R, Clampet J (eds) Iceland and the trials of 21-century tourism, *Skift*. New York

Allgood B, Hofberg M, Musikanski L et al (2019) Assessing Community-Based Wildlife Conservation Programs with the Gross National Happiness Framework. Int J Com WB 2:301–337. https://doi.org/10.1007/s42413-019-00045-7

Azara I, Michopoulou E, Niccolini F, Taff BD, Clarke A (2018) Tourism, health, well-being and protected areas. CABI

Benur AM, Bramwell B (2015) Tourism product development and product diversification in destinations. Tourism Manag 50:213–224. https://doi.org/10.1016/j.tourman.2015.02.005

Berkes F, Folke C (eds) (1998) Linking Social and Ecological Systems: Management Practices and Social Mechanisms for Building Resilience. Cambridge University Press, Cambridge, UK

Boluk KA, Cavaliere CT, Higgins-Desbiolles F (2019) A critical framework for interrogating the United Nations sustainable development goals 2030 agenda in tourism. J Sustain Tourism 27(7), 847–864. https://doi.org/10.1080/09669582.2019.1619748

Calderwood LU, Soshkin M (2019) The Travel & Tourism Competitiveness Report 2019. World Economic Forum. http://www3.weforum.org/docs/WEF_TTCR_2019.pdf

Capocchi A, Vallone C, Pierotti M, Amaduzzi A (2019) Overtourism: A literature review to assess implications and future perspectives. Sustainability 11(12):3303. https://doi.org/10.3390/su11123303

Center for Responsible Travel—CREST (2018) The case for responsible travel: Trends & Statistics 2018. Retrieved from https://www.responsibletravel.org

Center for Responsible Travel (2019) The Case for Responsible Travel: Trends & Statistics 2019 (7). https://www.responsibletravel.org/docs/CaseforResponsibleTravel_2019_Web.pdf

Cheer JM, Milano C, Novelli M (2019) Tourism and community resilience in the Anthropocene: Accentuating temporal overtourism. J Sustain Tourism 27:554–572. https://doi.org/10.1080/09669582.2019.1578363

Cheer JM, Milano C, Novelli M (2020) Tourism and community resilience in the Anthropocene: Accentuating temporal overtourism. Anthropocene Ecologies 27(4):133–151. https://doi.org/10.4324/9781003000099-10

Cheung KS, Li L (2019) Understanding visitor–resident relations in overtourism: Developing resilience for sustainable tourism. J Sustain Tourism 27(8):1197–1216. https://doi.org/10.1080/09669582.2019.1606815

Choi N, Kim J, Lee SJ (2020) The Usefulness of Intersubjective Community Wellbeing as a Community Development Indicator: Evidence from Comparing Three Approaches to Measuring Community Wellbeing. Int J Com WB 3:173–192. https://doi.org/10.1007/s42413-020-00059-6

Chung MG, Dietz T, Liu J (2018) Global relationships between biodiversity and nature-based tourism in protected areas. Ecosyst Serv 34:11–23. https://doi.org/10.1016/j.ecoser.2018.09.004

Claudet J, Loiseau C, Sostres M, Zupan M (2020) Underprotected marine protected areas in a global biodiversity hotspot. One Earth 2(4):380–384. https://doi.org/10.1016/j.oneear.2020.03.008

Cloutier S, Ehlenz MM, Afinowich R (2019) Cultivating Community Wellbeing: Guiding Principles for Research and Practice. Int J Com WB 2:277–299. https://doi.org/10.1007/s42413-019-00033-x

Colding J, Barthel S (2019) Exploring the social-ecological systems discourse 20 years later. Ecology Soc 24(1):2. https://doi.org/10.5751/ES-10598-240102

Croes R, Manuel A, Rivera KJ, Semrad KJ, Khalilzadeh J (2017) Happiness and Tourism: Evidence from Aruba. Retrieved from The Dick Pope Sr. Institute for Tourism Studies website: http://doi.org/10.13140/RG.2.2.29257.85602

Damnjanović I (2020) Overtourism Effects: Positive and Negative Impacts for Sustainable Development. In: Leal Filho W, Azul AM, Brandli L, Lange Salvia A, Wall T (eds) Industry, Innovation and Infrastructure. Encyclopedia of the UN Sustainable Development Goals. Springer, Cham. https://doi.org/10.1007/978-3-319-71059-4_112-1

Dudley N (ed) (2008/2013) Guidelines for Applying Protected Area Management Categories. Gland, Switzerland: IUCN. x + 86 pp. WITH Stolton S, Shadie P Dudley N (2013) IUCN

WCPA Best Practice Guidance on Recognising Protected Areas and Assigning Management Categories and Governance Types, Best Practice Protected Area Guidelines Series No. 21, Gland, Switzerland: IUCN. xxpp. https://portals.iucn.org/library/sites/library/files/documents/PAG-021.pdf

Dudley N, Stolton S, Belokurov A, Krueger L, Lopoukhine N, MacKinnon K, Sandwith T, Sekhran N (eds) (2010) Natural solutions: protected areas helping people cope with climate change. IUCN-WCPA, TNC, UNDP, WCS, The World Bank, WWF, Gland, Switzerland, Washington, DC, and New York, NY

Dwyer L (2020) Tourism development and sustainable well-being: A beyond GDP perspective. J Sustain Tourism:1–18. https://doi.org/10.1080/09669582.2020.1825457

Epler Wood, M., Milstein, M., Ahamed-Broadhurst, K., 2019. *Destinations at Risk: The Invisible Burden of Tourism.* The Travel Foundation

EUROPARC Federation (2018, March 29) European day of parks. Retrieved June 13, 2020, from https://www.europarc.org/nature/european-day-of-parks/

Fiorello A, Bo D (2012) Community-based ecotourism to meet the new tourist's expectations: An exploratory study. J Hospitality Mark Manag 21(7):758–778. https://doi.org/10.1080/19368623.2012.624293

Fletcher R, Mas IM, Blanco-Romero A, Blázquez-Salom M (2020) Tourism and degrowth: An emerging agenda for research and praxis. Tourism Degrowth 27(12):1–19. https://doi.org/10.4324/9781003017257-1

Global Sustainable Tourism Council (GSTC) (2017, February 8). Retrieved June 10, 2020, from https://www.gstcouncil.org/

Global Wellness Institute (2018) Global Wellness Tourism Economy: Europe (November 2018). https://globalwellnessinstitute.org/

Global Wellness Summit (2019) 2019 Global Wellness Trends Report. https://www.globalwellnesssummit.com/2019-global-wellness-trends/download-report/

Goodwin H (2017) The challenge of overtourism (Working Paper). Responsible Tourism Partnership. Available at https://haroldgoodwin.info/pubs/RTP'WP4Overtourism01'2017.pdf

Gowreesunkar VG, Vo Thanh T (2020) Between Overtourism and Under-Tourism: Impacts, Implications, and Probable Solutions. In: Séraphin H, Gladkikh T, Vo Thanh T (eds) Overtourism. Palgrave Macmillan, Cham. https://doi.org/10.1007/978-3-030-42458-9_4

Green Destinations (2020) Sustainable destination awards 2020, April 27. Retrieved Oct 4, 2020, from https://greendestinations.org/2020-top-100-awards-itb-berlin/

Guthey GT, Whiteman G, Elmes M (2014) Place and sense of place. J Manag Inq 23(3):254–265. https://doi.org/10.1177/1056492613517511

Hall CM (2019) Constructing sustainable tourism development: the 2030 agenda and the managerial ecology of sustainable tourism. J Sustain Tourism 27(7):1044–1060. https://doi.org/10.1080/09669582.2018.1560456

Hockings M, Dudley N, Elliott W, Napolitano Ferreira M, MacKinnon K, Pasha MK, Phillips A, Stolton S, Woodley S, Appleton M, Chassot O, Fitzsimons J, Galliers C, Kroner RG, Goodrich J, Hopkins J, Jackson W, Jonas H, Long B, et al (2020) Editorial essay: COVID-19 and protected and conserved areas. Parks 26(1):7–24. https://parksjournal.com/wp-content/uploads/2020/06/Hockings-et-al-10.2305-IUCN.CH_.2020.PARKS-26-1MH.en_-1.pdf

Interreg CENTRAL EUROPE Programme—CEETO (2018) Handbook of successful and innovative practices for a sustainable tourism inside protected areas, November. https://www.interreg-central.eu/Content.Node/Handook-Sustainable-Tourism-EN-CEETO-Interreg.pdf

Iriarte L, Musikanski L (2019) Bridging the Gap between the Sustainable Development Goals and Happiness Metrics. Int J Com WB 1:115–135. https://doi.org/10.1007/s42413-018-0012-2

IUCN (2015) Protected areas left short-changed by governments, March 11. Retrieved June 13, 2020, from https://www.iucn.org/content/protected-areas-left-short-changed-governments

IUCN (2019) Protected area categories, August 15. Retrieved June 13, 2020, from https://www.iucn.org/theme/protected-areas/about/protected-area-categories

Junot A, Paquet Y, Fenouillet F (2018) Place attachment influence on human well-being and general pro-environmental behaviors. J Theor Soc Psychol 2(2):49–57. https://doi.org/10.1002/jts5.18
Kianicka S, Buchecker M, Hunziker M, Müller-Böker U (2006) Locals' and tourists' sense of place. Mt Res Dev 26(1):55–63. https://doi.org/10.1659/0276-4741(2006)026%5b0055:latsop%5d2.0.co;2
Koens K, Postma A, Papp B (2018) Is Overtourism overused? Understanding the impact of tourism in a city context. Sustainability 10(12):4384. https://doi.org/10.3390/su10124384
Koščak M, Lapos P, Primožič T (2020) Handbook of successful and innovative practices for a sustainable tourism inside Protected Areas. Regional Development Centre Koper & CEETO Interreg CE Project partners
Lalicic L (2020) Solastalgia: An application in the overtourism context. Ann Tourism Res 82:102766. https://doi.org/10.1016/j.annals.2019.102766
Lange G (2015) Tourism in Zanzibar: Incentives for sustainable management of the coastal environment. Ecosyst Serv 11:5–11. https://doi.org/10.1016/j.ecoser.2014.11.009
Leung Y, Spenceley A, Hvenegaard G, Buckley R (eds) (2018) Tourism and visitor management in protected areas: Guidelines for sustainability. Best Practice Protected Area Guidelines Series No. 27, Gland, Switzerland: IUCN. xii + 120 pp
Lopes P, Pacheco S, Clauzet M, Silvano R, Begossi A (2015) Fisheries, tourism, and marine protected areas: conflicting or synergistic interactions? Ecosyst Serv 16:333–340. https://doi.org/10.1016/j.ecoser.2014.12.003
Mandić A (2019) Nature-based solutions for sustainable tourism development in protected natural areas: a review. Environ Syst Decis 39:249–268. https://doi.org/10.1007/s10669-019-09718-2
Mihalic T (2020) Conceptualising overtourism: A sustainability approach. Ann Tourism Res 84:103025. https://doi.org/10.1016/j.annals.2020.103025
Milano C (2017) Overtourism and tourismphobia: Global trends and local contexts. Retrieved from Barcelona: Ostelea School of Tourism and Hospitality website: https://doi.org/10.13140/rg.2.2.13463.88481
Milano C, Novelli M, Cheer JM (2019a) Overtourism and degrowth: a social movements perspective. J Sustain Tourism 27(12):1857–1875. https://doi.org/10.1080/09669582.2019.1650054
Milano C, Novelli M, Cheer JM (2019b) Overtourism and tourismphobia: A journey through four decades of tourism development, planning and local concerns. Tourism Plann Dev 16(4):353–357. Retrieved from https://doi.org/10.1080/21568316.2019.1599604
Muler Gonzalez V, Coromina L, Galí N (2018) Overtourism: Residents' perceptions of tourism impact as an indicator of resident social carrying capacity—case study of a Spanish heritage town. Tourism Rev 73(3):277–296. https://doi.org/10.1108/tr-08-2017-0138
Musikanski L, Rogers P, Smith S et al (2019) Planet Happiness: a Proposition to Address Overtourism and Guide Responsible Tourism, Happiness, Well-being and Sustainability in World Heritage Sites and Beyond. Int J Com WB 2:359–371. https://doi.org/10.1007/s42413-019-00038-6
Naidoo R, Gerkey D, Hole D, Pfaff A, Ellis AM, Golden CD, Herrera D, Johnson K, Mulligan M, Ricketts TH, Fisher B (2019) Evaluating the impacts of protected areas on human well-being across the developing world. Sci Adv 5(4):eaav3006. https://doi.org/10.1126/sciadv.aav3006
Negev M, Sagie H, Orenstein DE, Zemah Shamir S, Hassan Y, Amasha H, Raviv O, Fares N, Lotan A, Peled Y, Wittenberg L, Izhaki I (2019) Using the ecosystem services framework for defining diverse human-nature relationships in a multi-ethnic biosphere reserve. Ecosyst Serv 39:100989. https://doi.org/10.1016/j.ecoser.2019.100989
Organisation for Economic Cooperation and Development (OECD) (2019) Measuring well-being and progress. OECD Better Life Initiative. Retrieved from https://www.oecd.org/sdd/OECD-Better-Life-Initiative.pdf
Pappas I (2020) 2019 GSTC Destination Assessment—Dubrovnik. Global Sustainable Tourism Council. https://www.gstcouncil.org/wp-content/uploads/GSTC-Destination-Assessment-Dubrovnik-2019-Final-Report.pdf

Peeters P, Gössling S, Klijs J, Milano C, Novelli M, Dijkmans C, Eijgelaar E, Hartman S, Heslinga J, Isaac R, Mitas O, Moretti S, Nawijn J, Papp B, Postma A (2018) Research for TRAN Committee—Overtourism: impact and possible policy responses. European Parliament, Policy Department for Structural and Cohesion Policies, Brussels

Perkumienė D, Pranskūnienė R (2019) Overtourism: Between the right to travel and residents' rights. Sustainability 11(7):21–38. https://doi.org/10.3390/su11072138

Rasoolimanesh SM, Ramakrishna S, Hall CM, Esfandiar K, Seyfi S (2020) A systematic scoping review of sustainable tourism indicators in relation to the sustainable development goals. J Sustain Tourism: 1–21. https://doi.org/10.1080/09669582.2020.1775621

Sandifer PA, Sutton-Grier AE, Ward BP (2015) Exploring connections among nature, biodiversity, ecosystem services, and human health and well-being: Opportunities to enhance health and biodiversity conservation. Ecosyst Serv 12:1–15. https://doi.org/10.1016/j.ecoser.2014.12.007

Sari FO, Nazli M (2020) Exploring the effects of "excessive tourism growth" on public health and ecosystem. J Hospitality Tourism Insights, ahead-of-print (ahead-of-print). https://doi.org/10.1108/jhti-04-2020-0060

Sarkki S (2017) Governance services: Co-producing human well-being with ecosystem services. Ecosystem Services 27:82–91. https://doi.org/10.1016/j.ecoser.2017.08.003

Sharpley R (2020) Tourism, sustainable development and the theoretical divide: 20 years on. J Sustain Tourism 28(11):1932–1946. https://doi.org/10.1080/09669582.2020.1779732

Sirgy MJ (2018) What Types of Indicators Should Be Used to Capture Community Well-Being Comprehensively? Int J Com WB 1:3–9. https://doi.org/10.1007/s42413-018-0002-4

Skift (2019) Megatrends defining travel in 2019. Skift Magazine (12), January 16. https://skift.com/wp-content/themes/skift/img/megatrends-2019/Skift-Megatrends-2019.pdf

Skift (2020) Megatrends defining travel in 2020. Skift Magazine (15), February 3. https://skift.com/wp-content/themes/skift/img/megatrends-2020//Skift-Megatrends-2020.pdf

Slovenian Tourist Board (2020) I feel Slovenia. Retrieved Oct 4, 2020, from https://www.slovenia.info

Spenceley A, Snyman S, Rylance A (2017) Revenue sharing from tourism in terrestrial African protected areas. J Sustain Tourism 27(6):720–734. https://doi.org/10.1080/09669582.2017.1401632

Sung H, Phillips RG (2018) Indicators and Community Well-Being: Exploring a Relational Framework. Int J Com WB 1:63–79. https://doi.org/10.1007/s42413-018-0006-0

The International Ecotourism Society (2015) Retrieved June 13, 2020, from https://ecotourism.org/

The International Ecotourism Society (2019) Ecotourism is the solution to Overtourism, July 31. Retrieved Sept 10, 2019, from https://ecotourism.org/news/ecotourism-is-the-solution-to-overtourism/

UN World Tourism Organization; Centre of Expertise Leisure, Tourism & Hospitality; NHTV Breda University of Applied Sciences; and NHL Stenden University of Applied Sciences (2018) 'Overtourism'?—Understanding and Managing Urban Tourism Growth beyond Perceptions, Executive Summary. UNWTO, Madrid. https://doi.org/10.18111/9789284420070

UNWTO and UNEP (2005) Making Tourism More Sustainable: A Guide for Policy-Makers. UNWTO and UNEP, Madrid and Paris

UNWTO (2020a) "Sustainability as the new normal" a vision for the future of tourism. Retrieved June 9, 2020, from https://www.unwto.org/covid-19-oneplanet-responsible-recovery

UNWTO (2020b) Global and regional tourism performance. Retrieved June 8, 2020, from https://www.unwto.org/global-and-regional-tourism-performance

VanderWeele TJ (2019) Measures of Community Well-Being: a Template. Int J Com WB 2:253–275. https://doi.org/10.1007/s42413-019-00036-8

Villasante S, Lopes PF, Coll M (2016) The role of marine ecosystem services for human well-being: Disentangling synergies and trade-offs at multiple scales. Ecosystem Services 17:1–4. https://doi.org/10.1016/j.ecoser.2015.10.022

Weidenfeld A (2018) Tourism diversification and its implications for smart specialisation. Sustainability 10(2):319. https://doi.org/10.3390/su10020319

Wellness Summit, Global Spa & and International, SRI, "The Global Wellness Tourism Economy" (2017) Tourism Travel and Research Association: Advancing Tourism Research Globally. 15. http://scholarworks.umass.edu/ttra/2013marketing/White_Papers/15
Wellness Tourism Worldwide (2011) 4WR: Wellness, Who, Where, What? Wellness Travel 2020 Full Report. https://globalwellnesssummit.com/wp-content/uploads/Industry-Research/Global/2011-wellness-tourism-worldwide-wellness-for-whom.pdf
World Bank, Spenceley A (2020) Tools and resources for nature-based tourism. https://doi.org/10.1596/34433
World Tourism Organization (UNWTO) (2017) Measuring Sustainable Tourism: A call for Action—Report of the 6th International Conference on Tourism Statistics, Manila, Philippines, 21–23 June 2017. UNWTO, Madrid. https://doi.org/10.18111/9789284418954
World Tourism Organization (UNWTO) (2019a) International Tourism Highlights, 2019 Edition. UNWTO, Madrid. https://doi.org/10.18111/9789284421152
World Tourism Organization (UNWTO) (2019b) Recommendations on Sustainable Development of Indigenous Tourism. UNWTO, Madrid. https://doi.org/10.18111/9789284421299
World Tourism Organization (UNWTO) and Guangdong Chimelong Group (2020) Sustainable Development of Wildlife Tourism in Asia and the Pacific. UNWTO, Madrid. https://doi.org/10.18111/9789284421572
World Tourism Organization (UNWTO) and United Nations Environment Programme (UNEP) (2019) Baseline Report on the Integration of Sustainable Consumption and Production Patterns into Tourism Policies. UNWTO, Madrid. https://doi.org/10.18111/9789284420605
World Travel & Tourism Council (WTTC) and McKinsey Company (2017) Coping with success: Managing overcrowding in tourism destinations. Retrieved from https://www.wttc.org/-/media/files/reports/policy-research/coping-with-success—managing-overcrowding-in-tourism-destinations-2017.pdf
World Travel & Tourism Council (WTTC) (2020) Economic impact | World travel & tourism council (WTTC). Retrieved June 8, 2020, from https://wttc.org/Research/Economic-Impact
World Travel and Tourism Council (WTTC) (2019) The economic impact of global wildlife tourism. August 2019. https://travesiasdigital.com/wp-content/uploads/2019/08/The-Economic-Impact-of-Global-Wildlife-Tourism-Final-19.pdf
World Wide Fund For Nature—WWF (2019) Blue Planet: Coasts. Retrieved Sept 12, 2019, from http://wwf.panda.org/our_work/oceans/coasts/
Zahra T (2017) Nautical and maritime tourism diversification strategies (Exploratory opinion) (INT/808-EESC-2016-05923-00-00-ac-tra). European Economic and Social Committee. https://www.eesc.europa.eu/en/our-work/opinions-information-reports/opinions/nautical-and-maritime-tourism-diversification-strategies
Zhang H, Gordon S, Buhalis D, Ding X (2017) Experience value cocreation on destination online platforms. J Travel Res 57(8):1093–1107. https://doi.org/10.1177/0047287517733557

Chapter 6
Stakeholder Management and the Imbalance of Power: A Central Mediterranean Perspective on Tourism in Marine Protected Areas

Karl Agius and Samantha Chaperon

Abstract Islands in the Mediterranean region are important tourist destinations, but overtourism has led to extensive challenges for island communities, not least due to its negative impact on fragile environments. As a result, attention is shifting towards alternative forms of tourism, such as ecotourism, which embrace sustainability principles. This involves the repositioning of protected areas in terrestrial and marine environments as not solely conservation instruments, but also as venues for ecotourism. Despite their small populations, stakeholder involvement in the management of these sites can be a challenging process which is fraught with conflict. Qualitative interviews with stakeholders were conducted in six islands in the central Mediterranean Region: The Aegadian Archipelago, the Pelagian Islands, and the Island of Pantelleria. Different levels of protection exist with the most peripheral islands allocated the highest level of protection and the most restrictions on tourism activity. Stakeholders fear that this will negatively impact their traditions and their livelihoods, and this has led to strong resistance against MPAs. Recommendations are made as to how genuine stakeholder involvement, better management and a rebalancing of power can lead to more competitive ecotourism destinations and improved well-being of local communities in an era of overtourism.

Keywords Ecotourism · Islands · Core–periphery · Stakeholder involvement · Central Mediterranean

---

K. Agius (✉)
Institute for Tourism, Travel and Culture, University of Malta, Msida, Malta
e-mail: karl.agius.05@um.edu.mt

S. Chaperon
Department of Marketing, Events and Tourism, University of Greenwich, London, UK
e-mail: S.A.Chaperon@greenwich.ac.uk

A. Mandić and L. Petrić (eds.), *Mediterranean Protected Areas in the Era of Overtourism*, https://doi.org/10.1007/978-3-030-69193-6_6

6.1 Introduction: Ecotourism and Marine Protected Areas (MPAs) in Islands

Islands are synonymous with tourism and are among the most visited tourist destinations in the world (Fotiou et al. 2002). The Mediterranean region, which is by far the world's most visited tourist destination, boasts hundreds of islands which tourists visit in their droves (Andolina et al. 2020). Most tourism in Mediterranean islands is for sun, sea, and sand (3S) holidays which take place in the peak summer months thus leading to seasonality (Cannas and Giudici 2015; Tovar-Sánchez et al. 2019; Alipour et al. 2020). Mass tourism has brought about a number of benefits to local communities, including employment (Bramwell 2004). However, for some island destinations, the number of tourists received far exceeds that of the local population at least throughout part of the year. Commonly known as 'overtourism', carrying capacity limits are breached, and this leads to various negative environmental, economic, and sociocultural consequences (Baldacchino 2008, 2015; Vogiatzakis et al. 2008; Said 2017; Briguglio and Avellino 2019). Facing significant pressures on their resources, attempts have been made to diversify the tourism product and to promote alternative, more sustainable forms of tourism which are smaller in scale, which draw on the distinctive features of the destination (Bramwell 2004), and can take place throughout the entire year. In this regard, ecotourism has been lauded as an alternative tourism product to mass tourism on islands (Weaver 1993) including those in the Mediterranean (Cidalia Tojeiro 2011; Said 2017; Agius et al. 2019).

It has been argued that if managed appropriately and sustainably, ecotourism can not only overcome seasonality by extending the tourist season (Garrod and Wilson 2004) but can also reduce pressure on the destination during peak times (Buckley 2009). Some islands have already rebranded themselves from 3S destinations to 'nature islands' (Weaver 2017). Ecotourism is a form of tourism which takes place in natural settings, is educational and interpretative in disposition, and embraces sustainability (Weaver and Lawton 2007). Ecotourism mostly takes place in peripheral and insular places. Indeed, such areas boast a great richness of species, including charismatic megafauna, and an abundance of ecotourism venues which include MPAs (Hoyt 2005).

Various benefits have been associated with MPAs associated with ecotourism. MPAs have the potential to raise the environmental and socioeconomic profile of a coastal or insular region, to promote sustainable tourism (López Ornat 2006; Dalias et al. 2007) and to offer opportunities to local fishermen to supplement their income (Pham 2020). On the other hand, MPA regulations can lead to negative impacts on locals' everyday lives, cause inequalities between communities, and create conflicts (Neva 2020). Achieving a balance between the protection of habitats and allowing for the development of marine ecotourism opportunities is challenging (Hoyt 2005). One approach has been to delineate zones where tourism can be encouraged while actively trying to minimise conflicts with other uses (Salm et al. 2000). In the case of island MPAs, zoning can maximise protection while minimising restrictions against anthropogenic use such as small-scale commercial fishing and ecotourism activities such as wildlife-watching (Schofield et al. 2013).

For MPA management and zoning to be a success, the involvement of stakeholders in decision-making processes and good governance is key (Bustamante et al. 2014). Early consultation with all stakeholders is considered crucial to reducing conflicts in an MPA in the long term (Francour et al. 2001). While stakeholder involvement in MPAs has received much attention in the academic literature, for central Mediterranean islands, this best practice approach has not been without its challenges. For islands, the small size and limited economic opportunities can result in extensive pressure on coastal and marine areas and conflicts between the various users. Stakeholders include artisanal and recreational fishermen, conventional and ecotourism operators, environmental non-governmental organisations (NGOs), tourists that visit islands for a variety of reasons (e.g. free-diving, boat trips, diving, swimming, wildlife-watching, etc.), as well as the residents. Since the presence of MPAs can have a particularly notable impact on the fishing industry, the success of an MPA will often be hugely dependent on the industry's attitude towards it (Pita et al. 2011).

In the case of archipelagos, the different cultural and community interests on each island makes stakeholder involvement an even more complicated and difficult process (Sheehan and Ritchie 2005). The smaller, more peripheral islands of an archipelago are usually dependent on the larger neighbouring islands which function as economic and political centres (Cross and Nutley 1999; Karampela et al. 2015). In tourism terms, this means decisions about the nature of the industry's development are often taken by leaders at the centre, with varying levels of stakeholder involvement from the periphery (Chaperon and Bramwell 2013). A characteristic of tourism governance in archipelagos is dominance and subordination (Baldacchino 2015) whereby the power dynamics between the islands are unequal, with the core having more power than the islands at the periphery. The core is often accused of controlling and exploiting the periphery, and this leads to 'core-periphery conflict' (Keller 1987; Jordan 2004; Chaperon and Bramwell 2013).

6.2 Ecotourism and MPAs: The Central Mediterranean Context

The economies of islands in the Mediterranean are increasingly dependent on the tourism sector (Mazzola et al. 2019). This has led to an increase in the urban footprint at the expense of the natural environment to the extent that landscapes have at times been considered to be more cultural than natural (Cassar et al. 2008). For islands where the natural terrain and coastline have been heavily impacted by tourism development, but where there are still rich marine environments, it is feasible to develop marine ecotourism products.

In the case of archipelagos, the peripheral islands have greater potential in this respect. These islands may be more difficult to reach for international visitors due to their remoteness, may not have been marketed as strongly as the core islands, and so have not experienced the same levels of tourism development as 'core' islands. Thanks to this, they have more pristine environments and extensive terrestrial and marine protected areas (Agius et al. 2019). Therefore, while these islands often face many challenges with their economic development because of their geographical peripherality (and often political too), these features also offer opportunities in tourism terms (Garrod and Wilson 2004; Chaperon and Theuma 2015; Weaver 2017).

The Mediterranean region has huge marine ecotourism potential due to its rich marine ecosystems (Fotiou et al. 2002; Agius et al. 2018; Andolina et al. 2020). Although the Mediterranean Sea only represents 0.82% of the World's ocean area and 0.3% of its volume, the richness of its species corresponds to 4–18% of all marine species, depending on the phylum taken into consideration (Lejeusne et al. 2010). A total of 12,000 marine species have been recorded (Gržetic et al. 2013). Marine ecotourism—a category of ecotourism which is practised in marine and coastal settings (Cater 2003)—is one of the fastest-growing segments (Sakellariadou 2014) and is supported by the blue growth strategy (Neva 2020). Islands have been widely regarded as ideal venues for marine ecotourism (Cater and Cater 2007; Fotiou et al. 2002; Halpenny 2001; Sakellariadou 2014), and this is often due to the presence of MPAs which act as conservation tools for marine biodiversity (Pham 2020) and ideal sites for marine ecotourism (Agardy 1993; Gerovassileiou et al. 2009; Petrić and Mandić 2014; Agius et al. 2019). In the case of the Mediterranean, these are mostly located around or adjacent to islands (Francour et al. 2001) increasing such potential. With the increasing awareness and demand for sustainable tourism (UNWTO 2019), these islands have the opportunity to develop, or further develop ecotourism products.

While various studies have been conducted on MPAs in the central Mediterranean in the field of natural sciences, there is limited literature in terms of island studies, ecotourism, and its role in addressing overtourism in this region. Furthermore, studies of the socioeconomic impacts of MPAs are scarce, and research into the involvement of stakeholders in the design, implementation, and management of MPAs is limited to a small number of geographic areas (Pascual et al. 2016). This chapter will discuss the importance of stakeholder involvement in the process of creating and managing a protected area as well as some of the challenges in doing this. In archipelagos, the challenges are enhanced due to the core–periphery relations that are present and the power imbalances between islands. Thus recommendations are made as to how this can be addressed. If well managed, MPAs can serve as an opportunity to attract tourists seeking nature-based experiences away from crowds and can offer islands and archipelagos that are suffering overtourism the chance to redress the balance.

6.3 Areas of Study: Aegadian Islands, Pelagian Islands, and Pantelleria

The area of study consists of six Italian islands, all located in the central Mediterranean (see Fig. 6.1). This includes the Pelagian archipelago, which comprises of Lampedusa and Linosa; the Aegadian archipelago which comprises of Favignana, Levanzo, and Marettimo, and the island of Pantelleria (see Table 6.1).

Established in 1991, the MPA in the Aegadian islands is the largest MPA in Italy and encompasses the entire archipelago (Mannino et al. 2016). It was considered to have been ineffectively managed by the Italian Coast Guard between 1991 and 2000, and its management was transferred to the Municipality of Favignana. The MPA is managed by a director, appointed by the mayor of the municipality, and an advisory board. The MPA stretches westward off the coast of Trapani and covers approximately 540 km^2, including 22 km of protected coastline. It is divided into four zones (Zones A, B, C, and D) each with varying levels of protection and restrictions (Himes 2007a). The MPA includes the largest and best-preserved meadow of Posidonia oceanica in the Mediterranean Sea, serving as a vital nursery for hundreds of species. The institutional mission of the MPA includes the protection and enhancement of the marine environment, environmental education and the promotion of sustainable development, with particular reference to the eco-compatibility of tourism (Donati 2015).

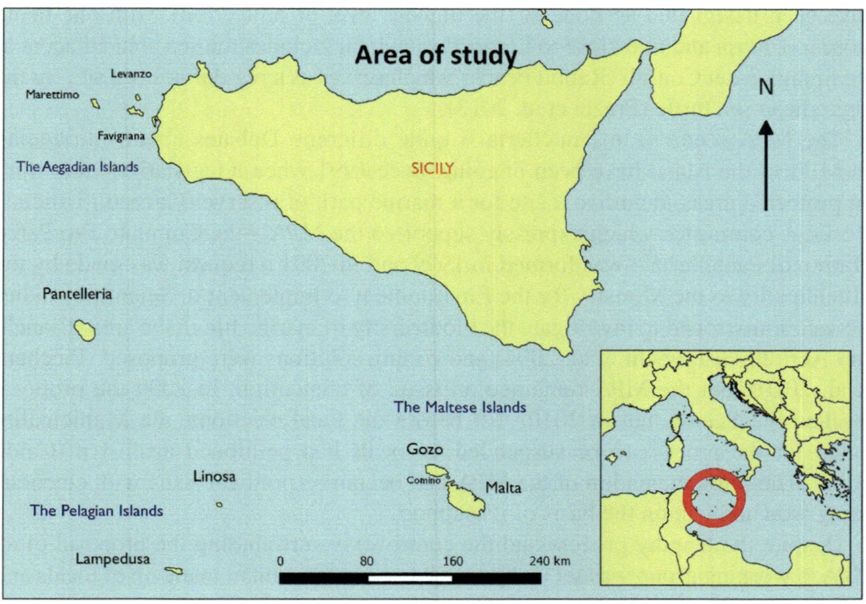

Fig. 6.1 Map showing Islands under study *Source* Prepared by the authors

Table 6.1 Characteristics of the Islands under study

Islands	Aegadian Islands			Pelagian Islands		Pantelleria
Factor	Favignana	Levanzo	Marettimo	Lampedusa	Linosa	
Density habitants/km^2	157	39	48	261	81	93
Permanent population	4500	220	820	5703	438	7736
Area (km^2)	19.8	5.8	12.4	20.2	5.4	84
Coastline (km)	33	15	18	33.3	11	51.5
Annual tourist arrivals	207,843			253,710		151,917
Size of MPA km^2	540			41.36		–

Sources Himes (2007a), Gallia (2012), Giardina (2012), ISTAT (2017), ENAC (2018) and Libero Consorzio Comunale di Trapani (2019)

The MPA in the Pelagian Islands was instituted in 2002 by the Italian Ministry for the Environment (La Manna et al. 2014). In 2003, its management was assigned to the Municipality of Lampedusa and Linosa with the aim to protect the marine vegetation and fauna, biological resources, and geomorphology of the area (Cooperativa Sesto Continente 2012). The entire area includes 41.36 km^2 and 46.28 km of coastline (Giardina 2012). The MPA is managed through a system of zones. Three key areas have been designated as Zone A (the highest level of protection), with one being close to Linosa and two close to Lampedusa which includes the sea area adjacent to the Spiaggia dei Conigli (Rabbit beach) which serves as a regular nesting site for the loggerhead sea turtle (Prazzi et al. 2013).

The MPA scenario in Pantelleria is quite different. Debates about introducing an MPA in the island have been ongoing since 1991 when a legislative framework on protected areas identified a site for a marine park or reserve (Gazzetta Ufficiale 1991). A committee which expressly supported the MPA—the Comitato Pro Parco Marino di Pantelleria—was formed in 1999 and in 2001 a request was made by the Municipality to the Ministry for the Environment to implement it. Scientific studies were commissioned to investigate the biodiversity of marine life in the area (Bianchi and Acri 2003; Bianchi et al. 2004) and zoning solutions were proposed (Picchetti et al. 2010), but the MPA remained an issue of contention. In 2009 the proposal reached the Senate, but in 2010, just before the local elections, the Municipality asked for the process to be suspended as locals had petitioned against it (Caldo 2019). The implementation of the MPA had become a political issue with elections being won and lost on the basis of its support.

Despite the lengthy process and the controversy surrounding the proposal of an MPA, a working group was set up by a civil society movement made up of locals and NGOs to continue working on its design and development. A participatory approach was taken, which involved identifying key stakeholder groups and nominating a representative from each to contribute to discussions. This allowed a bottom-up approach which took into consideration views from local community members and prioritised what was important to them, such as the protection of their local traditions (Rampini 2016). To date, the MPA has still not been implemented and has been described as a valid initiative approved through laws but hindered by politics and bureaucracy (Picchetti 2015).

6.4 Research Approach

A phase of multisited ethnography was carried out which involved 'sojourning' in the islands under study (Falzon 2009). Three study visits were made to each island between 2013 and 2016 and phases of observation were carried out to examine the issues related to the management of the MPA and to gain insights into the various perspectives on them. Between 2015 and 2020, semi-structured interviews were held with various stakeholders to explore these issues further. Two subtypes of strategic informant sampling technique were used to recruit interviewees. The first is expert sampling which involves the selection of 'typical' and 'representative' individuals. The second technique used, also known as snowball sampling, involves asking an initial set of informants to propose other potential sample members (Finn et al. 2000). The stakeholder interviewees included representatives from the local tourism industry, conventional and ecotourism operators, fishermen (both artisanal and recreational), NGOs, academics (in the field of natural science), policymakers, MPA management bodies, and also tourists. Interviews were held face to face in person and online, and also by phone. The combination of face-to-face and online interviews has frequently been used in tourism research (Power et al. 2017) with both methods permitting valid and high-quality interviews (Suryani 2013). The interviews were not recorded so as not to deter participants from active participation (Parker-Jenkins 2018). Instead, comprehensive field notes were taken during and immediately after the interviews (e.g. Decrop 1999). The field notes were coded and sustained through content analysis of existing legal framework related to the management of MPAs in the area of study using the method adopted by Stoffelen (2019). The following sections will present some of the key findings from this research, exploring some of the reasons for supporting and opposing the implementation of an MPA, and the various challenges facing MPA management.

6.5 Case Study: Pantelleria

The implementation of an MPA around the island of Pantelleria has been debated for several years. A terrestrial National Park is already in place, but the same protection for the waters that surround the island has raised many concerns, and it is yet to come to fruition. The Municipality and the Ministry for the Environment have recently expressed their commitment to reactivating the process (Caldo 2019), but according to policymakers interviewed this has already sparked opposition. Pantelleria has a limited tourism season, mainly in the summer months where seaside tourism is the major activity; Pantelleria does not have sandy beaches and so does not precisely offer the 3S product. However, the rich natural environment, including volcanic landscapes and impressive seascapes, offer opportunities for other types of tourism. There is agreement among policymakers, and other stakeholders, including locals and tourism operators, interviewed that Pantelleria should not become a mass tourism destination. Due to recent travel bans associated with COVID-19, Pantelleria has experienced a surge in young domestic tourists with little interest in natural and cultural attractions, and this has raised concerns among local residents and local ecotourism operators interviewed. The implementation of an MPA is again being proposed as a way of controlling tourism activity and attracting more nature-based tourists. This has the potential to build on the promotional efforts which have already been made to attract tourists to the terrestrial National Park. There have been suggestions made by NGOs and policymakers to follow the Cinque Terre model (mainland Italy) where the National Park includes both the terrestrial and marine protected areas. The Cinque Terre management body has recently become involved in the Pantelleria case to try to create awareness and generate support for the implementation of a similar model (Ente Parco Nazionale Isola di Pantelleria 2020).

6.5.1 Support for MPA Implementation

In support of the implementation of an MPA for Pantelleria are the small number of artisanal fishermen who see this as an opportunity to obtain exclusive fishing rights for locals and to deter outsiders; this is mainly aimed at fishermen from Sicily that fish in what the locals consider to be their waters. Another perceived benefit is that an increase in controls would prevent the illegal activity of recreational fishermen and offer new income opportunities through a permitted and appropriately managed tourism offer. A fisherman commented positively that the organisation of tours on fishing boats has already offered them a supplementary income. An NGO representative agreed that the positive perception of the MPA among professional fishermen is thanks to work undertaken by a civil society movement composed of locals and NGOs supporting the implementation of the MPA. Marine ecotourism operators and diving centres see the implementation of an MPA as a great opportunity to attract more tourists. For example, managers of diving centres based on Pantelleria commented that the underwater archaeological sites and the subsequent protection of the marine resources have already attracted more tourists to the area that have an interest in the natural environment. As a result, their businesses have benefited, they appreciate the importance of protecting such sites, and this has created a greater awareness to others working in the tourism industry of the positive impacts that greater protection and management can bring. This reflects Russ and Alcala's (1999) assertion that only when stakeholders fully understand the needs and benefits of an MPA will the support increase. Environmental NGOs have also supported the implementation of the MPA due to concerns related to the issuing of permits for oil exploration close to Pantelleria, fearing oil slicks and other hazards for the island.

6.5.2 Opposition to MPA Implementation

The main area of concern regarding the implementation of the MPA seems to be around zoning and the sense of disquietude about the restrictions that would be in place. Recreational fishermen have been particularly vocal against the implementation of the MPA. Their argument is that while there is no MPA on the island of Pantelleria, there are five sites along the coast (Cala Gadir, Cala Tramontana, Punta Li Marsi, Punta Tracino, and Punta Tre Pietre) that are designated as protected areas due to the presence of underwater archaeological artefacts. These areas already face restrictions that are equivalent to the most heavily protected MPA 'A' zones, and they are not willing to face further restrictions. Other recreational users, such as free-divers, are opposed to the MPA because this would prohibit their activities. Companies that operate excursions to the island are also opposed to the proposed MPA as they are convinced it would impact their visits and limit their activities.

According to NGO representatives, initial attempts to set up an MPA around the island were made by using a top-down approach, without any genuine involvement of the local community. More recent attempts to actively involve community stakeholders in the planning process have also been problematic. The process has been criticised for involving too many 'external' stakeholders. This includes non-native residents, several of which have only lived in the islands for a few years, and also environmental NGOs which have no direct link to the island. Some local community and tourism stakeholder respondents felt like they had no ownership over the proposal. If the MPA were to be implemented, there are also opposing views on who would be responsible for its management. This question has been turned into a political issue, with supporters of one party arguing that it should be managed in the same way as the National Park, and supporters of another party arguing that it must be managed by a separate body.

The case of Pantelleria shows how lack of stakeholder management in the initial phase, as well as lack of political commitment, derailed the setting up of an MPA. The absence of a bottom-up approach and foreign elements created doubt leading to failure to gain public support. Strong lobby groups such as recreational fishermen and conventional tourism operators have put pressure on successive administrations and influenced their political agenda to counteract the establishment of an MPA which they believe will have an impact on their traditions. Furthermore, there is fear for their livelihood as they believe that their tourism activity will be further controlled and limited. Protection has been associated with restrictions rather than with conservation, the well-being of local communities and added value to the competitiveness of the tourism product. There is a clear power imbalance between on one side recreational fishermen, tourism operators, and other resource users such as free-divers and on the other hand the few remaining professional fishermen, ecotourism operators including diving centres and NGOs. Stakeholders in favour represent a small minority, and thus politicians have considered the political repercussions. While Pantelleria has a terrestrial national park, the absence of an MPA implies that Pantelleria has missed out on the opportunity to strengthen the image of the island as an ecotourism destination remaining off the radar, for tourists seeking destinations away from the crowds in summer or seeking an ecotourism experience off season. This has contributed to the existing problem of seasonality which leads to a peak of tourism in the summer months and extensive economic losses beyond these months. In the long term, this might affect the tourism sector of the island as more tourists seek new destinations where they can immerse themselves in nature.

6.6 Case Study: Pelagian Islands

Unlike the MPA in the Aegadian Islands (established in 1991), much less academic research has been carried out on the implementation and subsequent management of the MPA in the Pelagian Islands (established in 2002). The MPA in the Pelagian Islands faced much less opposition, and this was agreed and explained by a variety of different stakeholder interviewees.

Artisanal and professional fishermen from the Pelagian Islands commented positively about the implementation of the MPA, albeit from two different perspectives. The majority of respondents viewed the MPA as having no direct impact on them or on their livelihoods, while a smaller group believed that the presence of the MPA was actually beneficial for them. The MPA in the Pelagian Islands does not extend very far beyond the shore (it covers an area of 41.36 km^2 compared to the 540 km^2 of the MPA in the Aegadian Islands), meaning that trawling—the main fishing activity in the archipelago (Celona and Comparetto 2009)—is conducted beyond the designated MPA zones, and did not impact the professional fishermen at all. Fishing was also an activity and industry in decline as more people were choosing to work in the tourism sector. These factors meant that there was little opposition to the MPA from the local community. A small number of respondents complained that environmental protection was being prioritised over the needs of the fishing industry with fishing not permitted in zone A, but in fact, this only covers a very small area of the MPA, and fishing is still permitted in the larger zones B and C.

Several respondents, including academics and NGOs attributed a lack of opposition to the MPA as being due to the weak level of its enforcement. According to regulations, certain limitations must be in place in the MPA. These limitations include the number of anchorage permissions, the number of boats permitted for rental, the number of operators organising boat tours, and also the number of diving centres (Ministero dell'ambiente e della tutela del territorio e del mare 2008) but these regulations have been rarely enforced. According to Orsini (2015), the local communities have gradually abandoned traditional industries, such as fishing, and have moved towards working in the tourism sector which today serves as a major economic activity for the local community. As a result, local authorities have shied away from high levels of enforcement as it causes conflict with the resource users. Instead, in collaboration with environmental NGOs, they are prioritising the need to raise awareness of the importance of the MPA to the local community. They are also working with universities that are carrying out academic research on the MPA. Studies conducted on the impact of vessel traffic in the MPAs and the effects on the bottlenose dolphin has also concluded that better enforcement is needed (Papale et al. 2012). Furthermore, it has been noted that the intense boat traffic during the summer period (especially between May and October) leads to several collisions with turtles (Prazzi et al. 2010). This also causes disturbance to the bottlenose dolphin (*Tursiops truncatus*), which in return causes their displacement from coastal areas (La Manna et al. 2013, 2014). Unlicensed and uncontrolled anchoring has also been reported as having a damaging impact on the seabed. According to Guidetti et al. (2008), MPAs in Italy are usually underfunded and understaffed, and this further affects their ability to successfully manage an MPA. The Coast Guard's mission includes surveillance in the MPA to identify any illegal activity, but tourism operators explained that the Coast Guard's main activities recently have focused on search and rescue in response to the ongoing migration crisis faced by Lampedusa. The lack of enforcement of MPA regulations has meant the area is not sufficiently protected, it has led to a reduction in the potential for wildlife sightings, and is detrimental to the success of ecotourism in the archipelagos.

In the case of the Pelagian islands, the limited size of the MPA and lack of enforcement have camouflaged challenges faced in other MPAs, but this does not mean that stakeholder involvement and management of the MPA is taking place effectively. Similarly to Pantelleria, its management is influenced by local authorities who are aware of the importance of tourism activity for the livelihood of local communities and have thus shied away from full enforcement. This is coupled with the continuing immigration crisis, which means resources for enforcement are stretched. Meanwhile, due to a lack of understanding of its potential, most operators still work in nautical and 3S tourism with limited involvement in ecotourism. The islands are branded by operators as the 'Caribbean of the Mediterranean', leading to peak season overcrowding on the beaches and deserted spaces for the rest of the year.

6.7 Case Study: Aegadian Islands

The importance of meaningful stakeholder involvement in successfully achieving an MPA has been widely discussed (Himes 2003, 2007b), but this was found to be lacking in the initial stages of MPA implementation in the Aegadian Islands. This is not surprising as a lack of stakeholder involvement is common in Italian MPAs, where only a few have been able to effectively manage natural resources through the collaboration of managers and interested stakeholders (Guidetti and Claudet 2010).

The implementation of the MPA in the Aegadian Islands did not receive much support initially as stakeholders were unclear about the potential impacts. In particular, this led to antagonism between the fishermen and the MPA management team (Badalamenti et al. 2000). Furthermore, initial management of the MPA was based on a very top-down approach, failing to encourage stakeholder participation in the management process, hindering support for it (D'Anna et al. 2015) and consequently bringing about a general lack of public trust in its governance (Guidetti et al. 2008). The MPA continued to face opposition to the way it was managed. Fishermen argued that they wanted to be more involved in the management of the MPA as they believed decisions were mainly influenced by academic research on environmental issues in the MPA and did not sufficiently take into consideration the needs of the local economy.

A new management body in 2010 addressed the issue 'head on' and stakeholders acknowledged that the situation did improve with their increased involvement (D'Anna et al. 2015). Representatives of local authorities claimed that the presence of the MPA led to various socioeconomic benefits for the fishing industry due to an increase in tourism. Fish stocks, including threatened species, have increased. Destructive fishing techniques, such as spear-fishing, have been prohibited while local fishermen have been given exclusivity to fish in the area. Diving centres and operators organising boat tours to observe wildlife remarked that MPAs are essential for their activity as adequate protection and management meant that visitors could observe more marine life. They claimed that marine protection boosted their activity, and thus, they had a personal interest in complying and supporting such initiatives. Operators and tourists attributed the abundant marine life and the extremely high visibility in the sea to the presence of MPAs making the islands ideal for diving, snorkelling, and wildlife-watching.

Unsurprisingly, the conflict between local authorities, conservationists, fishermen, tourism operators, and the local community still exists (D'Anna et al. 2016). Whereas protected areas should be seen as beneficial for local economies and communities, in several cases, this is not understood or viewed as such by local communities (McNeely 1994). In fact, representatives from the fishing and tourism industries interviewed claimed not to have experienced any of the promised benefits, such as new job opportunities related to tourism or higher fish stocks. On the contrary, they believe that fishing was a dying trade and that the presence of MPAs led to stricter controls which impacted their livelihood. Operators renting boats to conventional tourists also remarked that due to the MPA there were too many restrictions in place, particularly in relation to the reduced number of permitted anchoring sites, imposed to avoid damage to *Posidonia* meadows. There were also limitations on the use of motorised boats, arguing that this impacted tourism activity and that more flexibility was required. These stakeholders argued for the scaling down of zone A protection.

In the Aegadian Islands, most MPA conflicts are found on the main island of Favignana. This island, the largest in the archipelago, has remained geared up for more conventional mass tourism and is very seasonal in nature. Despite policies to promote alternative tourism (e.g. Guerra 2015) and the promised allure of the MPA to tourists in off-peak months, this has failed to materialise. Favignana has a heavily developed internal terrain and coastline which historically was due to industrial rock (tuff) extraction, and more recently the tourism sector (Groppi et al. 2018). The sea remains the only real alternative for nature tourism on the island, especially for 'hard ecotourists'. The use of charismatic species such as the Mediterranean monk seal (*Monachus monachus*) that has been sighted close to the island can serve to promote activity that supports both conservation and tourism, and successful management of the MPA is extremely important in this respect.

The main, recurring conflict between stakeholder groups centres around the balance between the fishing industry and conservation. Different stakeholder groups award different weighting to biological, economic, and sociocultural performance indicators when assessing MPA performance (Himes 2007b). Studies on MPA zoning in southern Europe have called for increased dialogue between stakeholders such as scientists, MPA managers, and fishermen so that the benefits of, and problems with, MPAs can be better understood (Mangi and Austen 2008). It is important to take into consideration both economic and cultural factors in all decisions regarding MPAs, and this should be done from the initial planning phase through to its daily management (Himes 2003). As this case suggests, failure to do so will result in low levels of support for the MPA and dissatisfaction in the local community (Badalamenti et al. 2000)

In the case of archipelagos, stakeholder management may further highlight imbalances in power and cause rivalry not only within an island but between islands, reflecting conflictual core–periphery relations (Jordan 2004; Chaperon and Bramwell 2013) and characterised by domination and subordination (Baldacchino 2015). At the crux of the problem is the lack of genuine stakeholder involvement, which has been discussed previously, but in this case, there is the argument that stakeholder interests from the larger, 'core' island are given priority. According to fishermen in Marettimo, the most remote and peripheral island of the Aegedian archipelago, attempts to encourage their participation in the MPA management process were 'fake' and although they were asked for their views, it was only the interests of the local authorities on the main island (Favignana) that were acted upon. Core–periphery relations are also exhibited in relation to the MPA zoning. Locals and tour operators from Marettimo remarked that their island suffers from disproportionately high levels of protection. Zone A, which represents the highest level of protection and where least activity is permitted, is situated very close to Marettimo's shoreline, limiting what tourism and fishing activities can take place, and negatively impacting the local economy. Locals argued that in reality, the entire archipelago merits the same level of protection, and it was unfair to attribute the MPA to the whole archipelago when the highest grade of protection was enforced solely around one island. Locals believe that such decisions were taken by authorities based on the core island to prioritise tourism activity on the larger island. Thus, the smaller island was placed at a disadvantage by shifting the 'burden of protection'. Meanwhile, academics, representatives of the management body of the MPA, politicians, and NGOs argued that a higher level of protection could serve as an advantage for the islands as they have the potential to attract more marine ecotourism. This could, in fact, help to address the inequality that exists in terms of tourism between the core and the periphery. Therefore, through the use of the MPA and its higher level of protection, Marettimo can attract the marine ecotourists seeking to avoid the crowds on Favignana or the day-trippers on Levanzo.

6.8 Conclusion and Recommendations

In each case study presented here, the stakeholder involvement in the implementation and management of MPAs has been characterised by imbalances in power between stakeholder groups, both within and (in the case of archipelagos) across islands. Decision-making related to the MPAs tends to be taken by bodies appointed by local authorities, themselves elected by the local community, but seemingly without sufficient consultation with the local community, and this has led to widespread dissatisfaction. The implementation of an MPA on Pantelleria is still being debated, mainly due to the vociferous recreational fishermen, free-divers, and tourism industry stakeholders who are vociferously opposed to it, fearing the impact it would have on their traditions and livelihoods, and who seem to have the power in this respect. For the Pelagian case, the power again of the tourism industry stakeholders, in particular, and the stretched resources of the coast guard due to the immigration crisis seems to prevent the local authorities from adequately managing the MPA and operating full enforcement. In the Aegedian Islands, core–periphery relations are more prominent, and the perceived power imbalance between the larger, 'core' island and the smaller, more peripheral islands is strongly felt by the communities based on the latter.

This study has contributed to the limited knowledge on the involvement of stakeholders in the design, implementation, and management of MPAs in small islands in the central Mediterranean, in particular in the case of the Pelagian Islands and Pantelleria where research on the matter has been neglected. It has also shown how characteristics of archipelagos, with core–periphery relations, in particular, have an impact on the implementation and management of MPAs and how this can be mitigated through genuine involvement of stakeholders. Acknowledging the inequalities between the islands, it has also been argued that marine ecotourism could take full advantage of the higher levels of protection at the more peripheral islands, creating appropriate settings for ecotourism products which could draw tourists away from the overcrowded coastal areas, contribute to the local economy and redress the balance. In general, islands in the central Mediterranean region have failed to take advantage of the opportunities presented by protected areas to shift from mass tourism on crowded beaches/coasts to more sustainable tourism within the MPAs. This has been due to lack of enforcement, the dominant role of 3S and nautical tourism operators, and lack of incentive given for the change. This needs to be addressed through genuine stakeholder involvement; endorsement of the MPAs, and a better understanding of their potential is key. By doing so, MPAs can not only reduce tourism pressures on the main islands of archipelagos but also reduce inequalities faced by peripheral islands. MPAs can offer the opportunity to attract ecotourists and to tackle seasonality since several activities extend beyond the 3S tourism season.

For all cases, clear recommendations can be made. It is imperative that all MPA implementation and management processes need to include equitable and meaningful participation of all stakeholder groups. Without this, there may be widespread dissatisfaction with the zoning decisions, and (as for Pantelleria) it may never exist. Resistance to MPA proposals can be addressed through effective participatory processes, coupled with consistent engagement of stakeholders over time, and transparency of the decision-making process. Genuine participation entails empowerment for engagement, and this calls for education and capacity building of local communities to get involved in the process of planning, implementing, and managing MPAs (Gaymer et al. 2014). This is key to shift from a top-down (government-led) to a bottom-up (community and user-led) approach.

In order to ensure benefit-sharing, mechanisms must be introduced by management bodies to assess the economic and sociocultural impacts and benefits arising from the establishment of MPAs and strive to share these equally, in particular between main and peripheral islands. In the case of archipelagos, policy benefits must compensate for costs incurred by peripheral islands due to higher levels of protection. Wider awareness is needed of the benefits of MPAs, and this can be achieved by using success stories from locals, and recruiting local stakeholders as 'MPA ambassadors'.

Regulations associated with MPAs need to be better enforced through further investment to ensure regular patrols and controls. Training must be provided to all resource users operating within MPAs to encourage compliance. In order to reduce the influence of powerful stakeholders and local politicians in decision-making about the management of the MPA, collaboration with local NGOs and research institutes is encouraged. The MPA alone is not enough to encourage the development of marine ecotourism; promotional efforts are also needed to attract these tourists, with the aim of eventually reducing the reliance on conventional 3S tourism to support the economy, and a more sustainable tourist destination which meets all stakeholders needs.

References

Agardy MT (1993) Accommodating ecotourism in multiple use planning of coastal and marine protected areas. Ocean Coast Manage 20(3):219–239

Agius K, Theuma N, Deidun A (2018) Does absence of charismatic species impact the ecotourism potential of Central Mediterranean islands? Xjenza Online 6:153–164

Agius K, Theuma N, Deidun A, Camilleri L (2019) Small islands as ecotourism destinations: a central Mediterranean perspective. Island Stud J 14(2):115–136

Alipour H, Olya HG, Maleki P, Dalir S (2020) Behavioral responses of 3S tourism visitors: evidence from a Mediterranean Island destination. Tourism Manage Perspect 33:100624

Andolina C, Signa G, Tomasello A, Mazzola A, Vizzini S (2020) Environmental effects of tourism and its seasonality on Mediterranean islands: the contribution of the Interreg MED BLUEISLANDS project to build up an approach towards sustainable tourism. Environ Dev Sustain, 1–12

Badalamenti F, Ramos AA, Voultsiadou E, Sanchez Lizaso JL, D'Anna G, Pipitone C, Mas J, Ruiz Fernandez JA, Whitmarsh D, Riggio S (2000) Cultural and socio-economic impacts of Mediterranean marine protected areas. Environ Conserv 27(2):110–125
Baldacchino G (2008) Island tourism. In: Luck M (ed) Encyclopaedia of tourism and recreation in marine environments. CAB International, Wallingford, pp 254–255
Baldacchino G (2015) More than Island tourism: branding, marketing and logistics. In: Baldacchino G (ed) Archipelago tourism: policies and practices. Ashgate Publishing, pp 1–18
Bianchi F, Acri F (2003) The Island of Pantelleria (Sicily Strait, Italy): towards the establishment of a marine protected area. First oceanographic investigations. Bollettino di Geofisica Teorica ed Applicata 44:3–9
Bianchi F, Acri F, Aubry FB, Camatti E (2004) Contributo allo studio dell'idrologia e del plancton nelle acque costiere dell'isola di Pantelleria. Biologia Marina Mediterranea 11(2):683–686
Bramwell B (2004) Mass tourism, diversification and sustainability in Southern Europe's coastal regions. In: Bramwell B (ed) Coastal mass tourism diversification and sustainable development in Southern Europe. Channel View, Clevedon, pp 1–31
Briguglio L, Avellino M (2019) Has overtourism reached the Maltese islands? Occasional papers on islands and small states, 2019/01
Buckley R (2009) Ecotourism: principles and practices. CABI, Wallingford
Bustamante G, Canals P, Di Carlo G, Gomei M, Romani M, Souan H, Vanzella-Khouri A (2014) Marine protected areas management in the Caribbean and Mediterranean seas: making them more than paper parks. Aquat Conserv Marine Freshwater Ecosyst 24(S2):153–165
Caldo M (2019) Che fine ha fatto l'Area Marina Protetta di Pantelleria? Il Vicesindaco Maurizio Caldo ribadisce l'interesse dell'Amministrazione Comunale a riprendere l'iter istitutivo. Retrieved on 17 Nov 2020 from https://www.pantellerianotizie.it/2019/07/28/caldo-riguardo-lopportunita-di-creare-larea-marina-protetta-a-pantelleria/
Cannas R, Giudici E (2015) Tourism relationships between Sardinia and its Islands: collaborative or conflicting? In: Baldacchino G (ed) Archipelago tourism: policies and practices. Ashgate Publishing, pp 67–84
Cassar LF, Conrad E, Schembri PJ (2008) The Maltese Archipelago. In: Vogiatzakis I, Pungetti G, Mannion AM (eds) Mediterranean island landscapes. Springer, Dordrecht, pp 297–322
Cater C, Cater E (2007) Marine ecotourism: between the devil and the deep blue sea. CABI, Wallingford
Cater E (2003) Between the devil and the deep blue sea: Dilemmas for marine ecotourism. In: Garrod B, Wilson JC (eds) Marine ecotourism: issues and experiences. Channel View, Clevedon, pp 37–47
Celona A, Comparetto G (2009) Indagine sull'attività di pesca nell'arcipleago delle Pelagie. Boll Mus Civ Stor Nat Venezia 60(1):113–120
Chaperon S, Bramwell B (2013) Dependency and agency in peripheral tourism development. Ann Tourism Res 40:132–154
Chaperon S, Theuma N (2015) The Malta-Gozo-Comino story: implication of a Malta-Gozo fixed link on tourism activity. In: Baldacchino G (ed) Archipelago tourism: policies and practices. Ashgate, London, pp 51–66
Cidalia Tojeiro M (2011) The ecotourism sustainable strategy: engagement and overcoming. J Coastal Res, 221–225
Cross M, Nutley S (1999) Insularity and accessibility: the Small Island communities of Western Ireland. J Rural Stud 15(3):317–330
Dalias N, Lenfant P, Licari ML, Bardelletti C (2007) Assistance guide to the management of the Protected Marine Areas: management and follow-up of the diving activity. General Council of Pyrénées-Orientales: Interreg IIIC MEDPAN, pp 1–88
D'Anna G, Pipitone C, Fernández TV, Garofalo G, Badalamenti F (2015) Stakeholder perspectives about the governance of the Egadi Islands Marine Protected Area (W Sicily) MPA. Biol Mar Mediterr 22(1):38–39

D'Anna G, Fernández TV, Pipitone C, Garofalo G, Badalamenti F (2016) Governance analysis in the Egadi Islands marine protected area: a Mediterranean case study. Marine Policy 71:301–309
Decrop A (1999) Triangulation in qualitative tourism research. Tour Manage 20(1):157–161
Donati S (2015) Biodiversity protection and sustainable management of coastal areas: the Marine Protected Area of Egadi Islands. Energia Ambiente e Innovazione 61(4):4–8
ENAC (2018) Traffic data 2017. Retrieved on 13 Oct 2020 from https://www.enac.gov.it/sites/default/files/allegati/2018-Ago/ENAC_Traffic_data_2017_en.pdf
Ente Parco Nazionale Isola di Pantelleria (2020) "Coltivare la Bellezza" wokshop nazionale Rotary Club Pantelleria. Retrieved on 17 Nov 2020 from http://www.parconazionalepantelleria.it/news-dettaglio.php?id=56583
Falzon MA (2009) Multi-sited ethnography: theory, praxis and locality in contemporary research. Ashgate Publishing Ltd
Finn M, Walton M, Elliott-White M (2000) Tourism and leisure research methods: data collection, analysis, and interpretation. Pearson education
Fotiou S, Buhalis D, Vereczi G (2002) Sustainable development of ecotourism in small islands developing states (SIDS) and other small islands. Tourism and Hospitality Res 4(1):79–88
Francour P, Harmelin JG, Pollard D, Sartoretto S (2001) A review of marine protected areas in the northwestern Mediterranean region: siting, usage, zonation and management. Aquat Conserv Mar Freshwater Ecosyst 11(3):155–188
Gallia A (2012) La valorizzazione dei beni culturali e ambientali per lo sviluppo delle isole minori italiane. Rivista giuridica del Mezzogiorno 26(4):929–962
Garrod B, Wilson JC (2004) Nature on the edge? Marine ecotourism in coastal peripheral areas. J Sustain Tourism 12(2):95–120
Gaymer CF, Stadel AV, Ban NC, Cárcamo PF, Ierna J Jr, Lieberknecht LM (2014) Merging top-down and bottom-up approaches in marine protected areas planning: experiences from around the globe. Aquat Conserv Mar Freshwater Ecosyst 24(S2):128–144
Gazzetta Ufficiale (1991) Legge 6 dicembere 1991, no 394 Legge quadro sulle areee protette. Supplemento ordinario alla Gazzetta Ufficiale no 292 of 13 December 1991
Gerovassileiou V, Koutsoubas D, Sini M, Paikou K (2009) Marine protected areas and diving tourism in the Greek Seas: practices and perspectives. Tourismos 4(4):181–197
Giardina F (2012) Itinerari Sommersi nelle isole Pelagie: incontri tra natura, scienza e vita marina. Ministero dell' Ambiente e della Tutela del Territorio e del Mare
Groppi D, Garcia DA, Basso GL, Cumo F, De Santoli L (2018) Analysing economic and environmental sustainability related to the use of battery and hydrogen energy storages for increasing the energy independence of small islands. Energy Convers Manage 177:64–76
Gržetic Z, Lukovic T, Božic K (2013) Nautical tourism market suppliers in the Mediterranean. In: Lukovic T (ed), Nautical tourism. CABI, pp 47–53
Guerra A (2015) Isole Egadi: a pedali, a remi, a piedi. Dal turismo balneare ai nuovi turismi ambientali. Analisi della Marca e proposte di strategia verso nuovi prodotti turistici. Un focus sul cicloturismo. Vivilitalia, 64 p
Guidetti P, Claudet J (2010) Co-management practices enhance fisheries in marine protected areas. Conserv Biol 24:312–318
Guidetti P, Milazzo M, Bussotti S, Molinari A, Murenu M, Pais A, Carrada G (2008) Italian marine reserve effectiveness: does enforcement matter? Biol Conserv 141(3):699–709
Halpenny EA (2001) Islands and coasts. In: Weaver DB (ed) The encyclopaedia of ecotourism. CAB International, Wallingford, UK, pp 435–450
Himes AH (2003) Small-scale Sicilian fisheries: opinions of artisanal fishermen and socio-cultural effects in two MPA case studies. Coastal Manage 31(4):389–408
Himes AH (2007a) Fishermen's opinions of MPA performance in the Egadi Islands marine reserve. Mast 5(2):55–76
Himes AH (2007b) Performance indicators in MPA management: using questionnaires to analyse stakeholder preferences. Ocean Coast Manage 50(5):329–351

Hoyt E (2005) Sustainable ecotourism on Atlantic islands, with special reference to whale watching, marine protected areas and sanctuaries for cetaceans. Biol Environ Proc Roy Irish Acad 105B(3):141–154

ISTAT (2017) Personal communication—request for data

Jordan LA (2004) Institutional arrangements for tourism in small twin-island states of the Caribbean. In: Duval DT (ed) Tourism in the Caribbean: trends, development, prospects. Routlege, London, pp 99–118

Karampela S, Kizos T, Papatheodorou A (2015) Patterns of transportation for tourists and residents in the Aegean archipelago, Greece. Archipelago tourism: policies and practices. In: Baldacchino G (ed) Archipelago tourist destinations: Archipelago tourism: policies and practices. Ashgate, pp 35–50

Keller CP (1987) Stages in peripheral tourism development—Canada's Northwest Territories. Tourism Manage 8(1):20–32

La Manna G, Manghi M, Pavan G, Lo Mascolo F, Sarà G (2013) Behavioural strategy of common Bottlenose Dolphins (*Tursiops truncatus*) in response to different kinds of boats in the waters of Lampedusa Island (Italy). Aquat Conserv Mar Freshwater Ecosyst 23(5):745–757

La Manna G, Manghi M, Sara G (2014) Monitoring the habitat use of common Bottlenose Dolphins (*Tursiops truncatus*) using passive acoustics in a Mediterranean marine protected area. Mediterr Mar Sci 15(2):327–337

Lejeusne C, Chevaldonné P, Pergent-Martini C, Boudouresque CF, Pérez T (2010) Climate change effects on a miniature ocean: the highly diverse, highly impacted Mediterranean Sea. Trends Ecol Evol 25(4):250–260

Libero Consorzio Comunale di Trapani (2019) TAV.4- Favignana: Movimento Turisito Ricettivo Mensile Anno 2019 – Dati Provvisori. Retrieved on 13 Oct 2020 from http://www.consorziocomunale.trapani.it/provinciatp/images/FAVIGNANA_agg2019.pdf

López Ornat A (2006) Guidelines for the establishment and management of Mediterranean marine and coastal protected areas. Tunis, MedMPA project, UNEP-MAP RAC\SPA, pp 1–158

Mangi SC, Austen MC (2008) Perceptions of stakeholders towards objectives and zoning of marine-protected areas in southern Europe. J Nat Conserv 16(4):271–280

Mannino AM, Parasporo M, Crocetta F, Balistreri P (2016) An updated overview of the marine alien and cryptogenic species from the Egadi Islands Marine Protected Area (Italy). Mar Biodivers 47(2):469–480

Mazzola F, Pizzuto P, Ruggieri G (2019) The role of tourism in island economic growth and resilience. J Econ Stud 46(7):1418–1436

McNeely JA (1994) Protected areas for the 21st century: working to provide benefits to society. Biodivers Conserv 3(5):390–405

Ministero dell'Ambiente e della tutela del territorio e del mare (2008) Approvazione del regolamento di disciplina dell' area marine protetta Isole Pelagie. Decreto 4 febbraio 2008. Retrieved on October 14 from http://www.ampisolepelagie.it/pdf/Regolamento%20Esecutivo%20delle%20discipline_04_02_2008_Isole%20Pelagie.pdf

Neva L (2020) Problematic blue growth: a thematic synthesis of social sustainability problems related to growth in the marine and coastal tourism. Sustain Sci 15(4):1233–1244

Orsini G (2015) Lampedusa: from a fishing island in the middle of the Mediterranean to a tourist destination in the middle of Europe's external border. Italian Stud 70(4):521–536

Papale E, Azzolin M, Giacoma C (2012) Vessel traffic affects Bottlenose Dolphin (*Tursiops truncatus*) behaviour in waters surrounding Lampedusa Island, south Italy. J Mar Biol Assoc United Kingdom 92(8):1877–1885

Parker-Jenkins M (2018) Problematising ethnography and case study: reflections on using ethnographic techniques and researcher positioning. Ethnography Educ 13(1):18–33

Pascual M, Rossetto M, Ojea E, Milchakova N, Giakoumi S, Kark S, Korolesova D, Melià P (2016) Socioeconomic impacts of marine protected areas in the Mediterranean and Black Seas. Ocean Coast Manage 133:1–10

Petrić L, Mandić A (2014) Visitor management tools for protected areas focused on sustainable tourism development: the Croatian experience. Environ Eng Manage J 13(6):1483–1495
Pham TTT (2020) Tourism in marine protected areas: can it be considered as an alternative livelihood for local communities? Mar Policy 115:103891
Picchetti G (2015) A futura memoria... l' AMP di Pantelleria: note di una sconfitta ... in 22 note, 84 p
Picchetti G, Caravello A, Ghelia M, Di Martino V (2010) Proposal for the national park area of Sicily channel: the Pantelleria Marine Protected Area. Biologia Marina Mediterranea 17(1):74–75
Pita C, Pierce GJ, Theodossiou I, Macpherson K (2011) An overview of commercial fishermen' attitudes towards marine protected areas. Hydrobiologia 670(1):289
Power S, Di Domenico M, Miller G (2017) The nature of ethical entrepreneurship in tourism. Ann Tourism Res 65:36–48
Prazzi E, Nicolini G, Piovano S, Giacoma C (2010) Protezione di *Caretta caretta* (Reptilia Chelonia) nella Riserva Naturale di Lampedusa. Naturalista Sicil S 4(34):265–294
Prazzi E, Nicolini G, Piovano S, Giacoma C (2013) La spiaggia dei Conigli a Lampedusa: un modello di gestione per la conservazione di *Caretta caretta*. In: Di Tizio L, Brugnola L, Cameli A, Di Francesco N (eds) Atti II Congresso SHI Abruzzo Molise "Testuggini e Tartarughe" (Chieti, 27–29 settembre 2013). Ianieri Edizoni, Pescara, pp 127–133
Rampini G (2016) Pantelleria Island: a testing ground for a participatory approach aimed at fostering the creation of an MPA proposed and developed by local inhabitants. Paper presented at the CIESM Congress, Kiel, Germany—Rapp. Comm. int. Mer Médit., vol 41, p 550
Russ GR, Alcala AC (1999) Management histories of Sumilon and Apo Marine Reserves, Philippines, and their influence on national marine resource policy. Coral Reefs 18(4):307–319
Said N (2017) From mass tourism to ecotourism: the case of the Balearic Islands, the Maltese Islands and the republic of Cyprus. Occasional papers on islands and small states, 03/2017
Sakellariadou F (2014) The concept of marine ecotourism: case study in a Mediterranean Island. Int J Clim Change Impacts Responses 6(1):33–39
Salm RV, Clark JR, Siirila E (2000) Marine and coastal protected areas: a guide for planners and managers. IUCN
Schofield G, Scott R, Dimadi A, Fossette S, Katselidis KA, Koutsoubas D, Lilley MKS, Pantis JD, Karagouni AD, Hays GC (2013) Evidence-based marine protected area planning for a highly mobile endangered marine vertebrate. Biol Conserv 161:101–109
Sheehan L, Ritchie B (2005) Destination stakeholders: exploring identity and salience. Ann Tourism Res 32:711–734
Societa Cooperativa Sesto Continente (2012) Marine protected area "Pelagie Islands"
Stoffelen A (2019) Disentangling the tourism sector's fragmentation: a hands-on coding/post-coding guide for interview and policy document analysis in tourism. Curr Issues Tourism 22(18):2197–2210
Suryani A (2013) Comparing case study and ethnography as qualitative research approaches. Jurnal Ilmu Komunikasi 5(1):117–127
Tovar-Sánchez A, Sánchez-Quiles D, Rodríguez-Romero A (2019) Massive coastal tourism influx to the Mediterranean Sea: the environmental risk of sunscreens. Sci Total Environ 656:316–321
UNWTO (2019) International tourism highlights. Retrieved on 14 Oct 2020 from https://www.e-unwto.org/doi/pdf/10.18111/9789284421152
Vogiatzakis IN, Mannion AM, Pungetti G (2008) Introduction to the Mediterranean Islands landscapes. In: Vogiatzakis IN, Pungetti G, Mannion AM (eds) Mediterranean Island landscapes. Springer, Dordrecht, pp 3–14
Weaver DB (1993) Ecotourism in the small island Caribbean. GeoJournal 31(4):457–465
Weaver DB (2017) Core–periphery relationships and the sustainability paradox of small island tourism. Tourism Recreation Res 42(1):11–21
Weaver DB, Lawton LJ (2007) Twenty years on: the state of contemporary ecotourism research. Tourism Manage 28(5):1168–1179

Chapter 7
Geoeducation and Tourism in Estrela UNESCO Global Geopark (Portugal) and Its Contributions to the Construction of a Sustainable Destination

Gonçalo Fernandes, Rosa Branca Tracana, Emanuel Castro, and Magda Fernandes

Abstract Geoparks have the potential to stimulate the development of local communities, among others through geotourism, while conserving resources and ensuring education. With this chapter, we discuss how the Estrela UNESCO Global Geopark through its research network, in an articulated and synergistic way involving the local community, research centres, and governments, promotes in situ experiences and the transmission of knowledge and values about the territory, its heritage, and its sustainability. The relationship between geoconservation, science, education, and tourism must be established to facilitate an integrated approach to the territory, in the natural and anthropocentric dimensions, fostering it as a pedagogical resource, and ensuring its sustainability. Geoeducation and geotourism have gained relevance in tourism development projects, especially in low-density and least developed territories, as the Estrela Geopark. Education is the foundation of geotourism and a pillar for heritage conservation, supporting the sustainable development of communities.

Keywords Estrela UNESCO Global Geopark · Tourism · Geoeducation · Heritage · Sustainability

G. Fernandes (✉)
Instituto Politécnico da Guarda, CITUR, CICS.NOVA, Guarda, Portugal
e-mail: goncalopoeta@ipg.pt

R. B. Tracana
Escola Superior de Educação, Comunicação e Desporto, Ci&DEI, Instituto Politécnico da Guarda, Guarda, Portugal
e-mail: rtracana@ipg.pt

E. Castro · M. Fernandes
Associação Geopark Estrela (AGE), Guarda, Portugal
e-mail: emanuelcastro@geoparkestrela.pt

M. Fernandes
e-mail: magdafernandes@geoparkestrela.pt

A. Mandić and L. Petrić (eds.), *Mediterranean Protected Areas in the Era of Overtourism*, https://doi.org/10.1007/978-3-030-69193-6_7

7.1 Introduction

Tourism plays an important role in the management and conservation of natural areas, so the link between protected areas and tourism activity is vital for their sustainability and enhancement as a territorial asset. This relationship is complicated due to the different interests involved. Tourism is a critical and strategic component in the process of planning and management of conservation areas, such as the UNESCO Global Geoparks. Sustainable tourism in protected areas ensures financial resources which are often essential to implement and improve conservation processes, stimulate learning, and to valorise the territory. However, the designation of the protected area, in many cases, limits communities to develop specific economic activities, including tourism which might deteriorate the state of the protected resources.

Portugal mountain areas of high environmental/natural value are classified among others, like National Parks, Natural Parks, Natural Reserves, Biosphere Reserves and Geoparks, are established for the (i) conservation of the geological heritage; (ii) education for sustainability, and (iii) tourism and local development. In this sense, they introduce a great responsibility in the creation/development of economic, tourism, and social value. Furthermore, they help create territorial value by generating development strategies for the well-being of the communities, valuing the heritage, and promoting a sense of belonging and collaborative work.

This chapter seeks to explain the strategies recommended by the Estrela Geopark (EG), regarding the valorisation of a Serra da Estrela nature-based destination, promoting initiatives and developing geoeducation programs for sustainable tourism, committed to the territory and its communities. The process and the role of Geoparks in the valorisation of territories are explained. The EG is described, and the dynamics and interrelations between science, education, territory, and tourism are systematised and discussed in the context of the sustainable development goals.

The EG introduced responsibility in the development of the economic, tourist, and social value of Serra da Estrela as a Natural Park (PNSE) in Portugal. The integration of the EG in a worldwide network acknowledged its value. It promoted the conservation of heritage, and fostered the positioning of Serra da Estrela into a global market, allowing the emergence of new products and services based on nature and locals traditions. The Geopark is a reservoir of resources that are fundamental in the regulation of biological processes. It ensures the conservation of the environment and valorisation of cultural heritage and local traditions, as it serves as a museum of geomorphologic history and living laboratory dedicated to supporting the achievement of sustainable development goals.

7.2 UNESCO World Geoparks and Their Territorial and Educational Strategy

The UNESCO Global Geoparks Programme, approved in November 2015, constitutes a new paradigm of valorisation, promotion, and development of territories, anchored in the importance of geological heritage and geosciences. In this sense, a Geopark is a well-delimited territory, possessing a remarkable geological history which, due to its relevance, uniqueness and meaning, constitutes a shared legacy that must be safeguarded and valued for future generations (Zouros 2004). UNESCO Global Geoparks advocates a holistic vision of the territory, a concerted action among the different development agents, a strategy of conservation and valorisation of its sites of geological interest, and a policy of territorial development that is effectively integrated and participatory, focused on sustainability and valorisation of the endogenous resources of each territory. The first steps towards the establishment of UNESCO Global Geoparks were taken in 2000 with the creation of the European Geoparks Network (EGN), constituted by four Geoparks, Haute-Provence Geological Reserve, (France), Lesvos Petrified Forest (Greece), Geopark Vulkaneifel, (Germany), and Maestrazgo Cultural Park (Spain). They were made official in 2001 through the signature of a formal agreement between EGN and UNESCO Division for Earth Sciences (Zouros 2004). In the next few years, the list included several territories that saw the potential of this designation to improve the development of their communities. The interest in this strategy took a new scale with international relevance, and in 2004 the Global Geoparks Network (GGN) was created under the auspices of UNESCO. The International Geosciences and Geoparks Programme (IGGP) was created in 2015 in a joint initiative by UNESCO Global Geoparks Programme and the International Union of Geosciences (IUGS). This facilitated the creation of international cooperation mechanisms, based on geological heritage, through a bottom-up approach, which allowed joint work on heritage conservation with the involvement of communities, also focused on the sustainable development of territories (UNESCO 2015). Currently, the program involves 161 classified territories, spread over 44 countries around the world.

The central objective of a Geopark is education, not only as a strategy of geoconservation, but also as a way of valorisation of heritage, in the promotion of the sense of belonging, and societal development itself. Geoparks are excellent to support sustainable development as they reconcile conservation and development goals while raising environmental awareness of communities. They play an essential role in promoting education, particularly regarding the geological heritage, promoting geoconservation strategies, ensuring future generations access to testimonies of the geological history of our planet, thus contributing to the progression of scientific knowledge and education in earth sciences. Geoparks as UNESCO territories are real living laboratories. The education within geoparks is promoted through activities with well-defined objectives, based on the contents addressed in formal education, thus complementing what is taught in schools and other formal education institutions (Cascais and Terán 2011). The interpretation of the natural and cultural heritage, done in geoparks, contributes to a better understanding and acquisition of knowledge, providing significant and fruitful learning. Pinto and Pereira (2008) state that non-formal education, as an intentional, systematic, structured, and specific educational practice, favours the development of personal and social skills that the school itself has difficulty in developing. Thus, non-formal education carried out in the geoparks, through study visits, has several advantages for the education and training of individuals, taking into account the previous knowledge, the personal experience of each one, and the feeling of the place. Therefore, all UNESCO Global Geoparks must develop and operate educational activities for all ages to spread greater awareness of geological heritage and its relationship with other non-geological, cultural, and intangible heritage elements. Geoparks should be able to promote educational programs for schools, or special activities for children through "Kids Clubs" or "Fossil Fun Days". In parallel, geoparks should be able to provide training for adults and the local population, empowering them with greater resources and knowledge (UNESCO 2015). The interpretation and dissemination of heritage value represent a strategy of conservation and tourism practices that tend to be more sustainable and protective of existing heritage resources.

7.3 Estrela Geopark, Brief Life for a Long History

Serra da Estrela is a territory with geological and geomorphological resources, consisting of glacial, periglacial, and granite landforms whose singularity and heritage value bring to this territory an eco-cultural dimension of great relevance. The geopark constitutes a "living laboratory" in which the natural resources, the ways of life of the population and the biological and geological diversity blend into cultural landscapes. The Serra da Estrela with 1993 m is the highest mountain range in mainland Portugal and is part of the Iberian Central Cordillera (Fernandes et al. 2016; Gomes et al. 2017). Bounded by fault scarps, a granite massif occupies the central area forming a summit plateau between ci. 1400 and 2000 m, while in its surrounding the core area, there is an interplay between schists and greywackes. During the Last Glacial, a plateau ice-field and five radiating valley glaciers occupied the highest parts of the mountain with an estimated equilibrium line altitude at 1,650 m asl (Vieira 2004). The plateau style of the glaciation and the ELA just below the plateau edge made Estrela very sensitive to climate fluctuations, which has resulted in several terminal moraine complexes that reveal several glacial stages. The central plateau area shows widespread glacial erosion features and almost complete stripping of the Cenozoic weathering mantle (Vieira 2004). The non-glaciated plateaus show a rich landscape dominated by granite weathering landforms. The remarkable glacial landscape of Serra da Estrela when considering its setting in SW Europe, together with other relevant geoheritage such as periglacial, weathering and mass wasting phenomena, tectonic, petrological and hydrogeological features, is at the core of the Estrela UNESCO Global Geopark (Castro et al. 2017) (Fig. 7.1).

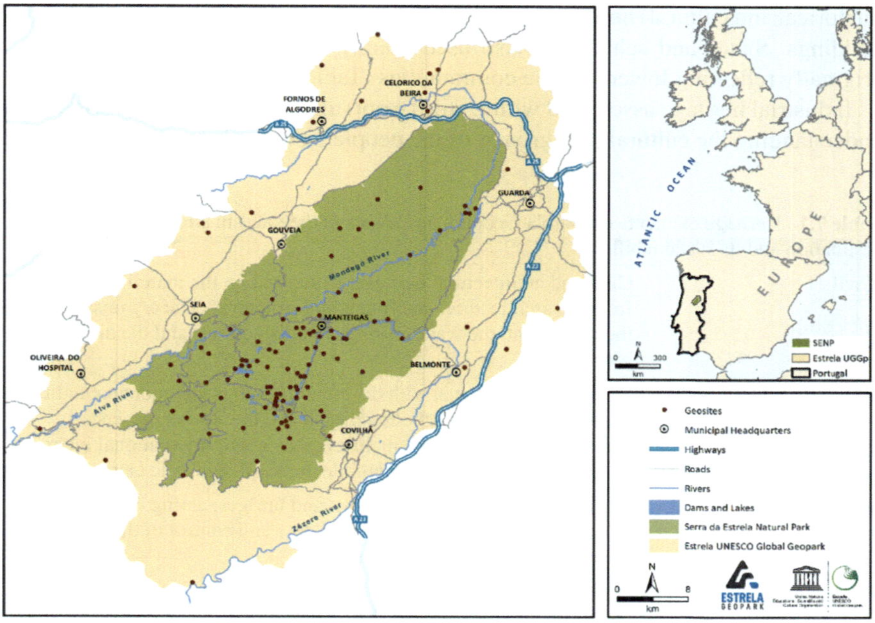

Fig. 7.1 Estela Geopark location (*Source* AGE -Associação Geopark Estrela)

The EG integrates the Serra da Estrela, from its southwestern limit on the border with the Açor Mountains to the northeast contact with the Iberian Meseta surface, including also the northwest and southeast foothills, where for ages communities live in close relationship with the natural environment.

The EG initiated the development of a collective strategy, reconciling environmental, social, and economic goals, which would involve for the first time all municipalities of Serra da Estrela into the value creation, conservation, and territorial promotion (Castro et al. 2017; Fernandes et al. 2020).

The richness of its natural and cultural heritage and the local community's way of life make Serra da Estrela a unique site and potentially appealing nature-based destination. The EG intends to enhance natural spaces by developing a Science and Education Network for Sustainability, promoting environmental education projects, strengthening relations with the community, improving nature-based tourism, and promoting actions that increase the well-being of local communities. Geopark takes a collaborative approach to support the transfer of knowledge and education, through processes and actions, supported by the natural environment as didactic resources. The development of materials and programs that allow structured knowledge of the territory's history, georesources, geoconservation practices, and the promotion of sustainable tourism strategies is highly prioritised.

The cultural heritage of the EG, especially the architectonic, has developed in close connection with geological traits of the territory. Granite marks the architectonic (historical and cultural) heritage, especially outstanding in castles, civil and religious buildings. Slates and schists are also used mainly in the metasedimentary areas, originally reflecting lower-income communities (Table 7.1).

Industrial heritage associated with textile (wool) and dairy activities is critical for understanding the cultural background of the people from the Estrela Mountain and

Table 7.1 Heritage elements of Estrela Geopark and assets in the tourism strategy to promote the destination and its sustainability

Civil	Civil and architectural heritage, evidenced by the materials used, frontage architecture, and functional structures. The organisation of the settlement, the water management structures, and the division of properties
Religious	Diversity of elements, with different meanings and monumentalities associated with the Catholic religion and the Jewish presence, generating in some municipalities elements of international interest, materialised in the existing architecture and symbolic elements
Military	Composed of forts, castles, walls, and bridges, dating from the Roman Empire, as a defensive structure and control of the territory of the mountains over the adjacent areas
Archaeologic	A diverse set of elements that show the human presence in these territories and the forms of their appropriation since prehistory, Rome, and Medieval times. Some elements have outstanding preservation and historical and scientific value in understanding the occupations and forms of human presence
Mining	Mining complexes, which represent different extractive moments and historical contributions, with significant and specific importance for the local economy, technological evolution, and historic preservation
Textile	Sets of factories and wool processing plants, revealing the strong link between agropastoral, the abundance of water, and the holding of specialised knowledge in the production of woollen yarns and fabrics. Some of these units are still active, and others of historical relevance turned into Museums (Real Fábrica de Panos- Covilhã)
Rural	Composed of traditional activities linked to agropastoral, structures for dividing the property and respective walls, ponds and dams, set of utensils and mechanism that last over time and represent the specific mountain ways of life
Festivities and Traditions	Religious and pagan festivities, many associated with vegetative cycles, or solar moments (solstices and equinoxes). Promotion of local products, crafts and mountain-specific traditions (sacred and profane)
Pastoralism	A striking element of Serra da Estrela's economic activity, giving meaning to two regional symbols, Cheese and Dog. The cheese corresponding to a valued gastronomic product and recognised for its organoleptic characteristics. The dog is a symbol for the protection of herds, their conduct, and an element of struggle and support for shepherds and the most isolated communities on the mountain

Source Authors elaboration

the evolution of the landscape, due to the links between economic activities and land-use dynamics. On the other hand, mining played a huge role in the economy of the region and is still of significance in some areas. The mining heritage, which has also been included in some geosites, shows the direct relation between society and geological resources, being especially significant to promote education and discussion on sustainability.

Most visitors are however limited to roadside stops for photos or shopping (daily visitors), concentrating more time in the villages, but not spending more than one or two nights in Estrela Mountain, with most of them still not searching for nature-related activities. Hence, despite the good existing trail network, it is rare to meet more than two to four people in a daylong walk in the plateau. This fact is essentially cultural, and relates to the lack of walking tradition by the Portuguese, and is also due to the relatively small number of foreign tourists. Therefore, the potential for growth of geotourism, nature-based, and adventure tourism in the region is considerable. The potential increase in visitor numbers after obtaining the UNESCO designation will also pose significant challenges for managing geosites and limiting their degradation. Therefore, the EG is providing to the municipalities and the Nature Park, adequate management and monitoring plans, especially for those geosites with more potential for visitation.

Education is the basis of geotourism and a pillar for the sustainability of local ecosystems. Geoparks provide an opportunity for education of visitors through the interpretation of geologically and geomorphologically significant resources. Geoeducation is focused on the knowledge of the history of our planet, as well as the influence of abiotic nature on human development (Farsani et al. 2012; Hose and Vasiljevic 2012). The educational strategies involving regional heritage are stimulating the development of educational resources for the teaching outside the classroom, generating experiences and learning in situ. The possibility of extending knowledge outside the school environment is of particular importance to the development of children and adolescents (Świercz and Smorzewska 2015). Thus, the cooperation between schools, universities, and Geoparks is highly encouraged to improve existing and develop experimental models based on cross-sectional learning and territorial valorisation.

7.4 Geotourism, Heritage Enhancement, and the Promotion of the Well-Being of Local Communities

The territories face permanent interactions between humans, culture, and the environment, generating different forms of organisations and networks. The transformation of these assets into leisure, entertainment, and knowledge is increasingly relevant to the socio-economic inclusion of the local population in tourism-related activities (Eder and Patzak 2004; Berman et al. 2008). Geological heritage plays an educational role in society by raising awareness about the extent of the influence of geological forms and processes on the history of the earth and humanity (Miśkiewicz 2016). It contributes to the identity of territories, by bringing didactic and scientific value, fostering the self-esteem of communities, and by promoting geotourism activities. The current dynamics suggest the creation of collaborative networks involving various stakeholders, which fosters an increase in territorial, educational, and organisational capacities. The increased capacities favour the holistic valorisation of the geological heritage and promote social welfare and local community well-being. The geological heritage and the geoeducation associated with it (Fig. 7.2), allows an in-depth knowledge of resources, increasing participation of communities in geoconservation, increasing capacities and skills in the development of sustainable nature-based tourism activities, and associated entrepreneurship strategies.

Geoeducation plays, as a constructive process and diffuser of knowledge, a relevant educational role in society focused on raising awareness about the importance of geological sites and their value as a cultural, social, and tourist asset. In the context of the sustainable use of natural and cultural heritage, the UNESCO Geoparks are expected to implement projects to preserve the quality of the environment, promote science, education, and the welfare of communities as well as to promote tourism (Hose 2005; Zouros 2004).

Geoeducation and geotourism gained relevance in tourism development projects, especially in low-density and least developed territories. This new paradigm of "looking" at the territory has allowed the construction of a new narrative on the valorisation

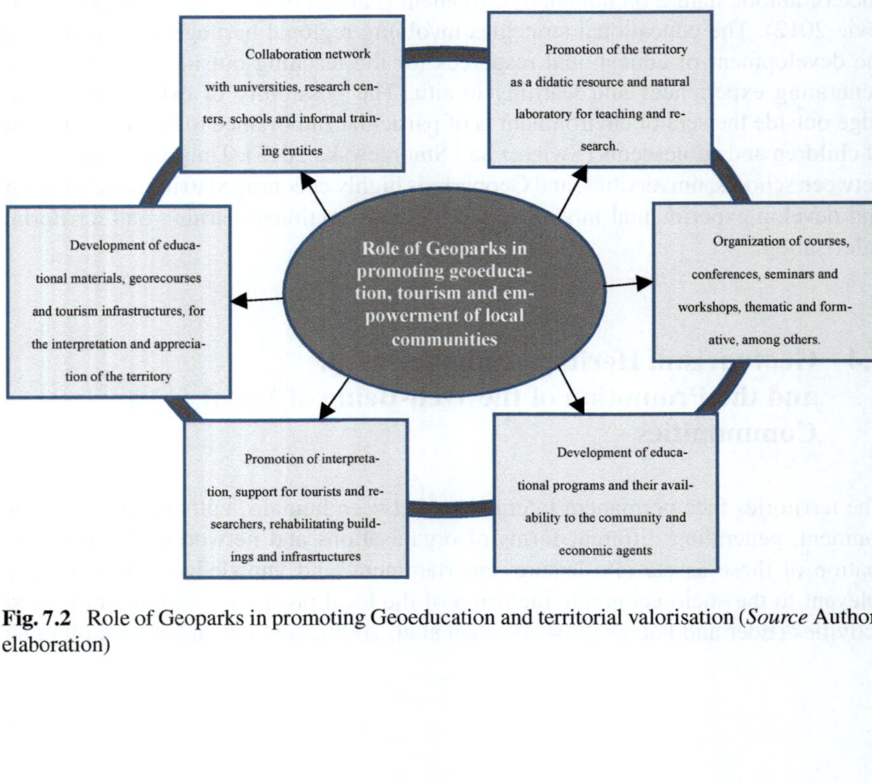

Fig. 7.2 Role of Geoparks in promoting Geoeducation and territorial valorisation (*Source* Authors elaboration)

of endogenous resources, the landscape, and territorial tourism development policies (Dowling 2015; Frey et al. 2016). Education and science play a vital role in the conservation and maintenance of geological and cultural diversity, stimulating research in the territory and actively contributing to the existence of a network of joint initiatives among other projects, publications, information exchange and conferences, participated by various actors. The interaction of the Estrela Geopark we propose, with the territorial system, allows the identification of strategies and actions that promote new synergies between crucial stakeholders to reconcile the conservation and economic development goals (Fig. 7.3). It embodies a collaborative approach to the sustainable development supported by geological heritage, science, and education fostering the appreciation of natural and cultural assets.

This approach to the dissemination of heritage values has an educational value in itself and constitutes a relevant contribution to the teaching-learning process, facilitating didactic activities, and the development of fieldwork (Brilha 2005; Fernandes et al. 2016). The development of educational promotion strategies focused on learning experiences in situ fosters an integrated analysis of the landscape and contributed to raising awareness about the environmental challenges that places of significant heritage value continuously face. The referenced educational programs aim to develop the awareness of communities regarding the preservation and valorisation of eco-cultural resources and sustainable tourism practices, given their fragility. In this context, particular attention should be paid to pedagogical approaches which

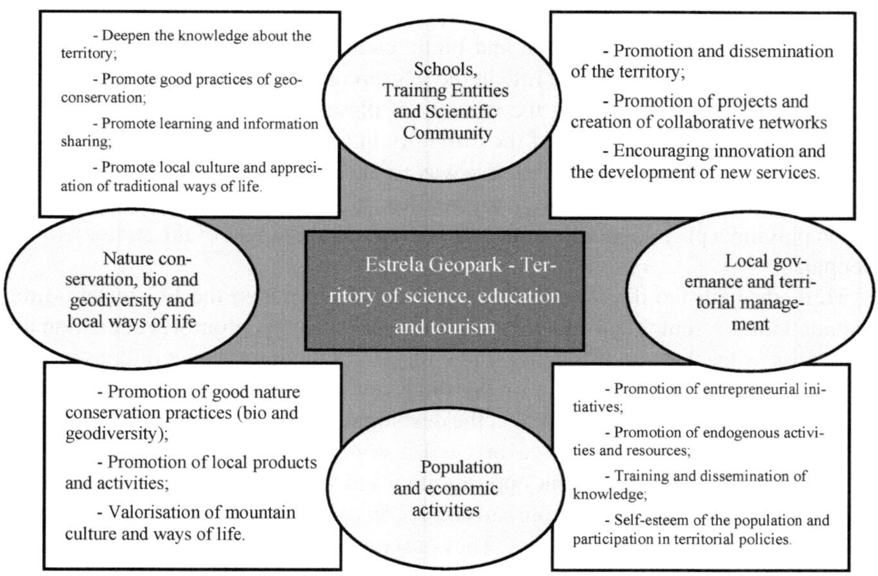

Fig. 7.3 Interaction of the Geopark star in the territorial system and promotion of science, education, and tourism (*Source* Authors elaboration)

consider different levels of education, aiming to stimulate curiosity and interest not only in geosciences but also the conservation and enhancement of local heritage.

7.5 Geoeducation in the Estrela Geopark

7.5.1 Heritage Enhancement and Geoconservation

Geoconservation is one of the critical areas of EG action since the preservation of the geological heritage allows the creation of sustainable development strategies in these territories classified by UNESCO. As such, it is part of its daily work to promote actions aimed at conservation and valorisation in an integrated and holistic way. It can be said that for the success of a Geopark's strategy, its various areas of operation require the geological heritage to be preserved and valued, and thus contribute to the achievement of this goal. Thus, the goal of geoconservation in the EG is to define an approach that allows the protection, preservation, interpretation, and valorisation of the unique geological heritage of the territory. Interpretation is one of the great tools for promoting the preservation and enhancement of geological heritage. Therefore, to promote knowledge on the territory and raise awareness among local communities and visitors, the EG has been implementing interpretative structures throughout the territory. These allow the visitor to understand not only geological and geomorphological issues but also the biodiversity and culture of the various sites, showing the clear link between the abiotic and biotic elements. The Estrela Geopark Torre Interpretation Centre has a vital role in the interpretation of the natural, cultural, and landscape heritage, reinforcing the strategy of dissemination and promotion of the territory and its values. Aware of the difficulty in transmitting scientific knowledge in an accessible, exciting, and attractive way to the public, it is also important to rely on interpretation as a vital strategy for the dissemination purposes, with communication playing a prominent role in the transversal development of the strategy of this Geopark.

Thus, through the development of educational programs, the EG fosters direct contact with the geological and geomorphological heritage, promotes education and awareness of teachers and students, and emphasises the importance of conservation of this heritage. Tourism is one of the pillars of the UNESCO Geoparks since it promotes the valorisation of heritage, the development of new products and services, and encourages the practice of local arts and customs, fostering economic growth and the creation of new employment opportunities and value for the territory. From this point of view, the focus on geotourism creates an opportunity to enhance and further develop the territory of this Geopark. The conservation and valorisation of the geological heritage are present in a transversal way in the several areas of Geopark's performance, either in the context of science, education, culture, tourism, and, inevitably, in the communication itself.

7.5.2 Promotion of Geosciences and Outdoor Learning

Knowledge and education are the way to promote a more environmentally conscious and sustainable society. Since 2016 and its constitution, the EG has been working hard to promote research and education as a strategic area for the dissemination of knowledge according to the UNESCO Program recommendations. The realisation of outdoor educational programs focused on all levels of education transforms the EG into a living laboratory for the non-formal education, where tutors use adequate methodologies to transfer the knowledge, promoting the discovery, conservation, and valorisation of this heritage. According to Salvador and Vasconcelos (2007), outdoor learning activities are performed outside the classroom, although not necessarily in a natural environment. Thus, outdoor activities can take place in any of the learning environments mentioned by Orion (2001), including the outdoor learning environment (natural environment, for example, natural areas); outdoor/indoor learning environment (semi-natural environment, for example, zoos, natural parks, biological farms) and indoor environment (human-made environment, for example, science museums, science and technology centres, interpretive centres, among others). The outdoor activities, such as field trips, allow the development of a set of interdisciplinary skills related to the promotion of citizenship and the development of knowledge-being (Silva 2014; Tracana et al. 2018). The EG encourages the visitation of the 124 geosites. Being aware that it is not always easy to leave school with the students due to curriculum and financial reasons, the EG has created an indoor educational program, called "The Estrela Goes to School". Within this program the Geopark technicians go to the Schools to carry out several activities, working with the students on transversal themes, related to geo and biodiversity, the cultural heritage, as well as to develop projects related to the Sustainable Development Goals (SDGs).

Taking into account the specificities of the Geopark's mountain areas, in the 2019/2020 school year, particular emphasis was placed on SDG 13 "Climate Action, taking urgent measures to combat climate change and its impacts", through the implementation of an annual thematic project. These projects were structured in four stages throughout the year, in which students from the adherent schools worked on the main concepts associated with climate change, moving from awareness-raising to active and responsible participation in combating and adapting to this problem. The plan for 2020/2021, is to develop the initiatives related to SDG 4—"Quality Education: Ensuring Inclusive, Equitable and Quality Education and Promoting Lifelong Learning Opportunities for All". According to UNESCO (2015), embarking on the path of sustainable development will require a profound transformation in the way people think and act. Thus, to create a more sustainable world and address sustainability issues, as described in the SDG, individuals must become agents of change for sustainability. To do so, they need the knowledge, skills, values, and attitudes to contribute to sustainable development. Education is, therefore, crucial to achieving sustainable development and valuing heritage and territories (Fig. 7.4).

As part of the educational strategy that has been implemented in the Geopark, from 2016 to December 2019, more than 1500 students and 320 teachers have visited the

Fig. 7.4 Clarification sessions in the scope of the "Estrela vai à Escola" program (*Source* Associação Geopark Estrela (AGE), 2020)

territory, as a part of educational tourism, with the implementation of 50 outdoor educational programs (Fig. 7.5). The Estrela Geopark—Torre Interpretation Center, a crucial pedagogical tool, has received more than 3600 visitors since its opening in September 2018. The indoor educational program "An Estrela vai à Escola" hosted more than 40 conferences/workshops, with approximately 2600 participants.

During the year 2019, 2080 students and teachers from all over the country made indoor and outdoor activities with the Geopark, which highlights the importance of the effectively structured educational strategy. The example of the EG reaffirms education as the basic premise of Geopark development, which creates the foundation for the development of territory and community.

7.5.3 From Geoeducation to Scientific and Educational Tourism

Educational tourism should enable the coexistence of people of different cultures through a participative pedagogy, in which participants are encouraged to engage in an experiential interaction with the territory. This goal is achieved in the EG through the development of educational programs and the creation of a network of interpretative routes. These allow people not only to visit the territory but also to have access to scientific knowledge and eco-cultural information about the geosites

Fig. 7.5 Outdoor educational program in Estrela Geopark (*Source* Associação Geopark Estrela [AGE] 2020)

among others through interpreted tours and interpretative panels. In the pilot project implemented in 2019/2020, the EG worked with partners to promote Sustainable Development Goals through educational tourism. The project consisted of integrating the Estrela Geopark in the schools' Educational Projects. The aim was to produce a set of concrete activities, i.e. joint preparation of projects and initiatives, or mutual proposals according to the challenges of each activity plan. The project yielded several successful outputs, including the workshop; "Healthy Living Camp", which involves outdoor activities during a weekend, and outdoor pedagogical routes, which include visits to different sites of geological interest, museums, and interpretation centres, structured based on the programmatic contents covered in each year of schooling. The activities are organised within the Torre Interpretation Centre, which is located at the highest point in mainland Portugal, and which is the most visited site in EG. In the centre, guided tours and interpretation are carried out, supported by graphic material and illustrative panels, not only in the field of tourism but also in the field of educational activities (Fig. 7.6).

The Science and Education Network for Sustainability of Estrela Geopark (RCES) was created in 2019, to support and promote research applied to the territory of the EG. This network aims to promote science and education as catalysts for territorial development, enable cooperation between science and citizens; promote research applied to the reality of territories and the needs of populations; strengthen the involvement of different partners and institutions; foster the development of networks (museums, interpretation centres, associations, communities, tourism industry); and

Fig. 7.6 Visit to the Geopark Estrela tower interpretation center (*Source* Associação Geopark Estrela [AGE] 2020)

enable the dissemination of knowledge through events and training actions in the community. The promotion of science and knowledge through RCES, educational tourism, and the implementation of joint projects with the educational institutions contribute to the promotion of *Education for Sustainable Development Goal* in this territory. These strategies meet the objectives of UNESCO, to encourage changes in the way of obtaining knowledge, the importance of strengthening values and attitudes, enabling a more sustainable and just society for all.

7.6 Conclusions

The Estrela UNESCO Global Geopark aims to contribute to the protection and enhancement of natural and cultural heritage, with particular emphasis on geological heritage. The Geopark seeks to develop and disseminate knowledge and promote geoeducation and community-based geotourism sustainably. The relationship between geoconservation, science, education, and tourism must be established to allow an integrated approach to the territory, promoting knowledge, awareness of heritage, and the development of an appreciation of the environment. The educational programs, the interpretative routes, and the RCES of the EG promote the cooperation and networking between all stakeholders involved, including community, local government, educational institutions at all levels, tourism industry and EG managing agency. It allows the development of new approaches for the sustainable valorisation of geology, capable of enhancing the endogenous value of the territory and its enjoyment by tourism and leisure activities.

The EG promotes an integrated development strategy, combining geoconservation with education and tourism, supporting the construction of development strategies for the well-being of the community and a participative collaboration in the promotion of the region as a heritage territory, didactic resource, and science place. The implementation of a geoconservation strategy based on established criteria and the creation of a science and education network for sustainability allows the deepening of knowledge of this territory in a transversal way by reinforcing the cooperation between key stakeholders and institutions. The active articulation with municipalities and the educational institutions allows for monitoring interventions, adapting joint strategies, and promoting sustainability initiatives based on UNESCO recommendations. The EG demonstrates that increasing attractiveness of the area supported by an integrated strategic approach can foster the sustainability of tourism development by reducing the effects of seasonality and crowding. It could create a feasible surrounding to promote geosciences and collaborative environment for the benefit of the entire ecosystem.

Acknowledgements This work is funded by National Funds through the FCT - Foundation for Science and Technology, IP, within the scope of the projects Refª UIDB/05507/2020 and Centre for Tourism Research, Development and Innovation. Furthermore, we would like to thank the Centre for Studies in Education and Innovation (CI&DEI) and the Polytechnic of Guarda for their support.

References

Berman M, Jonides J, Kaplan S (2008) The cognitive benefits of interacting with nature. Psychol Sci 19(12):1207–1212

Brilha JBR (2005) Património Geológico e Geoconservação: a conservação da Natureza na sua vertente geológica. Universidade do Minho

Cascais MGA, Terán AF (2011) Educação formal, informal e não formal em ciências: contribuições dos diversos espaços educativos. XX Encontro de Pesquisa Educacional Norte Nordeste, Universidade Federal do Amazonas – UFAM, 23 a 26 de agosto de 2011, Manaus-AM

Castro E, Vieira G, Loureiro F, Patrocínio F, Firmino G, Gomes H, Fernandes M (2017) Managing the Estrela aspiring Geopark: a framework for promoting geoheritage conservation. Conference Ciência

Dowling RK (2015) Geotourism's contribution to sustainable tourism. In: Hughes M, Weaver D, Pforr C (eds) The practice of sustainable tourism: resolving the paradox. Routledge, London, pp 207–227

Eder W, PatzaK M (2004) Geoparks—geological attractions: a tool for public education, recreation and sustainable economic development. Episodes 27(3):162–164. AUC Geographica, 51:2:235–246

Farsani NT, Coelho C, Costa C (2012) Geotourism and geoparks as gateways to socio-cultural sustainability in Qeshm rural areas. Iran. Asia Pac J Tourism Res 17:30–48

Fernandes G, Castro E, Firmino G, Ferreira A (2016) Geoparks and regional and local development: aspiring Estrela Geopark. European Geoparks Conference. https://doi.org/10.13140/rg.2.2.28770.94405

Fernandes GP, Castro E, Tracana RB (2020) Tourism as a strategy for Geo-Education and cultural valorization of territories: promoting eco-cultural values and Geoconservation in Estrela Geopark, Portugal. In Brandão F, Breda Z, Costa R, Costa C (eds) Handbook of research on the role of tourism in achieving sustainable development goals. IGI Global, pp 117–137. http://doi.org/10.4018/978-1-7998-5691-7.ch007

Frey ML, Schafer K, Buchel G, Patzak M (2016) Geoparks: a regional European and global policy. Geotourism. https://doi.org/10.1016/B978-0-7506-6215-4.50014-2
Gomes H, Fernandes M, Castro E, Vieira G (2017) The (Serra da) Estrela Aspir-ing Geopark (Portugal): preserving geoheritage, while promoting science and its links to local communities. Geophysical Research Abstracts vol 19, Conference Paper, EGU General Assembly
Hose T (2005) Geoturismo europeo. Interpretacíon geológica y promoción de la conservación geológica para turistas. In: Barretino D, Winbledon WAP, Gallego E (eds) Patrimonio geológico: conservación y gestión. Istituto Tecnológico Geomineiro de España, Madrid
Hose T, Vasiljevic D (2012) Defining the nature and purpose of modern geotourism with particular reference to the United Kingdom and South-East
Miśkiewicz K (2016) Promoting geoheritage in geoparks as an element of educational tourism. In Szponar A, Toczek-Werner S (eds) Geotourism: organisation of the tourism and education in the geoparks in the Middle-Europe Mountains. University of Business in Wrocław
Orion N (2001) A educação em Ciências da Terra. Da teoria à prática-implemantação de novas estraté-gias de ensino em diferentes ambientes de apren-dizagem. In: Marques L, Praia J. orgs. 2001. Geo-ciências nos currículos dos ensinos básico e secundário. University Aveiro, Aveiro, pp 93–114
Pinto L, Pereira S (2008) Educação Não Formal para uma Infância Real. Acedido em: http://www.inducar.pt/webpage/contents/pt/cad/Educacao_NaoFormal_para_uma_Infancia_Real.pdf
Salvador P, Vasconcelos CMS (2007) Actividades outdoor e a alfabetização científica de alunos de um clube de ciencias. Florianópolis, Linhas 8(2):76–90
Silva A (2014) A importância dos georrecursos para a divulgação e ensino das Geociências. Dissertação de Mestrado. Universidade Nova de Lisboa
Świercz A, Smorzewska E (2015) Selected examples of interactive teaching methods in the centre of Geoeducation in the city of Kielce (Poland). Procedia Soc Behav Sci 174:680–686. https://doi.org/10.2777/193011
Tracana RB, Lopes A, Direito AF, Ferreira ME (2018) À descoberta de recur-sos naturais: Uma atividade outdoor com alunos do Ensino Primário. Terræ Didatica 14(4):477–486. http://www.ige.uni-camp.br/terraedidatica/
UNESCO (2015) Statutes of the international Geoscience and Geoparks programme and operational guidelines for UNESCO global Geoparks: UNESCO, p 16
Viera G (2004) Geomorfologia dos Planaltos e Altos Vales da Serra da Estrela. Ambientes frios do Plistocénico Superior e Dinâmica Atual. PhD Tesis, Universidade de Lisboa, CEGOT, p 724
Zouros N (2004) The European Geoparks network. Episodes 27(3):165–171

Chapter 8
Investigating the Perception and Attitude of Business School Students Towards Overtourism at Marseille Calanques National Park

Hugues Seraphin, Simon Smith, and Dorra Yahiaoui

Abstract 'Sustainability is possibly the most important issue facing the tourism industry in the twenty-first century' (Edgell and Swanson in *Tourism policy and planning: Yesterday, today, and tomorrow*. Routledge, New York, p. 45, 2018). The study within this chapter focuses on the future leaders responsible for delivering a sustainable planet, i.e. higher education students. The context is set within the subject of overtourism; a phenomenon that is now irreversibly damaging the world's cultural and natural heritage. A case study focus on Marseille Calanques National Park in France provides a specific context for discussion. Data is collected from students at Kedge Business School (Marseille). Using a survey, they are asked questions related to overtourism and Marseille Calanques National Park. Kedge Business School was chosen because it is part of the Principles for Responsible Management Education (PRME) network. PRME is implemented into business to raise awareness of Sustainable Development Goals by adopting a holistic, interdisciplinary approach of education. PRME could also be viewed as a tool to inform the strategy of an organisation. However, the findings discussed raise questions over the impact PRME is making within this study context. From a practical point of view, there is a chance to really reflect on the impact of PRME versus overtourism and contemplate actions going forward.

Keywords Overtourism · Sustainability · Marseille Calanques National Park · Kedge Business School

---

H. Seraphin · S. Smith (✉)
University of Winchester Business School, Winchester, UK
e-mail: Simon.Smith@winchester.ac.uk

H. Seraphin
e-mail: Hugues.Seraphin@winchester.ac.uk

D. Yahiaoui
Kedge Business School, Talence, France
e-mail: dorra.yahiaoui@kedgebs.com

A. Mandić and L. Petrić (eds.), *Mediterranean Protected Areas in the Era of Overtourism*, https://doi.org/10.1007/978-3-030-69193-6_8

8.1 Introduction

> Sustainability is possibly the most important issue facing the tourism industry in the twenty-first century. (Edgell and Swanson 2018: 45).

This chapter focuses on this global challenge of achieving sustainability in tourism through aligning and connecting with recent debates surrounding 'overtourism' in particular. We, the authors, accept that overtourism is prominent within the global tourism industry, and we need to find more effective solutions, and fast. This strongly aligns with the United Nations Sustainable Development Goals (SDGs) (UNSDGs 2015a [online], 2015b [online]). The study presented here provides a particular focus on a French tourist destination using local higher education student participants. Thus, we have a particular emphasis on the following SDGs (UNSDGs 2015a [online]) because of the nature and context of the study:

1. SDG4: Quality education
2. SDG11: Sustainable cities and communities
3. SDG12: Responsible consumption and production.

A case study focus on Marseille Calanques National Park in France provides a specific context for discussion. We collect and present data from students at Kedge Business School (Marseille). Using a survey approach, they are asked questions related to overtourism and Marseille Calanques National Park. Kedge Business School was chosen in particular because it is part of the Principles for Responsible Management Education (PRME) network:

> The Principles for Responsible Management Education (PRME) is a United Nations-supported initiative founded in 2007 as a platform to raise the profile of sustainability in schools around the world, and to equip today's business students with the understanding and ability to deliver change tomorrow. (PRME 2020 [online])

The implementation of PRME would then hopefully raise awareness of SDGs by adopting a holistic, interdisciplinary approach of education. Thus, PRME could act as a tool to inform the sustainability strategy of an organisation.

The main aim of this study is to understand how overtourism is perceived and engaged with at Kedge Business School. We want to see if PRME is working in action by inspiring these students to become the future leaders and change agents for sustainability. Indeed, we want to get a sense of potential impacts on SDG4, 11 and 12 within Marseille Calanques National Park.

The chapter is structured as follows. The next sections outline literature surrounding overtourism and empowerment. Conceptual and contextual frameworks are then presented to further the position of this study. A methodology is briefly discussed. Results are discussed followed by an extended focus on empowerment. Conclusions are offered, including theoretical implications, practical implications, and a direction for future research.

8.2 Overview of Literature on Overtourism

8.2.1 What Is Overtourism?

Overtourism is defined by UNWTO (2018) as the excess number of visitors to a specific destination. In other words, overtourism occurs when a destination is accommodating visitors beyond its carrying capacity (Muler Gonzalez et al. 2018). Both definitions are in line with Richardson (2017), who defines overtourism destinations as those suffering the strain of tourism, and Singh (2018), who defines overtourism destinations as those where the number of tourists are higher than the number of locals. That said, the term 'overtourism' became an official word when it entered the Collins dictionary in 2018 (Dickinson 2018; Singh 2018). Venice is an excellent example of a destination that epitomises this phenomenon (Seraphin et al. 2018). As a phenomenon, overtourism is associated with a list of negative impacts, and these include overcrowding spaces; inappropriate behaviour of visitors; touristification of destinations (i.e. mass tourism impact that gears commercial and societal aspects of a destination towards a tourist rather than a local); displacement of local populations; and, pressure on the environment (Koens et al. 2018; Muler Gonzalez et al. 2018). The phenomenon is also a further indication that achieving sustainability in tourism is still falling short. Yet, the large number of existing academic publications and initiatives provide evidence that sustainability is an important topic for the industry and academia (Edgell and Swanson 2018; Jacobsen et al. 2019).

8.2.2 Existing Research on Overtourism

Thus far, research on overtourism has covered the topic from three different angles:

- The origins of the phenomenon of overtourism
- The impacts of the phenomenon of overtourism
- The solutions to tackle the phenomenon of overtourism.

The typology of research on this subject can be summarised in Fig. 8.1.

It is worth highlighting the fact that Fig. 8.1 does not provide an exhaustive list of references published so far on the topic of overtourism, but instead a comprehensive (and representative) sample of existing research. That said, when looking at Fig. 8.1, the topics of research on overtourism could be categorised as follows (based on the number of outputs):

- Responses to overtourism
- Impacts of overtourism
- Origins of overtourism.

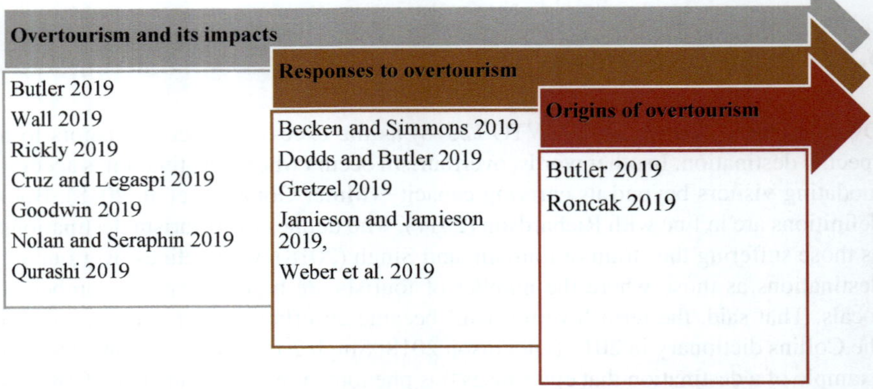

Fig. 8.1 Typology of research on overtourism (*Source* The authors)

8.2.3 *Gap in Literature*

When it comes to stakeholders in overtourism research, existing research has largely considered locals, local authorities, those working in the industry, and the tourists and visitors of affected destinations. However, we could not find research that has considered Business School students, and yet, Business Schools are central hubs for training future leaders in this industry and beyond (Don Keough cited in Favre 2017). This is an important element to take into consideration as 'businesses are the product and the extension of the personal characteristics of their leaders' (Don Keough cited in Favre 2017: 558). Thus, business leaders are potentially key stakeholders in dealing with overtouirsm. Indeed, Visser (2015) argues that greater sustainability could be achieved, among other dimensions, by unlocking change through transformational leadership and through integrated value. A transformational leader could potentially and critically instil pride, build a sense of mission and effectively teach and coach employees towards the required sustainable values and organisational practice (Smith 2016; Hater and Bass 1988). As a result, the study within this chapter is contributing towards addressing this gap in the literature and offering a Business School angle through the student's voice.

8.3 Empowerment of Students as Future Leaders and Agents of Change for the Tourism Industry

Higher education institutions (HEIs), particularly business schools, have an important role to play in education for the concept of sustainability through effective integration of sustainability (environmental, social, and economic dimensions) into their curriculum. This is despite the fact the latter of the dimensions, i.e. economic, can sometimes be perceived as the anti-thesis of sustainability. Because of a change of paradigm and initiative in teaching towards more sustainable practices, there is a desire for Business Schools to do better. For example, by teaching students how to design sustainable products, and ethically promoting and distributing those sustainable products (Kemper et al. 2019). HEIs can be considered as agents of change or transformational agents (Kemper et al. 2019).

By encouraging students at Kedge Business School to think about the impacts of overtourism on Marseille Calanques National Park, this study is encouraging students to be sustainable thinkers; that is to say, individuals with critical thinking and a questioning attitude. The study is also encouraging students to be sustainable actioners. In other words, these students become individuals involved in communities and sustainability projects. Finally, the involvement of students in this activity is turning them into sustainability transformers or activists, i.e. individuals who are enjoying initiating and observing the change of attitude and perception (Kemper et al. 2019).

The study presented here is also empowering students as stakeholders of the tourism industry, as there is a link between knowledge and empowerment. This is all the more important as empowerment is widely recognised as a prerequisite for sustainable tourism development (Joo et al. 2020). Indeed, the empowerment that happens as a result of an enhancement of competencies enables members of a community to have a better understanding of what is going on in their surrounding environment and to take action and control. In the tourism industry, this translates into: an increased involvement and sense of ownership of development plans; pride and self-esteem of the neighbourhood and inhabitants; and, development of social capital among fellow residents (Joo et al. 2020). Empowerment and responsible tourism are closely related as one of the pillars towards the new ideology of responsible tourism that involves the inclusion of multiple stakeholders, including the academic world alongside practitioners (Burrai et al. 2019).

8.4 Conceptual Framework

With the above research in mind, this study hypothesises relationships between students at Kedge Marseille Business School and tourism development in the city (and particularly, the impacts of the industry on the Calanques National Park). The conceptual framework of the study can be summarised in Fig. 8.2.

Objective suggestions from students are all about suggestions or recommendations that are applicable in the industry (due to an expertise developed via prior experience), while metaphorical suggestions are less applicable (due to the students lack of understanding of the industry). These approaches are based on Rakic and Chambers (2012) research on children whereby they argue that children are thinking both objectively (thoughts based on real-life facts) and metaphorically (thoughts based on their own perception of life). That said, this study is not to patronise students'

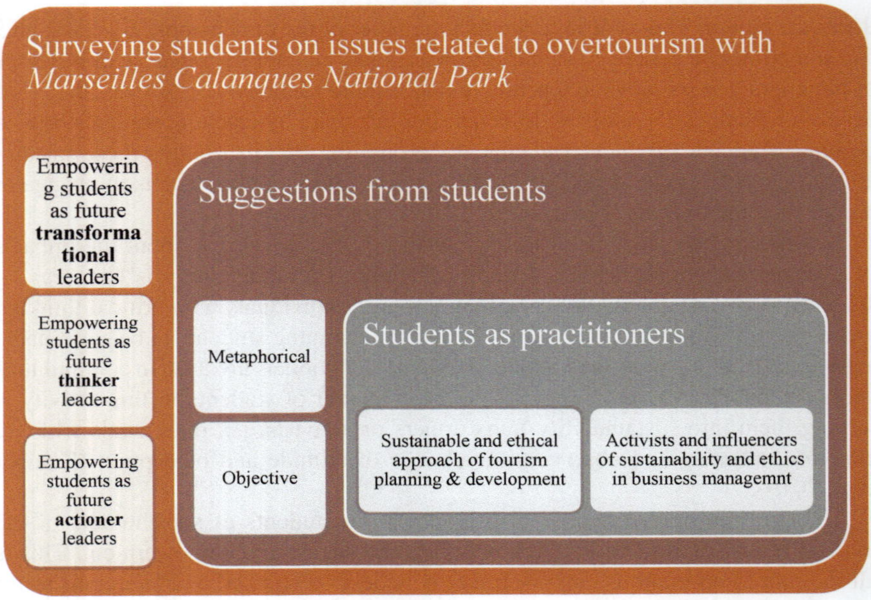

Fig. 8.2 Conceptual model of the study (*Source* The authors)

ability. It is adapting these two concepts, objective and metaphorical, to create a line of different thinking within a different segment of the population.

8.5 Contextual Framework

8.5.1 Marseille as a Destination

France can be classified as the leading tourism destination in the world. In 2017, for example, France received 86.9 million visitors (Table 8.1). Tourism was worth 7.2% of the France Growth Domestic Product (GDP) in 2017 (Ministere de l' Economie et des Finance, 2017 [Online]).

As for Marseille (Fig. 8.3), tourism and cognate sectors are central to the economic development of the city. Marseille, being the fourth biggest port of the Mediterranean area (and the largest in France), has helped with the development of the cruise sector. In 2018, some 31 cruise companies represented just over 1.7 million stopover passengers who visited Marseille. Awards and recognitions have helped Marseille to attract such a wealth of visitors; these include being the European Capital of Culture (2013), the Capital of Sport (2017), the Capital of Love (2018), and so on. Marseille is due to host the Olympic Games in 2024, so future prospects for tourism look

Table 8.1 International tourist arrivals by county of destination

Rank in 1980	Rank in 2017	Destination	Arrivals in 2017 (millions)	Change 2017/2016 (%)
1	**1**	**France**	**86.9**	**5.1**
3	2	Spain	81.9	8.8
2	3	United States	76.9	0.7
18	4	China[a]	60.7	2.4
4	5	Italy	58.2	11.1
8	6	Mexico	39.3	12.0
7	7	United Kingdom	37.7	5.3
52	8	Turkey	37.6	24.1
9	9	Germany	37.5	5.3
27	10	Thailand	35.4	8.9
Total World			**1,326.4**	**7.0**

[a]Excluding Hong Kong and Macao
Source World Tourism Organisation, 2017

Fig. 8.3 Marseille and Calanques National Park (*Source* Calanque National Park [Online])

fruitful too. From an economic point of view, tourism benefits the city thanks to taxes collected (6.085.405 euros in 2018) and job creation (15.211 people were working in the industry in 2017). Other assets of the city include a wide range of accommodation and facilities, events, easy access, its nature assets such as the Calanques National Park, etc. (Office de Tourism et des Congres Marseille 2018 [Online]).

8.5.2 *Calanques National Park and Related Perverse Impacts of Overtourism*

The visit of the Calanques (Fig. 8.3) is a very popular attraction in Marseille. Here are the recent numbers for tourists visiting the Calanques for water sport activities:

- 2016: 126,410 tourists
- 2017: 145,237 tourists
- 2018: 149,000 tourists.

That said, the number of visitors to this national park (created in 2012) is increasing year on year. Altogether (for sea- and land-related activities), the national park receives 2 million visitors per year. In summer 2018, some 51,724 tourists visited the park on a daily basis. This causes major environmental issues and highlights the reasons why ecology rangers were hired to raise tourists awareness of such issues (Office de Tourism et des Congres Marseille 2018 [Online]). These aspects and impacts of overtourism within this destination are central to the study conducted here. As the figures continue to rise, the pressure to find solutions also becomes more intensified.

8.5.3 *Kedge (Marseille) Business School as a PRME Institution*

Kedge Business School is the result of a merger between BEM, and Euromed Management in 2013. The school has 10 campuses (including Kedge Marseille), spread across three continents. A wide variety of programmes are offered, among these are: Marketing; Supply Chain; Corporate Social Responsibility; Wine and Spirits; Innovation and Entrepreneurship; Finance; Creative Industries and Culture; and Health Management. Kedge is part of the Principles of Management Education (PRME) network (Kedge [Online]).

PRME was conceived in 2007, with the objective to foster some ethical values within future leaders, i.e. students in higher education (Annan-Diab and Molinari 2017; Parkes et al. 2017), who would then implement these values within the business they would lead or work for (Mayer and Hutton 2016). PRME was also created and launched in order to help with the achievement of the Sustainable Development Goals, for a more sustainable world (Annan-Diab and Molinari 2017; Parkes et al. 2017), and ultimately to fight poverty at local, national, and international levels (Rosenbloom et al. 2017). To have PRME fully imbedded in their curricula, higher education institutions have to review their curriculum design, teaching approach, research strategy and agenda, as well as, and equally important, work in partnership with all stakeholders of the sustainability ecosystem (Parkes et al. 2017).

8.6 Methodology

8.6.1 Previous Study

This study could be considered to be a continuation of the work of Kemper et al. (2019) who investigated the 'why' and 'how' of teaching sustainability. They used semi-structured interviews (face-to-face; telephone; skype interviews) to explore the views of authors (from Australasia, Europe, and North-America) concerning sustainability marketing papers on: non- and/or conflicts with marketing and sustainability; what sustainability looks like in the marketing curriculum; the current integration of sustainability within the marketing curriculum and scholarship; pressure and logics of the business school; and, active change/resistance. The results of their research revealed the existence of three main types of academics: 'The sustainability "transformer" wishes to engage in transformational learning, changing student mindsets, the "thinker" wants to encourage critical thinking to bring about the discussion of worldviews, while the "actioner" hopes learning by doing (community projects) will provide an appreciation for sustainability" (Kemper et al. 2019: 1).

8.6.2 Positioning of the Study

From the above section and the work of Kemper et al. (2019), we position ourselves as sustainability 'thinkers' and 'actioners'. This positioning has influenced the survey (online questionnaire) at the basis of this study. The students at Kedge were asked to share their view on:

1. Sustainability (generally)
2. Conflict or non-conflict between sustainability and tourism
3. Causes of overtourism in Marseille and the impacts on Calanques National Park and the overall city
4. Strategies to tackle overtourism and the related detrimental impacts on Calanques National Park and Marseille overall
5. The current integration of sustainability within the curriculum of the different programmes of Kedge Marseille Business School
6. Barriers and opportunities of Kedge Marseille being a PRME institution.

8.6.3 Study Site

Kedge Marseille Business School was selected as the site of this study to collect data. Indeed, a case study is 'a study that is bounded by a focus on a particular person, event, group, organisation, a town or a unit of analysis' (Hammond and Wellington 2013: 162). Additionally, as a method, 'case study shares with ethnography an understanding of local conditions' (Hammond and Wellington 2013: 162).

This method helps to understand the 'how' and 'why' of a phenomenon; and also helps to gain fresh insight and ideas about a topic (Hammond and Wellington 2013). Thus, Kedge students provide the local insights for studying the phenomenon of overtourism within the Calanques National Park context.

8.6.4 Approach of the Study

This study is centred around primary data collection. 'Empirical enquiry involves first-hand data collection by interviewing, observation, questionnaire, etc'. (Hammond and Wellington 2013: 166). In terms of our approach, we have adopted a deductive approach. This approach draws a general conclusion from individual instances or observations. It is a bottom-up approach (Hammond and Wellington 2013).

8.6.5 Survey Instrument and Analysis

Data collection was conducted between January and February 2020 using an online survey (questionnaire) function of Google (i.e. Google Forms). The survey (Appendix) contained a host of questions split into three sections:

1. Overtourism in Marseille
2. Strategies to tackle overtourism and related perverse impacts on Calanques National Park and Marseille overall
3. The current integration (challenge and opportunities) of sustainability within the curriculum of the different programmes of a PRME institution.

The version of the survey used in this study has been adapted from a survey developed by Joo et al. (2020), which aimed at identifying what residents think and feel about their community, interaction with tourists, and tourism in Fredericksburg. The data collected are analysed using the analytics function of Google Forms. SPSS is also used to operate some correlations. 83 Kedge Marseille Business School students responded to the survey.

8.7 Results

At this stage, it is important to mention two important points necessary for understanding the findings presented in this section:

– Keys to Likert scale adopted: 1 (agree strongly) to 5 (disagree strongly)

– The graphs below are from the analytics function of Google Forms. As for the tables, they were completed using the software package of SPSS.

Tourism in Marseille

The first noteworthy finding is that there are a significant number of students who believe that overtourism is an issue in Marseille. This extends to concerns regarding the development of the industry, mainly due to the negative impacts on the environment of the Calanques, and the life of locals (Fig. 8.4).

Strategies to Tackle Overtourism

Students are not particularly involved in local affairs surrounding tourism in Marseille, let alone involved in actions to protect the city against the detrimental impacts of overtourism. Figure 8.5 shows very little involvement in any active

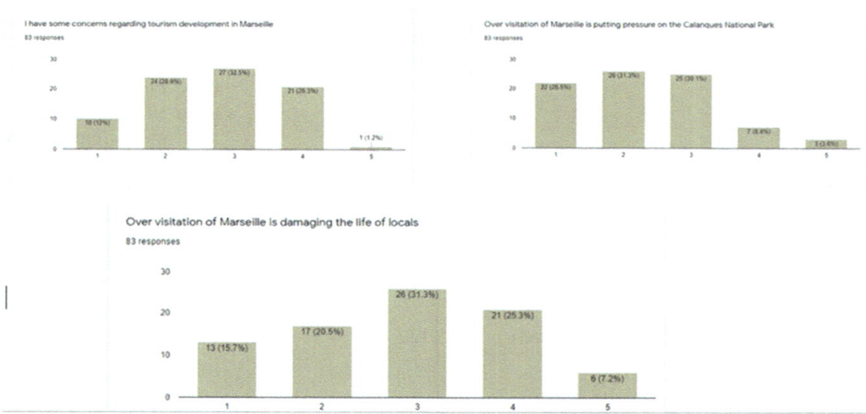

Fig. 8.4 Data results on overtourism in Marseille/*Calanques National Park* (*Source* The authors)

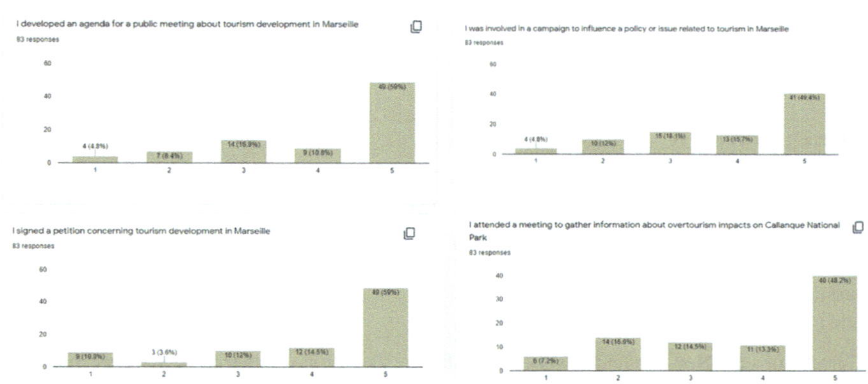

Fig. 8.5 Data results on student involvement in tourism issues (*Source* The authors)

or passive actions. Thus, these students have very little involvement in strategic developments for tackling overtourism.

Principles of Responsible Management Education (PRME)
Crucially, the fact that Kedge is a PRME institution does not appear to be translating into enhanced student understanding of sustainability, nor their increased involvement in sustainability initiatives (Table 8.2). In practice, PRME is failing in this institution, as one of the criteria to be a PRME member is the involvement with stakeholders in charge of sustainability (Parkes et al. 2017). Additionally, one of the objectives of PRME is to develop future leaders with a sustainability mindset

Table 8.2 Data correlations

Correlations (1)			
		I already knew Kedge was a PRIME institution	I feel I am knowledgeable enough about sustainability (environment) to be an agent of change about tourism development in Marseille
I already knew Kedge was a PRIME institution	Pearson Correlation	1	−.171
	Sig. (2-tailed)		.122
	N	83	83
I feel I am knowledgeable enough about sustainability (environment) to be an agent of change about tourism development in Marseille	Pearson Correlation	-.171	1
	Sig. (2-tailed)	.122	
	N	83	83
Correlations (2)			
		I already knew Kedge was a PRIME institution	I have a good understanding of sustainability (environment)
I already knew Kedge was a PRIME institution	Pearson Correlation	1	.136
	Sig. (2-tailed)		.220
	N	83	83
I have a good understanding of sustainability (environment)	Pearson Correlation	.136	1
	Sig. (2-tailed)	.220	
	N	83	83

[a]Correlations are Non-existent to very weak (.00 to .20 = Non-existent to very weak – according to: Silver et al. 2013)
Source The authors

(Annan-Diab and Molinari 2017; Parkes et al. 2017). It would therefore appear difficult to achieve as there is no correlation between being a PRME member and the students' understanding of sustainability issues and/or willingness to be involved in sustainability initiatives (Table 8.2).

8.8 The Importance of Empowerment

8.8.1 Empowering Sustainability: Tensions and Solutions

The most commonly accepted definition of 'sustainability' is the one developed by the World Commission on Environment and Development (WCED), namely a 'development that meets the needs of the present without compromising the ability of future generations to meet their own needs' (WCED 1987). Our discussion here in consideration of this definition will embrace Janusian thinking and ambidextrous management. In both contexts, this is to address difficulties in dealing with the tensions of present and future needs by considering them as opposites or potentially paradoxical (polar opposites) (Rothenberg 1996; Smith 2016; Vo-Thanh et al. 2020). In reference to Janus—the Roman god with two faces, who looked in opposite directions simultaneously—it appears that sustainability initiatives need to include children, teenagers, and younger adults (future generations), as well as the adults (present generations). For ambidextrous management, the potential paradox surrounds these tensions between the needs of the present versus the future.

To take the discussion one-step further, Seraphin and Vo-Thanh (2020) argue that, as children are neither passive nor powerless, they can play a significant role in the sustainability of the tourism industry, but only if empowered. As a result, Seraphin and Vo-Thanh (2020) suggested that empowerment fun activities could facilitate this endeavour. Indeed, they identify some activities in resort mini-clubs as having a potential to contribute towards Sustainable Development Goals (SDGs) 4 (Quality education), and 12 (Responsible consumption and production). Empowerment strategies and/or activities can be designed according to the age of the participants and the SDG targeted.

The Janusian thinking approach or ambidextrous approach suggested in this section is strongly connected with success since Janus has played an essential role in the creation of the world (Rothenberg 1996). Additionally, Vo-Thanh et al. (2020) explained that the application of ambidextrous management in organisations has contributed to the success, in terms of:

- A strategy towards sustainability (Martinez-Perez et al. 2016);
- Staff motivation and retention (Bouzari and Karatepe 2017; Ma et al. 2019)
- Innovation and performance (Cheng et al. 2016; Mihalache and Mihalache 2016).

Ultimately, all generations should be involved in sustainability initiatives. This view is further supported within the literature on sustainable tourism that highlights the importance of involving the present and future generations in initiatives (Hall et al. 2015). Other literature also supports the fact that the commitment of all stakeholders is a prerequisite for sustainability (Sloan et al. 2013).

Sustainability and Students at Kedge Business School
To build on the theoretical context above, the results of this study provide evidence that the younger generation (future) is not involved in sustainability actions. As sustainability requires the involvement of all stakeholders, actions must, therefore, be taken to motivate the younger members of communities to be involved. In other words, the younger generation needs to be empowered so that they can become a lot more involved. The results of this study are perhaps surprising, as well as disappointing, considering that Kedge is a PRME institution. More engagement was expected from students of this institution. As Janusian thinking and ambidexterity is filtered into this discussion, two factors could explain the non-engagement and lack of interest of students for sustainability (despite their acknowledgement that overtourism exists in Marseille and expressed some concerns). First, perhaps higher education institutions are not the best conveyor of sustainability messages; and second, if they are, it is the approach used by the school that is not suitable.

This failing of PRME at Kedge is all the more concern as, beyond the results of this study, sustainability is becoming more and more a mainstream way of thinking that underpins behaviour (Page 2019), and is largely influenced by the noticeable negative consequences of human activity on the planet (Sloan et al. 2013). Students in PRME institutions should be spearheading this change of perception. This, therefore, links to the topic of empowerment. Students from Kedge Business School should be more involved in tourism local affairs. As a result, sustainability and empowerment become related (Boley and McGehee 2014). Indeed, empowerment happens when individuals or groups are fully in control of their destiny and/or affairs, which has come through competencies developed as the result of learning (Joo et al. 2020). When empowered, individuals or groups contribute to the development of their community (Scheyvens 1999; Strzelecka et al. 2017). Indeed, knowledgeable individuals are more engaged in community affairs than others (Rocha 1997) as it gives them a sense of ownership (Joo et al. 2020; Strzelecka et al. 2017). Despite students at Kedge being educated people, they are not involved in sustainability initiatives overall, let alone initiatives to protect the Calanques. Education is therefore a criterion, but not the main criteria for empowerment. What would therefore engage students at Kedge to build towards sustainability actions for the protection of the Calanques against overtourism?

The Calanques and Students at Kedge Business School
The results of this study clearly acknowledge that Kedge and its students should be involved in tourism within the life of the city in one way or another (Fig. 8.6). One way suggested for more effective empowerment is to use more locally-based case studies.

Adams (2008) identified different approaches to achieve empowerment:

- a cathartic and facilitative strategy (enabling people to express their feelings)

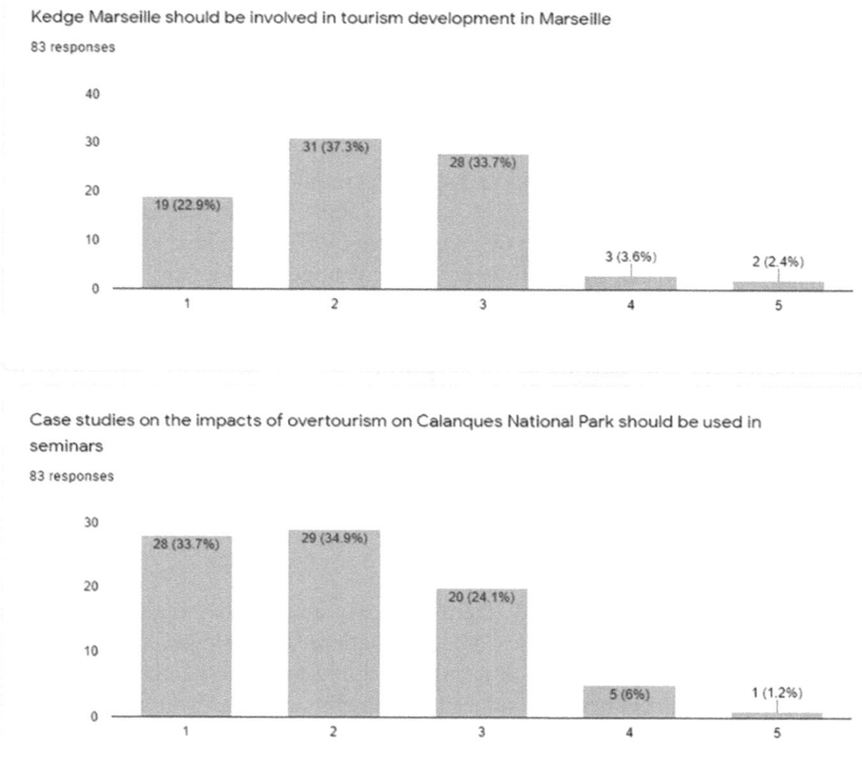

Fig. 8.6 Data results regarding Kedge Marseille's potential involvement in solving tourism issues (*Source* The authors)

- a catalytic strategy (enabling people to engage in self-discovery, self-directed living, and problem solving)
- supportive and catalytic strategies (enabling people to build self-confidence)
- self-advocacy strategy (enabling people to speak for themselves).

The empowerment strategy identified by students could be said to be a catalytic strategy. However, if Kedge was to implement the results of this survey and include case studies on Marseille Calanques in the curriculum, this could be considered as supportive and catalytic strategies. A more locally embedded curriculum could contribute, in the long-term, to the psychological, social, and political empowerment of students. Indeed, consider that Boley and Gard McGehee (2014), and Strzelecka et al. (2017) have also identified three types of empowerment in tourism:

- Psychological empowerment (is apparent when locals display some pride for their community and neighbourhood)

– Social empowerment (occurs when social capital is developed within the community); and finally
– Political empowerment (that occurs with the involvement of locals in decision-making).

Sometimes applying a twist to an already established way of doing things (in the case of this study, teaching methods and resources) could lead to great achievements. Daring to try a new approach is a prerequisite to those great achievements (Cardno et al. 2017). This study is supporting Callender (1997) that education (and more specifically, schools) can successfully empower younger individuals. That said, this study does highlight this will only happen if the right approach is adopted (i.e. as PRME is not reaching the standard required). More importantly, it is important to take students from the classroom to the field to interact with stakeholders, as the connection of people with their environment is central to an effective understanding of sustainability and empowerment (Camargo and Gretzel 2017).

8.9 Conclusion

Returning to the aim of this study, there is clearly a sense that students recognise overtourism issues within Marseille Calanques National Park. However, when it came to engagement with overtourism in terms of helping to solve the challenges, there was clearly a disconnect between values being taught and what was being exercised in practice.

Theoretical Implications
Worryingly, there is a gap between theory and practice within this study as students recognised the importance of overtourism in theory, but are not contributing much in terms of practice to help solve the problem. PRME, as a concept, is argued to be falling short in this regard. PRME may espouse the ideal values, but there is much work to do to convert this theory into effective practice—at least for the context of this study. Those currently engaged with, or wanting to be engaged with, PRME should perhaps not assume the badge will automatically lead to positive change. Instead, once PRME status is achieved, this is perhaps the crucial time to ensure activities in the university enable a practising of values—see the next section for expansion on practical implications. This is perhaps fundamental in terms of achieving true sustainability in the future (hopefully the near future).

Practical Implications
As the above indicates, more steps are required to convert ideology into action. Our study highlights and promotes empowerment in this regard. Students can be given more practical empowerment opportunities within their studies to enable bridging between theory and practice. In essence, students should become more actively involved within local issues to enable them to more effectively become the future

leaders we want them to be for sustainability in tourism. This will hopefully then feed more successfully into solving the challenges of overtourism.

Future Research

We would strongly suggest that Action Research is a way forward for research. This, in essence, could build into a curriculum the practical challenges to be focused upon (e.g. overtourism in Marseille Calanques National Park) which can then be tracked and assessed over time. Action Research can add the academic rigour needed to build and implement such a project, as well as providing the expertise to change as it progressed.

In addition to the above, there is clearly a need for greater research into the impact PRME is having on future leaders' business practice. Even if findings highlight limitations, this should only be a precursor for further action, as suggested within this study, to expand the impact PRME is having.

Appendix

See Table 8.3.

Table 8.3 Testing the empowerment theory in a tourism context

	Questions	Likert scale
Tourism in Marseille	Tourism makes me proud to be a Marseille resident/student	
	Tourism makes me feel special because people travel to see Marseille's city unique features	
	Tourism makes me want to tell others about what we have to offer in Marseille	
	Tourism makes me feel connected to Marseille's community	
	I have some concerns regarding tourism development in Marseille	
	As a destination Marseille is over visited	
	Over visitation of Marseille is damaging the unique features of the city	
	Over visitation of Marseille is putting pressure on the Calanques National Park	
	Over visitation of Marseille is damaging the life of locals	
Strategies to tackle overtourism	I am at the origin of a petition to influence a policy or issue related to tourism in Marseille	
	I signed a petition concerning tourism development in Marseille	
	I attended a meeting to pressure for change of Marseille's approach to tourism	
	I developed an agenda for a public meeting about tourism development in Marseille	
	I attended a meeting to gather information about overtourism impacts on Calanques National Park	
PRME institution	I already knew Kedge was a PRME institution	
	I am proud to be a student at a PRME institution	
	I have a good understanding of sustainability	
	I feel I am knowledgeable enough about sustainability to share my concerns about tourism development in Marseille	
	I feel I am knowledgeable enough about sustainability to be an agent of change about tourism development in Marseille	
	Kedge Marseille should be involved in tourism development in Marseille	

(continued)

Table 8.3 (continued)

	Questions	Likert scale
	Case studies on the impacts of overtourism on Calanques National Park should be used in seminars	

Source The authors (adapted from Joo et al. 2020)

References

Adams R (2008) Empowerment, participation and social work, 4th edn. Palgrave Macmillan, Basingstoke

Annan-Diab F, Molinari C (2017) Interdisciplinarity: practical approach to advancing education for sustainability and for the sustainable development goals. Int J Manage Educ 15(2):73–83

Boley BB, McGehee NG (2014) Measuring empowerment: developing and validating the resident empowerment through tourism scale (RETS). Tourism Manage 45:85–94

Bouzari M, Karatepe OM (2017) Test of a mediation model of psychological capital among hotel salespeople. Int J Contemp Hosp Manage 29(8):2178–2197

Burrai E, Buda DM, Stanford D (2019) Rethinking the ideology of responsible tourism. J Sustain Tourism. https://doi.org/10.1080/09669582.2019.1578365

Calanques National Park. http://www.calanques-parcnational.fr/fr/le-parc-national-des-calanques/identite-du-parc-national-des-calanques. Accessed 15 Apr 20

Callender C (1997) Education for empowerment. The practice and philosophies of black teachers. Trentham Print Design, Chester

Camargo BA, Gretzel U (2017) What do tourism students know about sustainability and sustainable tourism? An exploratory study of Latin American students. J Teach Travel Tourism 17(2):101–117

Cardno C, Rosales-Anderson N, McDonald M (2017) Documentary analysis hui: an emergent bricolage method for culturally responsive qualitative research. MAI J 6(2):143–152

Cheng J-S, Tang T-W, Shih H-Y, Wang T-C (2016) Designing lifestyle hotels. Int J Hosp Manage 58:95–106

Dickinson G (2018) Dear dictionaries, this is why 'overtourism' should be your 2018 word of the year. Retrieved April 20, 2018, from https://www.telegraph.co.uk/travel/comment/overtourism-word-of-the-year

Edgell DL, Swanson JR (2018) Tourism policy and planning: yesterday, today, and tomorrow, 3rd edn. Routledge, New York

Favre C (2017) The Small2Mighty tourism academy: growing business to grow women as a transformative strategy for emerging destinations. Worldwide Hosp Tourism Themes 9(5):555–563

Hall CM, Gossling S, Scott D (eds) (2015) The Routledge handbook of tourism and sustainability. Routledge, Abingdon

Hammond M, Wellington J (2013) Research methods: the key concepts. Routledge, London

Hater JJ, Bass BM (1988) Superior's evaluations and subordinate's perceptions of transformational and transactional leadership. J Appl Psychol 73(4):695–702

Jacobsen J, Kr S, Iversen NM, Hem LE (2019) Hotspot crowding and over-tourism: antecedents of destination attractiveness. Ann Tourism Res 76:53–66

Joo D, Woosnam KM, Strzelecka M, Boley B (2020) Knowledge, empowerment, and action: testing the empowerment theory in a tourism context. J Sustain Tourism 28(1):69–85

Kedge. https://student.kedge.edu/about-kedge. Accessed 15 Apr 20

Kemper JA, Ballantine PW, Hall M (2019) Combining the 'why' and 'how' of teaching sustainability: the case of business school academics. Environ Educ Res. https://doi.org/10.1080/13504622.2019.1667959

Koens K, Postma A, Papp B (2018) Is tourism overused? Understanding the impact of tourism in a city context. Sustainability 10:4384

Ma J, Zhou X, Chen R, Dong X (2019) Does ambidextrous leadership motivate work crafting? Int J Hosp Manage 77:159–168

Martínez-Pérez Á, García-Villaverde PM, Elche D (2016) The mediating effect of ambidextrous knowledge strategy between social capital and innovation of cultural tourism clusters firms. Int J Contemp Hosp Manage 28(7):1484–1507

Mayer D, Hutton B (2016) Global business education for management professionals. In: Sunley R, Leigh J (eds) Educating for responsible management. Putting theory into practice, Routledge, London, pp 185–210

Mihalache M, Mihalache OR (2016) Organisational ambidexterity and sustained performance in the tourism industry. Ann Tourism Res 56:142–144

Ministere de l'Economie et des Finance. https://www.entreprises.gouv.fr/etudes-et-statistiques/chiffres-cles-tourisme. Accessed 15 Apr 20

Muler Gonzalez V, Coromina L, Gali N (2018) Overtourism: residents' perception of tourism impact as an indicator of resident social carrying capacity—case study of a Spanish heritage town. Tourism Rev 73(3):277–296

Office de Tourism et des Congres Marseille (2018) http://www.marseille-tourisme.com/fr/espace-presse/chiffres-cles-du-tourisme/. Accessed 15 Apr 20

Page SJ (2019) Tourism Management, 6th edn. Routledge, Abingdon

Parkes C, Buono AF, Howaidy G (2017) The principles for responsible management education (PRME): The first decade—what has been achieved? The next decade—responsible management education's challenge for the sustainable development goals (SDGs). Int J Manage Educ 15(2):61–65

Rakik T, Chambers D (2012) An introduction to visual research methods in tourism. Routledge, Abingdon

Richardson D (2017) Suffering the strain of tourism. Travel Trade Gazette (special issue Word Travel Market 2017)

Rocha EM (1997) A ladder of empowerment. J Plan Educ Res 17(1):31–44

Rosenbloom A, Gudic M, Parkes C, Kronbach B (2017) A PRME response to the challenge of fighting poverty: how far have we come? where do we need to go now? Int J Manage Educ 15(2):104–120

Rothenberg A (1996) The Janusian process in scientific creativity. Creativity Res J 9(2):207–231

Scheyvens R (1999) Ecotourism and the empowerment of local communities. Tourism Manage 20(2):245–249

Séraphin H, Sheeran P, Pilato M (2018) Over-tourism and the fall of Venice as a destination. J Destination Mark Manage 9:374–376

Séraphin H, Vo-Than T (2020) Investigating the application of the principles for responsible management education to resort mini-clubs. Int J Manage Educ. https://doi.org/10.1016/j.ijme.2020.100377

Silver L, Stevens R, Wrenn B, Loudon D (2013) The essentials of marketing research. Routledge, Abingdon

Singh T (2018) Is over-tourism the downside of mass tourism? Tourism Recreation Res 43(4):415–416

Sloan P, Legrand W, Chen JS (2013) Sustainability in the hospitality industry: principles of sustainable operations, 2nd edn. Routledge, Abingdon

Smith SM (2016) Management and organisation—the 21st century global and international context. In: Stokes P, Moore N, Smith S, Rowland C, Scott P (eds) Organizational management: approaches and solutions. Kogan Page, London

Strzelecka M, Boley BB, Woosnam KM (2017) Place attachment and empowerment: do residents need to be attached to be empowered? Ann Tourism Res 66:61–73

UNSDGs (2015a) Sustainable development goals. https://sustainabledevelopment.un.org/?menu=1300. Accessed 29 Apr 20

UNSDGs (2015b) Transforming our world: the 2030 Agenda for sustainable development. https://sustainabledevelopment.un.org/post2015/transformingourworld. Accessed 29 Apr 20
UNWTO (2018) Overtourism? understanding and managing urban tourism growth beyond perceptions, Madrid
Visser W (2015) Sustainable frontiers: unlocking change through business, leadership, and innovation. Greenleaf Publishing, Sheffield
Vo-Thanh T, Séraphin H, Okumus F, Koseoglu MA (2020) Organisational ambidexterity in tourism research: a systematic review. Tourism Anal. https://doi.org/10.3727/108354220X15758301241701
WCED (1987) Our common future: Brundtland report. https://www.are.admin.ch/are/en/home/sustainable-development/international-cooperation/2030agenda/un-_-milestones-in-sustainable-development/1987--brundtland-report.html

Chapter 9
Collaborative Destination Management Based on Carrying Capacity Assessment from Resident and Visitor Perspectives: A Case Study of Crikvenica-Vinodol Riviera, Croatia

Neda Telišman-Košuta and Neven Ivandić

Abstract Within the context of the rapid worldwide growth of tourism frequently accompanied by overcrowding and overtourism issues, this paper juxtaposes resident and visitor perceptions of tourism impacts on Croatia's busy Crikvenica-Vinodol Riviera seeking to better understand destination social carrying capacity implications as these may inform improvement in collaborative destination management and governance. The analysis shows significant differences in perception of tourism, with predominantly positive views among visitors versus the more critical residents. As a means of overcoming the threat of diverging perceptions becoming a limiting factor of sustainable tourism development, the paper suggests destination management should be oriented towards building destination cohesion and advancing community values. This implies better monitoring of a destination's tourism carrying capacity parameters, further community capacity building in tourism, improving the quality of tourism planning and capacity for plan implementation, as well as promoting a collaborative culture among local stakeholders.

Keywords Social tourism carrying capacity · Collaborative destination management · Crikvenica-Vinodol Riviera · Croatia

9.1 Introduction

Over the past several decades tourism has proven to be a continuously growing, rapidly expanding and resilient global phenomenon. The current unprecedented medical crisis caused by the coronavirus pandemic will, if judging from past post-crisis performance, only temporarily cut short worldwide tourism growth. Fueled

N. Telišman-Košuta (✉) · N. Ivandić
Institute for Tourism, Zagreb, Croatia
e-mail: neda.kosuta@iztzg.hr

N. Ivandić
e-mail: neven.ivandic@iztzg.hr

A. Mandić and L. Petrić (eds.), *Mediterranean Protected Areas in the Era of Overtourism*, https://doi.org/10.1007/978-3-030-69193-6_9

by mature generating markets in Europe and North America, but particularly by new markets in the Asia-Pacific region, international tourist arrivals grew by 7% in 2017, followed by an additional 5% increase in 2018 (UNWTO 2019) and 4% in 2019 to reach total 1.5 billion overnight visitors worldwide (UNWTO 2020). All world regions report growth, including mature ones such as Europe (UNWTO 2020). Globally, tourism generates one in 10 jobs, 10% of GDP and 7% of total exports, being positioned as the third-largest export category in the world, behind fuels and chemicals (UNWTO 2019). With its low barriers to entry and the capacity to generate economic development in a relatively short time, tourism continues to be the sector of choice worldwide. Considering its attractiveness coupled with the world's growing population, affluence and mobility, the potential for the future growth of tourism seems unquestionable.

At the same time, these record growth rates are translating worldwide into experiences of overcrowding and overtourism to be coped with by destination residents and visitors alike, provoking increasing criticism of tourism, now spreading beyond academia to professional and public discourse, as the cause of overbuilt landscapes, loss of biodiversity and population displacement among other impacts. Evermore apparently, tourism's potential, within the dominantly pursued growth paradigm, for disruption of not only natural but also of social and cultural environments is unquestionable as well. In view of the tourism industry's high economic growth potential and at the same time its increasingly harmful pressures on natural and social structures, destination governance and management are becoming the critical issues of tourism sustainability, which, although not new notions, need to be tackled now with a new sense of urgency and sincerity. Improvement of destination governance and management models presupposes they are based on improved understanding of destination carrying capacity and collaborative management concepts. Drawing on an exploratory project assessing tourism carrying capacity of Croatia's Crikvenica-Vinodol Riviera (Ivandić and Telišman-Košuta 2019), a destination owning its appeal to its natural resources, initially, in the 1890s evolving as a climatic spa retreat and having turned today into one of the country's most developed seaside, summer holiday spots, this paper juxtaposes resident and visitor perceptions of tourism impacts searching for commonalities and discrepancies among them with the aim to improve the understanding of social carrying capacity thus leading to better, knowledge-based collaborative destination management and governance methodologies.

9.2 Guiding Research

Arising from past extensive research of tourism impacts focusing mostly on residents' attitudes and predominantly embedded in social exchange theory (Rasoolimanesh and Seyfi 2020) explaining interactions between parties as analyses of costs vs benefits (Emerson 1976), destination management and governance in today's circumstances of booming travel and faced with unprecedented issues of overcrowding and overtourism needs to be sharpened taking into account complex destination dynamics and possible carrying capacity, especially social carrying capacity, limitations.

Tourism destinations are in their nature unbalanced and conflicted systems as such difficult to manage (Telišman-Košuta and Ivandić 2020). They are fragmented entities made up of different stakeholder groups including local residents, tourism entrepreneurs and developers, political structures and various tourist segments who differ in knowledge, experience and worldviews, perceiving problems and solutions differently, as well as having different interests and expectations of the simultaneous use of local resources and tourism development (Buhalis 2000; Ritchie and Crouch 2003; Manente and Minghetti 2006; Wang and Pizam 2011; Bimonte and Punzo 2016; Boom et al. 2020). Leadership is often characterized by the more politically and/or economically influential actors dominating decision-making although they may not be the most qualified to do so, bypassing other partners and also weakening destination management organizations where the latter even exist (Boom et al. 2020). The private sector typically favours tourism growth, while the public sector is slow in formulating and implementing tourism strategies and for the most part, also hesitant in curbing growth (Dodds and Butler 2019). Local residents, many of whom are directly involved in the tourism industry, seem caught between personal economic benefits and unwanted social, but particularly environmental costs of tourism contributing to a sense of 'collective schizophrenia' (Telišman-Košuta et al. 2015). Overcoming in-destination conflicts is possible only with mutually beneficial development resulting from positive trade-offs between benefits and costs (Gursoy et al. 2018), yet fragmented as they are, destinations are likely to always be vulnerable to tourism-generated pressures.

Coupled with destination instability, recent years of high tourism growth rates had contributed to specific long researched impacts of visitor density, newly coined as overcrowding when looked at from visitor perspective and overtourism when seen from the resident viewpoint (Gossling et al. 2020), come to the forefront with a new level of acuteness (Koens et al. 2018; Dodds and Butler 2019; Capocchi et al. 2019; Perkumiene and Pranskuniene 2019). Overcrowding is acknowledged as a subjective, psychological response to visitor density which could be affected by nationality and cultural background, personal and situational characteristics including education, age and gender, environmental characteristics and activity types (Rasoolimanesh and Seyfi 2020; Gossling et al. 2020). Overtourism is a more complex and multifaceted phenomenon (Peeters et al. 2018) with a common thread running through several existing definitions being the notion of tourism impacts on a destination which result in perceived unacceptable deterioration of resident quality of life and/or visitor quality of experience (UNWTO 2018). Central to the concept are excessive numbers of tourists visiting a destination, which can be further aggravated by seasonal fluctuations typical for tourism and nowadays additionally exacerbated by low-cost flights, budget cruises and accommodation platforms such as Airbnb 'pumping' capacity and visitors (Dodds and Butler 2019). Overtourism is related to damaging effects of a broad scope of issues, ranging from overcrowding of iconic attractions, touristification and gentrification to growth-insistent mass tourism negating environmental, cultural and social sustainability of places (Capocchi et al. 2019; Koens et al. 2018;

Benner 2019). With UNWTO forecasting further growth of tourism, expecting international tourist arrivals to reach 1.8 billion by 2030[1] (UNWTO 2011), overtourism will, if not mitigated, continue being a problem in the future, undermining tourism itself by 'sawing off the branch it is sitting on'. Dealing with overtourism presupposes implementation of some kind of strategy, whether steering tourist flows within a destination, deterring tourist from even coming, increasing the capacity of existing destination systems (Dodds and Butler 2019) or other tailor-made management strategies being developed to cope with specific situations in different destinations (Seraphin et al. 2018). What is new in the ongoing discussion of overtourism are the levels of awareness it is raising and anti-tourism sentiment it is inspiring, intensifying the debate on the desirability of the continued growth-focused development model of tourism (Oklevik et al. 2019; Benner 2019). Concerns with growth and, as is argued, the destructive outcomes of growth strategies exemplified by overtourism, are leading to increasing interest in rethinking, even degrowing tourism (Higgins-Desbiolles et al. 2019).

In this sense tourism carrying capacity assessment, initially applied in tourism and recreation studies in the 1960s as one of the earliest attempts to define limits of tourism growth, has been receiving renewed attention since the 1990s as concerns with negative effects of tourism on destination sustainability began to grow (Kennell 2016). The concept has evolved over the years from emphasizing ecological limits of places, expressed as a numeric value of maximum visitor capacity, to include socio-cultural and economic aspects in an attempt to discern limits of acceptable change for communities (Carić and Klarić 2011; Mrđa et al. 2014). Common to a number of definitions of tourism carrying capacity assessment are notions of a destination's capacity to assimilate change without damaging its resources, reducing resident well-being or decreasing tourist satisfaction. As a multidimensional concept, tourism carrying capacity thus integrates different economic, environmental, cultural, social and perceptual aspects of destination tourism activity (Zelenka and Kacetl 2014). It can also be partially defined, focusing on a specific aspect (Lopez Bonilla and Lopez Bonilla 2008). The assessment of social carrying capacity is based, for example, on counterpointing resident and visitor experiences reflecting their willingness to enter into an exchange (Muler et al. 2018) as a result of perceived benefits and costs related to their understanding of how to share local resources (Bimonte and Punzo 2016). Their exchange, however, is not that of equals with residents being stable populations with limited choices and a long-term relationship with the destination as opposed to visitors who are a quickly adaptable population having a short-term relationship with destinations and a wide range of other choices (Bimonte 2008). Such partial assessment of carrying capacity from different perspectives cannot necessarily be expected to produce the same conclusions as a comprehensive assessment; namely, it is possible to expect that an assessment based on social factors result in a lower carrying capacity than an assessment based on economic and ecosystem

[1] To the best of the authors' knowledge, at the time of writing this paper the UNWTO has not published revised long-term forecast of international tourist arrivals as these may be impacted by the 2020 COVID-19 pandemic.

factors (Marsiglio 2017). In either case, tourism carrying capacity assessment is a knowledge-based judgement call on acceptable change and as such a management tool or a guiding framework for tourism development decisions (Coccossis and Mexa 2004). Given that sustainability inherently assumes living within limits, identifying the carrying capacity of destinations seems a logical priority (Butler 2020). Its implementation, however, whether through institutional guidelines, regulatory, economic or organizational measures, requires political support and the will of local residents to participate in the process (Coccossis and Mexa 2004).

As a data-based tool, tourism carrying capacity assessment can be a powerful facilitator of collaborative destination management and governance. Due to the complex character of tourism destinations, stakeholder cooperation is a necessary factor for planning, development and delivery of integrated tourism products (Beritelli 2010) and some form of cooperative action supporting interorganizational cohesion within destinations is needed in tourism probably more than in most other economic sectors (Scott et al. 2008). It is argued that destination competitiveness is based on internal cooperation or, in other words, that destinations should strive to first achieve collaborative advantage as a prerequisite of competitive advantage (Fyall et al. 2012). The usual enablers of collaborative practice within destinations such as teamwork, networks, awareness-raising campaigns, promotion of vision and brand concepts, although vital, do not in themselves seem sufficient to promote collaborative destination management when larger gross capital formation related to tourism is needed or planned. In such circumstances additional 'hard policy' interventions need to be introduced (Benner 2019) with tourism carrying capacity as a framework for tourism and spatial planning having the potential to encourage collaboration between a destination's public and private, inner and outer (e.g. visitors, tour-operators) stakeholders.

9.3 Juxtaposing Resident and Visitor Perspectives of Tourism Impacts: A Case Study of Croatia's Crikvenica-Vinodol Riviera

Keeping in mind that, by combining residents' views of destination characteristics and visitor satisfaction ratings of destination offer, social carrying capacity assessment is inevitably based on value judgements and attitude measurements (Severiades 2000) burdened by limitations inherent to value and attitude research (Rasoolimanesh and Seyfi 2020; Gossling et al. 2020), this paper juxtaposes resident and visitor perceptions of tourism and its impacts in a highly developed tourism destination. Specifically, the analysis focuses on Crikvenica-Vinodol Riviera and encompasses a side-by-side examination of resident and visitor views of tourism impacts on specific destination offer areas where highly developed destinations are typically most vulnerable, namely, perceptions of crowding and infrastructure services, perceptions of environmental and spatial characteristics and perceived overall satisfaction with tourism development.

9.3.1 The Location

Situated in northern Adriatic's Kvarner County, Crikvenica-Vinodol Riviera is a community of 19.900 residents (Croatian Bureau of Statistics 2011) encompassing three distinct territorial units stretching from the coast inland, into the Vinodol valley and the forested slopes of the Kapela mountains. It is by definition a nature-based destination, offering a variety of recreational outdoor tourism experiences (Newsome et al. 2002; Holden and Sparrowhawk 2002), owing its original appeal to a mild climate and healthy seaside aerosol, followed by a period of intense beach tourism development along 40 kilometres of coastline and aspiring to better realize the potential of its green hinterland for bicycling, walking, hiking and paragliding in the future.

Over the course of its 130-year tradition in tourism, the Riviera has always been a popular Adriatic vacation spot, being today one of Croatia's 'top ten' tourism destinations having registered 2.5 million overnights in commercial accommodation in 2019 (Croatian Bureau of Statistics 2020). Growth of overnights has been continuous over the past decade, reaching 9% during the 2016–2019 travel boom. It is traditionally oriented towards the domestic and close-by international markets, its market segments having shifted towards markedly laid-back family and 50 + guests focused on a limited set of interests and activities revolving around the beach, food and beverage offer and local entertainment events with visitor total daily expenditures in 2017 some 20% lower than in the surrounding regional destinations (Marušić 2018). Business is very seasonal and, characteristically, almost 70% of total yearly overnights in 2019 were realized in July and August. The Riviera offers 39 thousand beds in commercial accommodation, of which 60% are in, the so-called, household accommodation and around 13%, respectively, in camps and hotels (eVisitor 2020). Household accommodation has also been the fastest-growing, registering a 15% increase in capacity between 2016 and 2019, as opposed to a 10% increase in camps and an 8% in hotels over the same period.

Although various strategic documents pertaining to the Crikvenica-Vinodol Riviera unanimously acclaim sustainable tourism development, there is growing concern regarding aspects of overtourism, most notably spatial degradation, which have been becoming apparent for some time (Telišman-Košuta and Ivandić 2019). Beyond the rapid expansion of commercial household accommodation, as indicated above, research has shown there is substantial, also rapidly growing non-commercial capacity on the Riviera (e.g. second homes) which is a major source of statistically not registered tourism demand. In fact, estimates of total peak demand[2] indicate there are at least twice as many visitors on the Riviera as officially registered (Ivandić and

[2]For research purposes, peak demand is assumed to occur on the second Friday in August.

Telišman-Košuta 2019). Furthermore, occupancy rates, which in 2019 ranged from 21% in household accommodations to 45% in hotels, point to available capacity even at the height of the tourism season. The ensuing additional pressure on spatial resources, especially on the narrow coastal strip and particularly on the beaches is obvious. Existing spatial plans and building regulation allow for further expansion of building within urban zones and in planned tourism zones. This is the context in which the *Tourism development strategy of the Crikvenica-Vinodol Riviera 2019–2029* strongly recommended tourism carrying capacity of the Riviera be assessed as a priority action.

9.3.2 Research Method

Resident and visitor perspectives on selected destination attributes have been sourced from surveys of respective populations:

- **Resident survey:** *Attitudes of local residents on tourism development of Crikvenica-Vinodol Riviera* (Ipsos 2019) survey was conducted through computer-aided telephone interviews (CATI) in November 2019, on a representative sample of 330 residents of towns Crikvenica and Novi Vinodolski and the Vinodol municipality, aged 18+. Data analysis was based on descriptive statistics.
- **Visitor survey:** *TOMAS Summer 2017 – Attitudes and expenditures of tourists on Crikvenica-Vinodol Riviera* (Marušić 2018) is an excerpt of the longitudinal *TOMAS Summer 2017* survey, conducted through personal interviews, in July–October 2017, on a sample of 115 tourists in Crikvenica and Novi Vinodolski. Data analysis was based on descriptive statistics.

Both surveys are part of the Riviera's periodic tourism monitoring efforts with their respective questionnaires overlapping only to a certain degree.

9.3.3 Resident and Visitor Perspectives

Due to only partially coinciding questionnaires used in resident and visitor surveys on the Crikvenica-Vinodol Riviera, the analysis of their perceptions of tourism impacts and possible implications on social carrying capacity is based on different attributes making up, however, three thematically comparable destination offer domain sets: crowding and infrastructure services, environmental and spatial characteristics and overall satisfaction with tourism development.

Perception of crowding and destination infrastructure services

Resident and visitor perceptions of crowding and infrastructure services on Crikvenica-Vinodol Riviera are derived on the basis of respondents' views and experiences with congestion in public places, noise and public safety, traffic and parking,

Table 9.1 Resident and visitor experience with crowding and infrastructure services on Crikvenica-Vinodol Riviera

Share of residents expressing an opinion	Not experienced or did not negatively affect quality of life (%)	Experienced and has somewhat affected quality of life (%)	Share of visitors who did not experience or were not negatively affected by the situation	C-V Riviera (%)	Deviation from Kvarner average (%)
Crowding in public places	46	32	Crowding in public places	99	7
Disturbances of public order and safety during the tourist season	67	21	Improperly disposed waste	96	7
Shortages of water during the tourist season	93	5	Impossibility of separating waste	96	7
Shortages of electric power during the tourist season	92	5	Unpleasant odours	97	9
Availability of parking	28	26	Traffic congestion in destination	92	5

as well as with the functioning of public utility services such as water, electricity and waste management during the tourist season (Table 9.1 and Table 9.2).

The Riviera residents have mixed experiences with how tourism is affecting their quality of life. On the one hand, they are not experiencing water (93%) or electricity (92%) shortages during the tourist season and although not satisfied to the same extent with waste management, around three-quarters of residents perceive it to be either high or medium quality. Disturbances of public order and safety are by most (67%) also not seen as negatively affecting the community. On the other hand, crowding in public places, still somewhat ambivalently, but obviously becoming an issue, is perceived by half of the residents as affecting their quality of life. Lack of parking which the majority of residents (72%) have to deal with, as well as noise levels perceived by the majority (81%) to be either high or medium, are seen as the biggest irritants during the tourist season.

Visitors on the Crikvenica-Vinodol Riviera are not experiencing, nor being negatively affected by most typical tourism irritation factors being analysed. In fact, above 90% of visitors report not having negative experiences with crowding in public places, waste management or traffic congestion. Moreover, the quality of these elements of

Table 9.2 Resident and visitor perception of selected infrastructure services on Crikvenica-Vinodol Riviera

Share of residents expressing an opinion	High (%)	Medium (%)	Share of visitors expressing a high level of satisfaction[a]	C-V Riviera (%)	Deviation from Kvarner average (%)
Quality of waste disposal system	40	30	Destination accessibility	84	13
Quality of drainage system	43	34	Traffic organization in destination	35	−45
Noise level	46	35			

[a]Share of visitors expressing satisfaction levels of 6 and 7 on a scale from 1 (very poor) to 7 (excellent)

the destination offer is rated higher on the Riviera than in the surrounding Kvarner region. Even though the visitors are thus not seeing traffic congestion on the Riviera and its advantageous accessibility is broadly recognized (84%), traffic organization within the destination, namely, referring to parking availability, is for the majority of visitors (65%) unsatisfactory. This is also a destination attribute on which the Riviera lags significantly behind the regional Kvarner average.

Looking jointly at resident and visitor experiences with crowding and destination infrastructure services during periods of high tourism intensity, it is apparent that while the Riviera's basic communal infrastructure such as water, electricity, waste management and accessibility are considered satisfactory by all, noise, parking and crowding may be points of contention. Lack of parking, equally affecting both parties, can be presumed to generate competition and even resentment, not only between residents and visitors but also between different segments within each respective group. On the other hand, crowding in public places is the only aspect on which residents and visitors disagree, with visitors literally not perceiving it at all while the local community is almost evenly split between those who do and do not feel it is affecting their quality of life. Such divergent and even contrary experiences of crowding between residents and visitors, but also among residents can be additional causes of antagonism between stakeholder groups. Described tensions between and within various segments of resident and visitor populations competing for the same resources signal possible 'bottlenecks' in terms of Riviera's social carrying capacity.

Perception of destination environmental and spatial characteristics

Resident and visitor perception of Crikvenica-Vinodol Riviera's environmental and spatial characteristics is being assessed in terms of their views on environmental preservation and cleanliness of beaches, as well as, reflecting each group's particular focus, it is also based on residents' sentiments regarding various aspects in which tourism affects the use of space and visitors' evaluation of destination built and natural space attractiveness (Table 9.3).

Table 9.3 Resident and visitor perception of Crikvenica-Vinodol environmental and spatial characteristics

Share of residents expressing an opinion	Agree (%)	Partially agree (%)	Share of visitors expressing a high level of satisfaction[a]	C-V Riviera (%)	Deviation from Kvarner average (%)
Tourism facilities decrease the attractiveness of Riviera	41	29	Picturesqueness and tidiness of town	76	−10%
Building zones within urban areas need to be increased	36	21	Scenic and natural beauty	79	−9%
Coastline is overbuilt	46	27	Environmental preservation	66	−15%
	High	**Medium**	Cleanliness of beaches	75	−9%
Environmental preservation	53	36			
Cleanliness of sea/beaches	68	26			

[a]Share of visitors expressing satisfaction levels of 6 and 7 on a scale from 1 (very poor) to 7 (excellent)

Amid dissonant opinions, Riviera's residents are quite critical of tourism impacts on the destination's environmental and spatial characteristics. A large majority (70%) agree or partially agree tourism facilities are decreasing the attractiveness of the Riviera, while almost three-quarters of the respondents (73%) agree to some extent the coastline is overbuilt. At the same time, somewhat paradoxically, there is majority consent (57%) that building zones within urban areas, which are predominantly located along the already heavily developed coastline, can be increased. The pro-new construction majority is a tenuous one, however, indicating relative hesitance within the community towards the further building. The residents are, furthermore, also relatively critical of the levels of environmental preservation and beach cleanliness, which although seen by most to be high (53 and 68%, respectively) do not elicit consensual agreement, pointing to a pronounced perception of degradation of resources traditionally considered among the Riviera's strengths.

On the other hand, three-quarters of visitors or more express high satisfaction rates with most aspects of the Riviera's spatial and environmental characteristics. Specifically, a large majority applauds cleanliness of beaches (75%), picturesqueness and tidiness of towns (76%) and especially scenic and natural beauty (79%). The visitors are somewhat more critical only of environmental preservation, although two-thirds (66%) are highly satisfied with this aspect of the Riviera as well. These high satisfaction ratings are, nevertheless, significantly below the Kvarner region average, and particularly so in relation to environmental preservation, indicating that in terms of environmental and spatial qualities the Riviera lags behind its regional competitive circle.

Resident and visitor perceptions of the Riviera's spatial and environmental characteristics essentially differ, with the residents, although to varying degrees, tending to be more critical of the perceived environmental and spatial decline and the visitors, in comparison, showing higher tolerance for the sense of place as it is. Such divergence, compounded by the Riviera's perceived regionally inferior competitive position in relation to its environmental and spatial qualities, may result in mixed and confusing messages for the local community and possibly lead to the advocacy of conflicting development agendas by different local stakeholder groups. In such circumstances arising animosities not only between residents and visitors but even more so between residents themselves are generating 'bottlenecks' in terms of the Riviera's social carrying capacity.

Perceived overall satisfaction with tourism development

Overall satisfaction with tourism development on the Crikvenica-Vinodol Riviera can be inferred from residents' attitudes on the destination's tourism development model and visitors' views of the value for money of their trip (Table 9.4).

A significant majority of local respondents (70%) are critical of the existing mass tourism development model on the Riviera feeling it threatens their quality of life, a relative majority (41%) fully agreeing with the statement and slightly more than a quarter of the respondents (29%) partially agreeing. This is clearly a call for change. Interpreting this data, however, in view of almost unanimous community support for tourism as a course of future development (86%) and also seeing over a half of local residents (56%) derive a direct income from tourism, it should be primarily understood as a sign for change of direction away from mass tourism in favour of a different model of tourism development. Considering, nevertheless, the majority of residents either fully (70%) or partially (18%) agree the Riviera has become too

Table 9.4 Resident and visitor perceived overall satisfaction with tourism development on Crikvenica-Vinodol Riviera

Share of residents expressing an opinion	Agree (%)	Partially agree (%)	Share of visitors expressing a high level of satisfaction[a]	C-V Riviera (%)	Deviation from Kvarner average (%)
Mass tourism on the Riviera threatens the local quality of life	41	29	Value for money of entire stay	70	−14

[a]Share of visitors expressing satisfaction levels of 6 and 7 on a scale from 1 (very poor) to 7 (excellent)

dependent on tourism alone, this result could also imply support for diversification of economic activity reducing reliance on tourism.

At the same time, a substantial majority of visitors on the Riviera (70%) are highly satisfied with the value of money of their stay or, in other words, with the overall quality of the tourism product being delivered in the destination. Apparently, the tourism experience being offered is well adapted to the expectations of most visitors who, judging from their approximately 20% below regional average spend, likely represent the price-sensitive end of the family and 50+ market segments. Still, Riviera's perceived value for money is below the regional average indicating the visitors do recognize problems with the destination's tourism offer to some extent.

When looked at concurrently, it is obvious residents and visitors are expressing somewhat divergent perceptions of overall satisfaction with tourism development on the Riviera. Adhering to the visitors' in essence highly positive point of view and in keeping with the marketing axiom that 'the customer is always right' may be a source of frustration within the local community obviously showing signs of wanting to break away from the existing tourism development model. The fact that the community is, however, not unanimous in this 'change of paradigm' sentiment may be an additional source of internal tension. Moreover, changing the direction of tourism development in line with some of the residents' views could result in further frustration as this entails a change of market segments, which in turn entails substantial capital, marketing, organizational and mindset investment, as well as economic uncertainty inherent to repositioning efforts. Finally, a change in the Riviera's market segment mix may entail frustration on the part of visitors and particularly the travel organizers who would be displaced since such relocation would require an investment of time and financial resources. The resulting imbalances between and within stakeholder groups create tension points which may, in fact, be considered 'bottlenecks' in terms of the Riviera's social carrying capacity.

9.4 Implications for Destination Management

The above-presented analysis of resident and visitor perceptions of tourism impacts on the Crikvenica-Vinodol Riviera reveals relatively few commonalities but significant differences in the views of the two groups. Agreeing on the satisfactory state of basic communal infrastructure and the insufficiency of parking during the tourist season, residents tend to be more critical than visitors of all other tourism impacts looked at, ranging from their perception of crowding in public places to environmental and spatial qualities of the landscape and overall satisfaction with tourism development on the Riviera. The research further shows that visitors are a more homogenous group in their views, while there are noticeable discrepancies among residents, especially in relation to perceived tourism impacts on the Riviera's spatial and environmental characteristics and the overall direction of tourism development.

Differences in perception of tourism impacts on a destination, seen on the Crikvenica-Vinodol Riviera in the form of predominantly positive views among visitors versus the more critical residents, are to be expected between or within groups. Resident and visitor views are shaped by their different positions, namely, the visitors' experiences of leisure-time, their psychological need to validate destination choice and holiday expectations, a short-term and essentially superficial relationship with the destination as opposed to the residents' experience of personally working overtime or the destination as whole going into overdrive during the tourist season, as well as their long-term and more deeply felt emotional investment into a place to which they are bound by heritage, family, possessions and work. It is also a possibility that divergent viewpoints between the two groups can result from a mismatch between tourist market segments and local aspirations in tourism. Differences in opinion within the resident community can, among other reasons, stem from individuals' diverse levels of involvement in tourism or their particular interests, from different views of the destination's tourism potential or opposing beliefs of what should be the role of tourism in the local economy.

Although understandable and to a certain extent inevitable, diverging perceptions of tourism impacts and the possibly ensuing tensions between residents and visitors or simmering conflicting goals within either group have the potential of turning into 'bottlenecks' of a destination's social carrying capacity sphere which if not managed can become limiting factors of sustainable tourism development. It is, however, the residents, with their higher investment and stakes in the local tourism sector and the mandate to shape the local community, whose mutual relationships are more critical and need to be prioritized in destination management.

Stemming from Croatian experiences and focusing here on the management of developed tourism destinations with accrued expertise, but also bearing in mind their complexity and the usually still high potential for further tourism growth, further disruption and thus also for further generating 'bottlenecks' in terms of social carrying capacity, destination management[3] needs to become more inclusive, advocating local cooperation, building cohesion and advancing community values and priorities. Such collaborative destination management needs to increasingly focus especially on:

- **Improving tourism statistics available on destination level** in order to capture as wide a breadth of tourism activity as possible primarily by broadening the data scope to include information on non-commercial accommodation (e.g. second homes) and unregistered demand segments (e.g. second-home owners, VFR, not registered guests in commercial accommodation, one-day visitors), who represent a significant 'unseen' volume of users of destination services, as well as covering not only visitor, but also resident perceptions of tourism characteristics and impacts since the latter are key in shaping destinations.
- **Continuous monitoring of destination's tourism carrying capacity parameters** in order to keep abreast of tourism impacts entails formulating a set of

[3]The reference is primarily to destinations on city or municipality level.

'tourism pressure indicators' on the natural, sociocultural and economic environments, regular data collection, comparative analysis, report preparation with, if needed, a proposal of recommendations and corrective activities and report dissemination. The monitoring process also includes follow-up checks on the implementation of proposed corrective activities.

- **Community capacity building in tourism**, aspiring to advance general local knowledge of tourism thus increasing resident capabilities of responsibly partnering in the design of tourism offer, encompasses educational and information dissemination activities tailored for different segments of the local population (e.g. professionals in tourism and related fields, local government, NGOs, school children, the retired, etc.), covering a variety of topics in tourism (e.g. best practices, tourism trends, visitor profiles, local tourism plans, etc.) and being delivered through different platforms and programmes (e.g. tutorials, field trips, presentations, open discussions, etc.) specifically designed to be informative, immersive and applicable. A number of activities also target visitors promoting their responsible relationship with the destination (e.g. bringing them on board for certain destination initiatives, involving them in certain segments of tourism planning, etc.).
- **Improving the quality of tourism planning, capacity for plan implementation and monitoring**, so as to produce practical, community-supported and operationalized documents, by securing in the planning phase a multidisciplinary, professional and non-partisan strategy team, inputs from various tourism-related stakeholders (e.g. spatial planners, utilities, hospitality businesses, etc.) and meaningful community-wide discussion of plan proposals, to be followed in the implementation phase by support in organizing and functioning of workgroups taking over the execution of planned projects (e.g. staffing, workspace, administrative aid, etc.), assistance in securing project financing (e.g. information on financial sources, aid in the preparation of project bids, etc.) and monitoring of project execution.
- **Promoting a collaboration culture among local stakeholders** in order to attain and deliver a destination's unifying vision, by, among other initiatives, facilitating and/or supporting joint projects, regularly orchestrating and/or supporting interdisciplinary gatherings intended to promote joint goals and/or raise mutual awareness among stakeholders, organizing local 'expert advice teams' available for consultations in implementing joint destination-level projects, maintaining a 'referral database' with contacts of local service providers, but also by drawing in visitors, particularly those in a long-term relationship with the destination (e.g. loyal guests, second-home owners).

Part of a destination's tourism governance structure, collaborative destination management presupposes an independent organizational unit, inward-focused, devoted to coordinating tourism development within a community and separate from outward-focused destination marketing. It is responsible directly to the city or municipality government. It calls for a professional, apolitical team bolstered by a network of relevant partners.

9.5 Concluding Remarks and Future Steps

Increasing worldwide instances of overtourism and overcrowding, as well as shifting values reflecting concerns for the natural and sociocultural environments have led to increasing criticism of growth-driven mass tourism, at the same time also generating a renewed interest in the concepts and the application of tourism carrying capacity, destination governance and management. Several Croatian coastal destinations, heavily exposed to tourism induced pressures, have recently also started exploring issues of their tourism carrying capacity and, subsequently, of destination management. Drawing on data accumulated during the course of several related research projects on Crikvenica-Vinodol Riviera, this paper contrasts resident and visitor perceptions of tourism impacts exploring social carrying capacity issues and the ensuing implications for collaborative destination management and governance, being, however, limited by existing research design and particularly by survey structures and sample sizes.

In this context, priority next steps to be undertaken in researching destination social carrying capacity, as well as in improving the application of tourism carrying capacity and collaborative destination management concepts should encompass:

- Extending the research of resident and visitor perceptions of tourism impacts to a larger number of destinations differing in tourism volume and market segment structures in order to better understand visitor–resident dynamics in varying tourism development scenarios.
- Broadening research tools used in assessing visitor and resident perceptions of tourism impacts to include, beyond surveys, other methods allowing for better in-depth probing of views, thus securing more insightful research results.
- Designing complementary research tools administered to resident and visitor populations and securing proportionate sample sizes so as to ensure better comparability of obtained data.
- Further study and improvement of tourism pressure indicators reflecting destination diversity, also proposing minimal standard indicator sets according to destination type in order to facilitate the application of tourism carrying capacity.
- Further study and formulating of tourism carrying capacity benchmarks and/or criteria reflecting possible threshold levels of pressure indicators for different destination types.
- Further study and piloting of collaborative destination management organizations as regards their tasks and responsibilities, human and financial resources, links with stakeholders and local government, also controlling for possible alterations reflecting different destination types.

It is, however, awareness-raising about the need to manage tourism, doing it in a collaborative manner and using tourism carrying capacity methodology in the process, that appears to be a pressing and a continuous future task thus enabling destinations to take a pro-active stance towards tourism.

References

Benner M (2019) From overtourism to sustainability: a research agenda for qualitative tourism development in the Adriatic. MPRA Paper No. 92213. https://mpra.ub.uni-muenchen.de/92213
Beritelli P (2010) Cooperation among prominent actors in a tourism destination. Ann Tourism Res 38(2):607–629
Bimonte S (2008) The 'tragedy of tourism resources' as the outcome of a strategic game: a new analytical framework. Ecol Econ 67(3):457–464
Bimonte S, Punzo LF (2016) Tourist development and host-guest interaction: an economic exchange theory. Ann Tourism Res 58:128–139
Boom S, Weijschede J, Melissen F, Koens K, Mayer I (2020) Identifying stakeholder perspectives and worldviews on sustainable urban tourism development using Q-sort methodology. Curr Issues in Tourism. https://doi.org/10.1080/13683500.2020.1722076
Buhalis D (2000) Marketing the competitive destination of the future. Tourism Manage 21(1):97–116
Butler RW (2020) Tourism carrying capacity research: a perspective article. Tourism Rev 75(1):207–211
Capocchi A, Vallone C, Pierotti M, Amaduzzi A (2019) Overtourism: a literature review to assess implications and future perspectives. Sustain 11(3303):1–18
Carić H, Klarić Z (2011) Istraživanja prihvatnog kapaciteta – Međunarodna i hrvatska iskustva. In: Čorak S (ed) Izazovi upravljanja turizmom. Institut za turizam, Zagreb, pp 107–121
Coccossis H, Mexa A (eds) (2004) The challenge of tourism carrying capacity assessment: theory and practice. Ashgate, Farnham
Croatian Bureau of Statistics (2011) Census of Population, Households and Dwellings 2011, First results by Settlements. Statistical Reports, 1441. https://www.dzs.hr/
Croatian Bureau of Statistics (2020) Tourist arrivals and nights in 2019, First release, number 4.3.2., https://www.dzs.hr/
Dodds R, Butler R (2019) The phenomena of overtourism: a review. Int J Tourism Cities 5(4):519–528
Emerson R (1976) Social exchange theory. Ann Rev Sociol 2:335–362
eVisitor (2020) The Croatian national tourist information system eVisitor. Data from electronic database retrieved at 30 March 2020
Fyall A, Garrod B, Wang Y (2012) Destination collaboration: a critical review of theoretical approaches to a multidimensional phenomenon. J Destination Mark Manage 1(1):10–26
Gossling S, McCabe S, Chen N (2020) A socio-psychological conceptualization of overtourism. Ann Tourism Res 84. https://doi.org/10.1016/j.annals.2020.102976
Gursoy D, Ouyang Z, Nunkoo R, Wei W (2018) Residents' impact perceptions of and attitudes toward tourism development: A meta-analysis. J Hosp Mark Manage 28(3):306–333
Higgins-Desbiolles F, Carnicelli S, Krolikowski C, Wijesinghe G, Boluk K (2019) Degrowing tourism: rethinking tourism. J Sustain Tourism 27(12):1926–1944
Holden A, Sparrowhawk J (2002) Understanding the motivations of ecotourists: the case of trekkers in Annapurna, Nepal. Int J Tourism Res 4(6):435–446
IPSOS (2019) Stavovi lokalnog stanovništva Crikveničko-vinodolske rivijere o razvoju turizma. IPSOS, Zagreb
Ivandić N, Telišman-Košuta N (2019) Sustainable tourism development of Primorje-Gorje County: methodological framework for tourism carrying capacity assessment, Phase I. Institute for Tourism, Zagreb
Kennell J (2016) Carrying capacity. In: Jafari J, Xiao H (eds) Encyclopedia of tourism. Springer International Publishing, New York, pp 133–135
Koens K, Postma A, Papp B (2018) Is overtourism overused? understanding the impact of tourism in a city context. Sustain 10(4384):1–15
Lopez Bonilla JM, Lopez Bonilla LM (2008) Measuring social carrying capacity: an exploratory study. Tourismos 3(1):116–134

Manente M, Minghetti V (2006) Destination management organizations and actors. In: Buhalis D, Costa C (eds) Tourism business frontiers: consumers, products and industry. Elsevier Butterworth-Heinemann, Oxford, pp 228–237
Marsiglio S (2017) On the carrying capacity and the optimal number of visitors in tourism destinations. Tourism Econ 23(3):632–646
Marušić Z (2018) TOMAS Ljeto 2017 – Obilježja turističke potražnje na Crikveničko-vinodolskoj rivijeri. Institut za turizam, Zagreb
Mrđa A, Carić H, Bojanić-Obad Šćitaroci B (2014) Značaj koncepta turističke nosivosti za prostorno planiranje. Prostor 22 (2) (48) 213–227
Muler V, Coromina L, Gali N (2018) Overtourism: residents' perceptions of tourism impacts as an indicator of resident social carrying capacity—a case study of a Spanish heritage town. Tourism Rev 73(3):277–296
Newsome D, Moore SA, Dowling RK (2002) Natural area tourism: ecology impacts and management. Channel View, Clevedon, UK
Oklevik O, Gossling S, Hall CM, Jacobsen JKS, Grotte IP, McCabe S (2019) Overtourism, optimization, and destination performance indicators: a case study of activities in Fjord Norway. J Sustain Tourism 27(12):1804–1824
Peeters P, Gossling S, Klijs J, Milano C, Novelli M, Dijkmans C, Eijgelaar E, Hartman S, Heslinga J, Isaak R, Mitas O, Moretti S, Nawijn J, Papp B, Postma A (2018) Research for TRAN Committee—overtourism: impact and possible policy responses. European Parliament, Policy Department for Structural and Cohesion Policies, Brussels
Perkumiene D, Pranskuniene R (2019) Overtourism: between the right to travel and resident's rights. Sustain 11(2138):1–17
Rasoolimanesh SM, Seyfi S (2020) Residents' perceptions and attitudes towards tourism development: a perspective article, Tourism Review, Vol. ahead-of-print, No. ahead-of-print. https://doi.org/10.1108/TR-11-2019-0461
Ritchie JB, Crouch GI (2003) The competitive destination: a sustainable tourism perspective. CABI, Cambridge, MA
Scott N, Cooper C, Biaggio R (2008) Destination networks. four Australian cases. Ann Tourism Res 35(1):169–188
Seraphin H, Sheeran P, Pilato M (2018) Over-tourism and the fall of Venice as a destination. J Destination Mark Manage 9:374–376
Severiades A (2000) Establishing the social tourism carrying capacity for the tourism resorts of the east coast of the Republic of Cyprus. Tourism Manage 21:147–156
Telišman-Košuta N, Ivandić N (2019) Strategija razvoja turizma Crikveničko-vinodolske rivijere 2019–2029. Institut za turizam, Zagreb
Telišman-Košuta N, Ivandić N (2020) Tourism and destination identities: policy implications. In: Marković Vukadin I, Krešić D (eds) Tourism in the VUCA world: towards the era of (ir)responsibility. Proceedings book, International Tourism Conference Dubrovnik 2019. Zagreb: Institute for tourism
Telišman-Košuta N, Ivandić N, Kunst I (2015) From a conflicted to a collaborative destination: a case study of marketing and management challenges in co-creating wellbeing in Novalja, Croatia. In: Pesonen J, Komppula R (eds) Tourism engagement: co-creating wellbeing. Proceedings of the 6th Advances in Tourism Marketing Conference. University of Eastern Finland, Joensuu, pp 316–320
UNWTO (2011) Tourism towards 2030/global overview. Madrid: UNWTO. https://www.e-unwto.org
UNWTO (2018) Overtourism? understanding and managing urban tourism growth beyond perceptions. UNWTO, Madrid. https://www.e-unwto.org
UNWTO (2019) International tourism highlights—2019 Edition. UNWTO, Madrid. https://www.e-unwto.org
UNWTO (2020) UNWTO world tourism barometer and statistical annex. UNWTO, Madrid. https://www.e-unwto.org

Wang Y, Pizam A (eds) (2011) Destination marketing and management: theories and applications. CABI, Cambridge, MA
Zelenka J, Kacetl J (2014) The concept of carrying capacity in tourism. Amfiteatru Econ J 1(36):641–654

Chapter 10
Limits of Acceptable Change (LAC) for Tourism Development in the Historic Centre of Porto (Portugal)

Ana Carolina Jordão, Zélia Breda, Medéia Veríssimo, Ivana Stevic, and Carlos Costa

Abstract The present study proposes to analyse the impacts of tourism in the historical area of the city of Porto in Portugal, according to the perception of the resident population, in order to define the limits of acceptable change (LAC) for tourism development in this area. Literature suggests that the impacts of tourism are felt more intensely by communities living in tourism destinations, and, because of this, they must be involved to play an active role in the development of local tourism policies. For that reason, the analysis of tourism impacts considering local perceptions is an essential part of adequate tourism planning and management process. The LAC method can be applied in the management of these tourism areas with the need to adapt and consider the changes caused by tourism, helping in the development of strategic plans to support sustainable development. It must be characterised by the active involvement of the community throughout the process, while decisions should be based on results that are acceptable to a variety of stakeholders. Based on the concepts mentioned above and the proposed objectives, the research methodology follows a qualitative approach. The data collection process was based on three focus groups with the participation of several stakeholders involved and/or affected by the development of tourism in the area. Those focus groups underpinned the LAC method and participatory planning, which contributed to the definition of the limits of acceptable change considering tourism development in the historic centre of Porto.

A. C. Jordão (✉) · Z. Breda · M. Veríssimo · I. Stevic · C. Costa
University of Aveiro, Aveiro, Portugal
e-mail: acjordao@ua.pt

Z. Breda
e-mail: zelia@ua.pt

M. Veríssimo
e-mail: medeiaverissimo@ua.pt

I. Stevic
e-mail: ivana.stevic87@ua.pt

C. Costa
e-mail: ccosta@ua.pt

A. Mandić and L. Petrić (eds.), *Mediterranean Protected Areas in the Era of Overtourism*, https://doi.org/10.1007/978-3-030-69193-6_10

Keywords Tourism impacts · Community involvement · Participatory planning · Local perceptions · Limits of Acceptable Change (LAC) · Historic city centre

10.1 Introduction

The tourism sector has affirmed, since 2018, its position as the largest economic export activity of the Portuguese economy, being responsible, in 2019, for 52.3% of service exports and 19.7% of global exports. The 336.8 thousand tourism-related jobs in 2019 represented a 6.9% portion of the national economy (Turismo de Portugal 2020). The World Travel Awards 2019 elected Portugal the best tourist destination in the world, for the third consecutive year. The city of Porto, the object of study in this research, was nominated the 'European Best Destination', in 2017, and is world-renowned for international tourism.

These impressive figures of tourism growth in Portugal, as well as in the city of Porto, have substantially affected the lives of its resident population. Within the historic centre of Porto, on the banks of the Douro River, lies a community that has seen, in recent years, its narrow and picturesque streets become crowded by visitors of various nationalities, its local produce stores be replaced by restaurants with terraces and menus in a foreign language. It has also seen the prices of services and leases rise considerably. Many have even been evicted from their homes to make way for short-term rentals for tourists.

The growth in tourism produces immense impacts, of different types and at different levels. In addition to the economic impacts, other effects are felt and perceived in the environment and societies, especially with regard to local communities in tourist destinations; tourism inevitably transforms and impacts—sometimes beneficially, sometimes harmfully—the lives of these communities. Due to the high economic benefits that tourism produces, its other impacts end up being sometimes neglected, and this can generate an uncontrolled growth of the activity, which will not be sustainable in the long run.

The World Tourism Organization (UNWTO 2018) recommends two main aspects for a correct and sustainable management of the growth of urban tourism: first, to monitor the perception of local communities towards tourism and promote the value of the sector among residents. This makes it possible to identify local communities' concerns early on and jointly develop management strategies to deal with perceived issues. Second, to determine the acceptable levels of impact of tourism on the city through a participatory process involving all relevant stakeholders. This will make it easier for local communities to benefit, create joint city experiences for visitors and residents and help in the communication with residents (World Tourism Organization, Centre of Expertise Leisure, Tourism and Hospitality; NHTV Breda University of Applied Sciences; NHL Stenden University of Applied Sciences 2018, p. 10).

In this context and in accordance with the recommendations of the UNWTO, this research organised focus group sessions as a 'laboratory' for the application of the Limits of Acceptable Change (LAC) planning system. These meetings reunited various tourism stakeholders from the historic centre of Porto in order to analyse the impacts of tourism activity and define the limits of acceptable change for the tourism development in that area, a protected historical urban destination, recently confronted with rapid tourism growth.

10.2 Tourism Planning—A Participatory and Sustainable Approach

Simpson (2001) defines a stakeholder as any individual or group of individuals that can affect, is affected or involved in an organisation, industry or phenomenon. The author identifies that the different groups of stakeholders involved and affected by tourism, normally found in an urban and heterogeneous society, are: government (composed of national, regional and local government bodies, regional and local tourism organisations and other government departments linked to tourism); non-governmental organisations; various associations (community, workers, environmental); professionals in the tourist industry and business owners in the sector; environmentalists; tourists/visitors; and the local residents of the region where tourism is developed, which even include people who have no interest, relationship or awareness of tourism.

Each of these stakeholders, for the most diverse reasons, tend to have different perceptions, attitudes and needs in relation to tourism and its development in their destinations. Even within groups, there is a conflict of opinion, and while some residents may rate tourism as a major factor in improving the quality of life in the region, others see tourism as an invasive force that acts against the identity of their community.

In addition to sociodemographic factors (such as age, level of education and income), other factors that influence the way in which residents of tourist destinations react, behave and assess the impacts of tourism are the community's economic dependence on tourism, the existence of a professional relationship with the sector, the distance from the residence to the tourism centres, the intensity of contact with tourists and the length of residence in the place (Almeida García et al. 2015; Beni 2006; Cooper et al. 2008; Frauman and Banks 2011).

Thus, the different groups of stakeholders are distinguished by their dependence on tourism, their interest in tourism; their perception of tourism; their view on the development of the activity; their influence and level of power (for decision making, for example) on tourism.

It is common to observe the presence and influence of market agents in the planning and management of tourist destinations. Visitors, although not present at the planning process, are usually prioritised and benefit from the decisions made. However, the other groups of stakeholders tend to be less represented in the planning process, and, even more rarely, have their interests, values and needs considered in the decision-making processes.

Gunn (1994) identifies that only through planning is it possible to avoid some negative impacts of tourism, however, he emphasises that, for it to be efficient in this way, it must be strategic and integrating, constituted by the ideas of all the involved participants. Hall (2004) adds that tourism planning must be pluralistic; involving the social, economic and physical dimensions of development and promoting a balance between them, as these can often be conflicting.

Participatory planning is a response to the challenge of developing more sustainable growth within the tourism sector. Meeting sessions between stakeholders, community public assemblies and surveys among residents about the impacts of tourism, for example, should be designed and carried out as a starting point in a planning process, to provide planners with information to enable them to develop plans and projects designed to address local concerns and issues (Lankford 2001).

Hall (2004) identifies, however, that there are limitations and difficulties to develop and maintain a participatory planning approach, such as the increase in financial and human resources that this participation requires, the prolongation of decision-making processes and the public's difficulty in understanding more technical and complex planning issues. This author also highlights the issue of managing conflicts of interest, one of the difficulties that planners who adopt this approach need to deal with and develop skills in, to better manage these moments. Simpson (2001) complements this list of limitations with the warning that multiple perspectives can result in low unanimity of opinion within and between stakeholder groups.

However, in most of the literature on tourism planning, the need to allow all relevant actors the chance to express their concerns and contribute to the planning process is considered to be a central role in the success of sustainable tourism development (Ahn et al. 2002; Frauman and Banks 2011; Gunn 1994; Horn and Simmons 2002).

Brida et al. (2011) confirm that tourism managers must have an in-depth knowledge of the characteristics of the destination that residents want to preserve and protect. For sustainable and successful management of tourist activity in any destination, it is of utmost importance that the local communities of the destinations are involved and have an active role in planning local tourism policy and that the decisions made regarding the development of tourism have their agreement and support.

McCool (1994, p. 52) indicates that the management of tourism destinations oriented towards sustainability requires two components: '(1) a technical planning system that deals with problems and forces explicit decision-making; and (2) a public involvement process oriented towards building consensus'. Considering the LAC planning system serves both these components, this study proposes its use in the context of aiming for a more sustainable tourism management and development.

10.3 The Limits of Acceptable Change (LAC)

The LAC planning system was developed and applied, for the first time, during the first half of the 1980s, for the management of visitors in the natural protected areas of Bob Marshall Wilderness, in Montana, by The National Wilderness Preservation System (NWPS), USA (McCool 1996; Stankey et al. 1985).

The development of the LAC method, according to its creators, also represented the reformulation of the method of determining pure and simple carrying capacity, which was only concerned with the limit of use and the number of people that an area could support without suffering major damage. However, it was noticed that this quantitative character of the carrying capacity failed in its objectives, since many of the problems caused by the recreational use of protected areas occurred not so much due to the number of people who were there, but (of the quality) of their behaviour. The concept of carrying capacity led managers to analyse 'how much' would be considered too much and to define quantitative limits, while the LAC system, on the other hand, started to lead them to a significantly different analysis, by asking what conditions are appropriate (or acceptable) and how these conditions are achieved (Stankey et al. 1985).

The LAC model proposes to evaluate the limits of acceptable changes from the impact of tourism activities on the destination; agrees on the degree of change that will be tolerated as well as the conditions desired after these changes; monitors the sector regularly and systematically and decides what actions will be taken if these limits are exceeded (Gonzalez et al. 2018).

McCool (2013) clarifies that the LAC system is based on the following propositions:

- Any level of human use (tourism, in this case) of an area results in some change in biophysical and social conditions;
- The character and amount of the resulting change will, at some point, become unacceptable for at least some members of society;
- Management is necessary to maintain such changes within parameters of acceptability or adequacy since it would not be possible to avoid or eliminate them.

10.3.1 The LAC Structure

The system structure is characterised by a cyclical and dynamic process, composed of nine stages, developed to structure the planning and, thus, lead to assist in the management and decision-making (Stankey et al. 1985), as shown in Fig. 10.1.

It is important to note that, although this structure presents a sequence of steps, its creators emphasise that it does not need to be strictly observed since these processes are largely interactive and circular rather than linear (Takahashi and Cegana 2005). McCool (1996) clarifies that 'it is important that planners understand the logic of each step and its sequence in the general process. By clearly understanding the logic, the steps can be modified as needed' (p. 7).

The complete application of the LAC method requires a detailed analysis of costs, political definitions and monitoring processes, which can only be effectively implemented in a real-life application. For that reason, and based on similar works (Ahn et al. 2002; Frauman and Banks 2011; Schetter and Schetter 2016, among others), this research focused on the application of steps 1, 3 and 5 of the LAC system, which

Fig. 10.1 The Limits of Acceptable Change planning system. *Source* United States Forest Service (2006)

are the first steps that demand collaborative processes, the objective of the focus groups carried out.

In the first step, the main problems and concerns are identified, such as resources and social aspects that need special attention or management problems that need to be solved. For this stage, the perception and knowledge of all interested, involved and affected parties must be considered. This helps identify and define the various issues and concerns associated with the perspective of tourism development in the area. Local residents, specialists and managers come together to identify the area's role and importance; the values, characteristics or special qualities that require attention; which management problems or concerns must be addressed; which issues in the management of the area are considered important by the public and what is the role of the area in a regional and national context. This dialogue between stakeholders helps to unify the agreement on important values and issues. The LAC is a very problem-oriented process, and the problems identified in this step will be addressed later (Bentz et al. 2016; McCool 2013).

In the third step, indicators are selected; they are the specific elements of the social environment or resources selected to represent—or be 'indicative of'—the conditions considered adequate and acceptable. This step is dedicated to the identification of the most important conditions of a scenario and the specific indicators that can better measure any change in these conditions. They must be easy to measure quantitatively.

Indicators are an essential part of the LAC system because their status reflects the general condition found in a scenario. This aspect of the selection of indicators is quite challenging and, even after the selection of relevant indicators, it can be difficult to measure and evaluate the actions necessary for the subsequent steps of the process (McCool 1994).

In the fifth step, the range of conditions for each indicator is defined, in measurable terms. The standards serve to define the 'limits of acceptable change'. These are the maximum allowable conditions that will be permitted in those specific indicators. They are not necessarily objectives to be achieved; the standards that define the range of acceptable conditions must be realistic, attainable and describe more than a simple reproduction of existing (unacceptable) conditions (McCool 2013).

These LAC steps allow not only to determine the desired conditions of an area but also to establish necessary indicators and standards to recognise when degradation or excess change has occurred (Frauman and Banks 2011).

10.3.2 LAC and Public Participation

Stankey et al. (1985) state the active community participation during decision-making processes as a fundamental component to the success of the planning process, as it was observed that the processes that involved the public presented a more complete work than those elaborated only by planners. After all, by combining the technical experience of specialists, with the knowledge of professionals in the sector and the valuable perception of the local community, the LAC process can result in more correct decisions and greater chances of success in its implementation (McCool 1996).

Defining what is acceptable to a variety of stakeholders is the essence of LAC's conceptual framework, as well as the means to seek some compromise between the different needs of these groups (Ahn et al. 2002; Stankey et al. 1985). The LAC system provides the necessary structure to assess the perceptions and feelings of the stakeholders participating in the processes in relation to the level of development and changes generated by the development of tourism. Through this participatory process, LAC is characterised as a management tool that contributes to preserve and reinforce local identity and values (Schetter and Schetter 2016).

10.3.3 The LAC System Applicability

As previously mentioned, the LAC system was developed in 1985, in the context of protected natural areas and, until now, most of the research using LAC has come from the areas of biology and ecology, applying the method for recreational use and conservation of natural parks and marine life.

In recent years, a few studies (with emphasis on Schetter and Schetter 2016) have started to apply this method as an instrument for planning and managing cultural heritage zones with a tourist influx. One study prior to this (Ahn et al. 2002) proposed an application of the LAC system for regional planning in urban communities.

McCool (2013), in his analysis of the LAC model applied to tourism, states that this planning system, although originally developed in the context of wild natural area management (Stankey et al. 1985), is suitable and should be tested for the tourism planning process in other contexts, at the local or regional level, especially if sustainable development is the main concern. Within this context, this study goes further and applies the LAC system to a protected historical urban destination.

10.4 Case Study: The Historic Centre of Porto

The city of Porto, characterised by a unique urban landscape with a history that goes back more than two thousand years, is the second-largest and most populous city in Portugal, located in the northwest of the country. The municipality has an official population of 237,591 inhabitants, according to the last Census (2011).

The Historic Centre of Porto is the oldest area and the heart of the city of Porto, the commercial, cultural and tourist centre of the region. Due to its great archaeological, historical, cultural, artistic and architectural value, the Historic Centre of Porto is, since 1996, an area classified as World Heritage by the United Nations Educational, Scientific and Cultural Organization (UNESCO). The entire area considered historic is constituted by an urban network, which includes houses, buildings, streets, churches and monuments.

Over the final decades of the twentieth century and up until the first decade of the twenty-first century, Porto's city centre had its period of decline and suffered a great process of 'desertification' (Freire 2015). However, over the last few years, this area has been the target of new economic dynamics. A new urban rehabilitation project is being developed, which has managed and promoted several strategies and operations to recover its heritage.

As a result of this rehabilitation, the historic centre of Porto has experienced a new era, with the valorisation and recovery of its architectural heritage, large financial investments, the arrival and installation of new businesses and new inhabitants. As Freire (2015) points out, the growth of tourism in the city of Porto has been one of the most important factors for these rapid interventions in the city.

10.4.1 Tourist Activity

The city of Porto was elected the European best travel destination in 2017, 2014 and 2012 (European Best Destinations 2019). In addition to this award, it has shown constant and exponential growth in the number of overnight stays, guests and revenues (INE 2019). The historic centre of Porto is the area where tourist activity is most concentrated. It stands out as a national and international tourist destination, in the segments of cultural and urban tourism, gastronomy, wine, events and short breaks (Associação de Turismo do Porto 2019).

The region's material and immaterial heritage, combined with the infrastructure for tourism that has seen heavy investments, has attracted millions of visitors in recent years (Turismo de Portugal 2020). A growth that has not only brought benefits, and despite being a relatively recent process, has already revealed negative impacts and some signs of discontent among some stakeholder groups in the city.

This current phase of extreme valorisation and urban rehabilitation, mainly due to the tourism 'boom', has, on the other hand, produced a significant increase in housing costs (also as a consequence of the rapid expansion of short-term rental accommodation), of goods and services. This process has caused much dissatisfaction among the residents since it has forced many inhabitants (mostly low-income families) to leave the homes they have lived in for decades, often without any warning, and move to other areas. This gentrification process has been taking place aggressively in the historical centre area over the past four years (Jornal de Notícias 2018; Lima 2018).

AirDNA (2019), a specialised company in short-term rentals, registered, in October 2019, 8580 'active' properties for rent in the city of Porto. This phenomenon of short-term rent has grown exponentially in the city, mainly in the historical centre area. On the map (Fig. 10.2), taken from the AirDNA website (2019), it is possible to observe a large number of these accommodations in the municipality, as well as their spatial concentration in the historic centre.

Since 2017, the residents of the region have created several associations, groups and movements in an attempt to solve the housing problem and circumvent other effects of the growth of tourist activity, such as precarious working conditions in the tourism sector and rising prices of essential goods and services, which often make them inaccessible to the local population (Pinto 2017). These associations claim to fight for a city with decent housing for all its residents and declare that this situation is almost leading to the extinction of unique customs and traditions of the local community. They argue that, while recognising that the growth of tourism and the expansion of short-term rentals have stimulated urban regeneration and economic

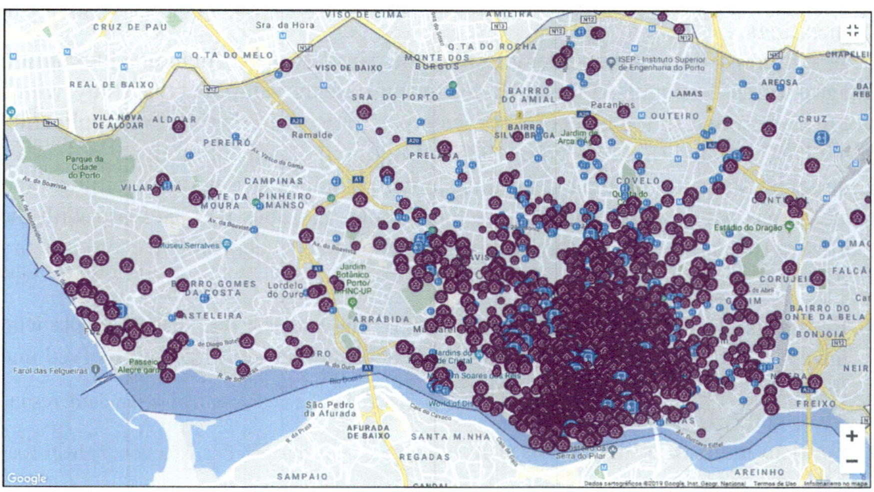

Fig. 10.2 Short-term rental accommodation in the city of Porto. *Source* AirDNA (2019)

dynamism, they are also generating a rise in prices and leading to a great loss of inhabitants and, crucially, the city's identity (Diario de Notícias 2018).

10.5 Methodology

The primary data for the empirical stage of this study were collected according to a qualitative approach, using the focus group technique as a way to promote a 'laboratory' of collaborative planning sessions among tourism stakeholders in the city of Porto, as defined by the LAC system. Three focus group sessions were held, bringing together twelve participants in total.

The selection of the focus groups' sample aimed to find social individuals who had an interest and a more significant link with the object of the investigation (Minayo 2004). Thus, the criteria for choosing participants aimed to bring together representatives from different sectors involved and/or directly impacted by the development of tourism in the historic centre of Porto, which can be summarised, according to the analysed literature (Beni 2006; Simpson 2001), in government officials, professionals and entrepreneurs in the tourism sector, traders, environmentalists, residents and tourists.

Among these stakeholder groups, tourists were not considered to compose the sample of this study based on other similar studies conducted (Ahn et al. 2002; Frauman and Banks 2011; Schetter and Schetter 2016). Also, in alignment with the research objectives, tourists were not included in the final sample because they usually do not have an in-depth view of management and/or tourism impacts and could disrupt the dynamics of the sessions with the community. The other groups of stakeholders were represented in the focus group sessions, with the exception of representatives of the local government, who although invited, were unable to attend the meetings.

In addition to these stakeholder groups defined by the tourism planning literature, a category called 'specialists' was included to compose the sample of this study, which would correspond to the category of 'scientists' present in the original application of the LAC method (Stankey et al. 1985). While in the application in natural areas, these scientists were biologists, for this study, specialists in tourism, sociology and related areas were invited to compose the sample. In Table 10.1, a board with the profile[1] of the participants of the focus groups conducted, according to each category of stakeholders represented:

- Residents of the historic centre, indicated by the letter R (including people who have some or no relationship with tourist activity, but are directly involved and impacted, since they live where tourism develops);
- Merchant (M);
- Tourism industry professionals (P);

[1] To preserve the identity and anonymity of the participants in this study, codes were defined and used to identify them.

Table 10.1 Category and profile of participants in the focus groups

Participant code	Gender	Age group	Nationality	Place and length of residence	Professional connection with tourism
R1	M	30–35	Brazilian	Porto historic centre; three years	Work not related to tourism
R2	F	36–40	Portuguese	Porto historic centre; +30 years	Works at a hostel in the historic centre
R3	F	30–35	Portuguese	Porto historic centre; nine years	Work not related to tourism
R4	F	30–35	Spanish	Porto historic centre; four years	Work not related to tourism
R5	M	36–40	Danish	Porto historic centre; 35 years	Work not related to tourism
M	F	30–35	Brazilian	Porto city centre (parish adjacent to the historic centre); four years	Owner of a commercial pastry shop in Porto city centre (indirect link to tourism)
P1	F	36–40	Portuguese	Historic centre; resident during childhood and returned four years ago	Manager of a short-term rental accommodation company in the historic centre of Porto
P2	F	36–40	Portuguese	Porto historic centre; seven years	Tour guide (receptive tours in Porto city centre mostly for Portuguese people)
P3	F	30–35	Portuguese	Porto city centre (parish adjacent to the historic centre); five years	Worked at the Porto Tourism Association for three years, working directly on the city's tourism promotion strategies. Currently works at a luxury hotel in the city
E	M	41–45	Portuguese	Matosinhos (adjacent municipality) for 40 years; works in the historic centre	Develops an 'alternative tourism' project (against mass tourism) in Porto historic centre

(continued)

Table 10.1 (continued)

Participant code	Gender	Age group	Nationality	Place and length of residence	Professional connection with tourism
S1	F	46–50	Portuguese	Vila Nova de Gaia (adjacent municipality)/has worked and frequented the historic centre for over 40 years	Sociologist/Ph.D. in Tourism; tourism professor at a university located in the historic centre of Porto. Has already coordinated research on the impacts of tourism on city residents
S2	F	30–35	Serbian	Porto city centre (parish adjacent to the historic centre); five years	Tourism PhD student; researching the impacts of tourism in the city of Porto according to the residents' perception

Source Own elaboration

- Environmentalist (E);
- Specialists in tourism studies, represented by the letter S.

The guide developed to conduct the focus groups was divided into four sections. The first section had an introductory nature. The second section was composed of questions about the impacts of tourism. In the third part, the application of the LAC system was initiated. This section was divided into three subsections, with reference to the first steps of the system that involve stakeholder participation (steps 1, 3 and 5):

- 3.1. Diagnosis of the area: LAC's step 1;
- 3.2. Definition of indicators of change, in reference to step 3 of the LAC system: observable and quantifiable signs, those that reflect the changes resulting from the presence of visitors and the development of tourism;
- 3.3. Definition of the acceptable limits for the change indicators: corresponding to step 5 of the LAC system. The central question for this stage was 'how much change is tolerable'? and the participants were encouraged to indicate, in a specific way, what they considered to be the limits, realistic and attainable, for each of the indicators previously defined.

The last part of the focus groups was about proposals for better management of the development of tourist activity in the city so that the limits previously established were reached and maintained.

10.5.1 Data Analysis

The technique used for the analysis of the data collected in the focus group sessions was content analysis. This technique was applied using both qualitative and quantitative approaches since it analysed the data by the textual citations of the participants' statements—qualitative approach—and by identifying and comparing the frequency and patterns of appearance of certain characteristics and terms in the contents of the analysed statements—quantitative approach (Kohlbacher 2006). This analysis was done using the software webQDA—Web Qualitative Data Analysis, version 3.0.

10.6 Discussion of Results

In this section, the data collected and the main results of this study will be reported, analysed and discussed, based on the methodology previously presented and in light of the theoretical framework analysed in this study. The following subsections are organised to present and discuss the results of each of the themes debated in the focus group sessions (the impacts of tourism, the application of steps 1, 3 and 5 of the LAC system and the stakeholders' proposals).

10.6.1 Perception of Tourism Impacts

The focus group sessions started with questions about the impacts of tourism perceived by the participants with regard to the city and to their lives and routines in particular. The initial questions were open and asked participants to identify the positive and negative impacts of tourism. As a result, there was a great predominance of negative impacts among the responses: amidst the 18 impacts most identified in the participants' statements, only one was positive (urban revitalisation). Regarding the dimensions, they mentioned five economic impacts, seven environmental/spatial impacts and six sociocultural ones.

Table 10.2 is a summary table of these impacts pointed out by the stakeholders, where the numbers show the frequency in which each one was identified in the participants' statements. In this matrix, the impacts are also related to the participants' descriptive codes of analysis: stakeholder group (residents/professionals/merchant/environmentalist/specialists); nationality (Portuguese or immigrants); place of residence (historic centre of Porto/adjacent parish/neighbouring municipality); length of residence (up to three years/four–nine years/more than 10 years).

A deeper analysis of this matrix on the relationship between the perceptions of impacts and each of the sociodemographic factors will not be carried out, since the

Table 10.2 Matrix tourism impacts × stakeholder classifications

Impacts	Groups					National		Place Resid.			Length Resid.		
	Resid.	Prof	Mer	Env	Spec	Port.	Immig.	HC	P	M	3y	4–9y	+10y
Real estate speculation	7	2	1	1	3	10	4	7	2	5	2	3	9
Precarious jobs	1	2	1	0	0	3	1	3	1	0	0	1	3
Increased cost of living	3	1	1	0	1	3	3	2	2	2	0	2	4
Increased price of goods and services	4	1	1	0	1	4	3	2	2	3	0	2	5
Rising rent cost	2	1	3	0	6	6	6	4	5	3	2	5	5
Parking problems	2	0	1	0	0	0	3	2	1	0	0	1	2
Overload of public spaces	3	1	0	0	1	3	2	2	1	2	0	2	3
Overload of leisure spaces	2	0	0	0	1	0	3	2	1	0	0	1	2
Urban mobility congestion	6	1	0	0	1	5	3	2	2	4	1	3	4
Overcrowding public transport	5	1	0	0	2	6	2	1	5	2	1	4	3
Air and water pollution	0	0	0	2	0	2	0	0	0	2	0	0	2
Urban revitalisation	2	0	1	0	1	2	2	2	2	0	0	3	1
Evictions	6	1	0	1	2	8	2	6	2	2	1	3	6
Noise, disorder	4	2	0	0	1	7	0	6	0	1	0	0	7
Loss of cultural identity	5	5	0	3	2	13	2	8	3	4	0	7	8
Loss of community life	5	0	0	1	0	6	0	5	0	1	0	2	4
Deprivation of passing/going to certain places	5	2	0	0	2	6	3	6	3	0	1	3	5
City centre's depopulation	3	0	0	2	1	4	2	2	1	3	1	2	3

Source Own elaboration, frequency data taken from the quantitative analysis using webQDA main objectives of the analysis are the general result of the discussion and construction of ideas as well as the opinions of the participants. However, some observations can be highlighted, such as:

- When analysing the participants' place of residence, most of the impacts related to housing problems (evictions, real estate speculation) and the disturbance of noise and bad behaviour of tourists were signalled by those residing in the historic centre, probably because they feel these effects more directly;
- Regarding the length of residence, almost all negative impacts were mostly highlighted by those who have lived in the historic centre and its surroundings for 10 years or more. This relationship between the length of residence and the perception of impacts is pointed out by several authors (Andereck et al. 2005; Beni 2006; Cardoso and Silva 2018; Cooper et al. 2008; Frauman and Banks 2011), who also relate this variable to the 'place attachment' of the resident, their memories and experiences lived over the years in that place, which ends up creating a nostalgic view of the past and making them more resistant and negative to new transformations.

The housing issue—encompassing several aspects, such as real estate speculation and rising rents, the prevalence of short-term rental accommodations, evictions and gentrification—was the most identified impact, addressed more than thirty times by all the participants, throughout the sessions.

The residents, especially those who demonstrated a great feeling of affection and attachment to their place of residence, pointed out tourism as a major cause of the community's housing problems. Some participants, however, mainly the immigrants, commented that the evictions had already occurred before the arrival of tourism, due to the terrible conditions in which the buildings were. One of the specialists (S2) said that it is important to clarify that 'Porto has not lost residents because of tourism', since this depopulation of the historic centre is a process that has been going on for years, due to the degradation conditions in which the streets and houses were. 'However, the effect of tourism is evident when observing the increased cost of rent' (S2).

What can be seen from this discussion is that there are different perspectives: the owners', the residents' and the market. But, in fact, as Silva (2017) points out, there is a great difficulty for residents to settle in the historic centre of Porto nowadays, as a result of the transformation of the space for tourism, since there is a low supply of permanent rentals and because the prices charged are impractical for a large part of the local population.

The second impact most highlighted by the participants concerns the question of cultural identity, and, in this regard, the opinion of residents was also, in general, very negative, pointing to a mischaracterisation and loss of the city's identity, both in material and immaterial ways. Among the physical aspects of this loss of local identity and traditional characteristics, the most mentioned were the new buildings that do not respect the surrounding architecture, the rehabilitation of buildings where only the facade is maintained (but the entire interior essence is destroyed) and the disappearance of traditional commercial establishments due to the arrival of new ventures, mostly foreigners. In addition, several immaterial aspects of culture and traditions that have been lost with the departure of the local population from the historic centre.

Residents who pointed out these impacts accused the public entities of being responsible for helping this process of dismantling the character of the historical centre, by allowing the 'destruction' of emblematic spaces, representative of the city's culture and history, such as taverns, restaurants and traditional stores that had been operating for decades, 'for financial interests that seek an immediate profit', according to one of the residents (R2). Participants reflected that the de-characterisation is building a tourist destination with no identity, just like many others in the world.

Another social impact mentioned several times in the focus group sessions, mostly by those who live in the historic centre, was in relation to the (unwanted) change in their habits and preferences, such as not going to or going through certain places due to the immense agglomeration of tourists or the effects of tourism in these places. Often, the territoriality of a community, which before the development of tourist activity obeyed other specific logics, is altered due to the production of spaces for tourism (Cruz 2009; Molina 2011). Thus, with these transformations, the community ends up redefining its spaces of circulation and leisure, often leaving parts of these spaces for the tourists.

The other impacts that showed greater expression among the participants were the increase in the prices of goods and services (and the consequent increased cost of living), the congestion in urban mobility, overcrowding of public transport and the noise and disorder on the streets of the historic centre.

10.6.2 LAC System—Step 1—Diagnosis of the Area

As seen, LAC is an issue-driven process and therefore begins with the identification of areas of concern, so the desired conditions can thus be determined. As Frauman and Banks (2011) point out, for this stage, the participation of all stakeholders is essential to help identify and define the various issues and concerns associated with the region and the perspective of tourism development in the area.

Based on the literature (Ahn et al. 2002; McCool 2013; Stankey et al. 1985; Takahashi and Cegana 2005), the following questions were discussed at this step:

- What is the role and importance of the historical centre in the regional and national context?
- What special values, characteristics or qualities of the area require attention? Which of these need to be maintained or achieved?
- What are the important issues and concerns in the area? What management problems are important and need addressing?

The role and importance of the historic centre of Porto were defined into four main groups: its historical value, its commercial importance, its central geographical position on the planet, and its economic importance—in the regional context, as a financial centre, which attracts investments, invigorates the economy and, consequently, also projects it at the national level, by positioning the city of Porto as the second most important city of Portugal.

As the most special characteristics of the area, the participants defined the feeling of pride, by the local residents, of their strong identity, their traditions, the security in the area and the hospitality of the community. They reflected that these qualities are being lost due to the disorderly growth of tourism and that they require attention in order to be preserved.

In relation to the management problems and concerns of the area regarding the growth of tourism, the stakeholders defined six main topics:

(1) The 'type' of tourism and tourists that has been attracted: this was the main concern of the area, mentioned twenty times by the participants. The concern stems from the impacts that a growing portion of tourists have caused: noise, dirt, bad behaviour and discomfort due to excessive alcohol consumption. Participants stressed that Porto has increasingly stood out as a cheap and good destination for parties. This has attracted a young public, which does not show much interest in local culture and causes many inconveniences and constraints for residents;

(2) Uncontrolled growth and unplanned tourism: This management problem was referred to nineteen times, by all stakeholder groups, who pointed to a lack of planning in government actions and policies, which, according to them, have not been adequate to the number of visitors the city has received in the past three years;
Horn and Simmons (2002) analysed, in their study, that in situations where development seems controlled or managed, less negative perceptions and unfavourable attitudes between the residents seem to exist. This theory is corroborated in this study, because the perception of lack of planning and management of the tourism development, among the participants, clearly caused a greater concern and rejection to the growth of the activity;

(3) Lack of public participation: one of the most mentioned management problems, pointed out ten times by almost all participants, who stressed the importance of participating in 'meetings' as the one promoted by the research. They complained about the lack of this initiative by the government and reflected that 'the local government makes decisions without respecting the opinion of those who live here' (P2) and 'public participation is the only way to transform things' (R3). As previously analysed, community participation in planning is an indispensable requirement for sustainable tourism development (Frauman and Banks 2011; Hall 2004);

(4) Economic growth 'above all': the environmentalist (E) highlighted that the tourism growth in the historic centre of Porto is based on a 'model of unsustainability, of constant growth'. The specialists defined it as 'unbalanced';

(5) Disrespect for the local population and prioritisation of the interests of tourists over the needs of residents: caused by the 'greed' of economic growth, some

participants pointed out that public management has shown a certain disrespect for the local population. The housing problems and the destruction of several traditional establishments are examples of this prioritisation. Residents pointed out that the city built and developed for its population became a 'visitation park'.

McCool (1996) stresses that the LAC system approach is aligned with the true sense of protecting an area, which is to protect the values for which this area was established. This principle can be applied to the protection desired by the community, of assuring adequate living conditions to its residents;

(6) The dominance of foreign capital in tourism enterprises and lack of support for the maintenance and development of local businesses: the problem of the excess of large foreign groups in the management of short-term rental, for example, is that it generates social disorders for the less favoured parts of the population and mischaracterises entire buildings for this use, which also represents a mischaracterisation of the local identity. In addition, the participants concluded that for the community to be strengthened and for its traditional characteristics to remain 'alive', it is essential that the local government supports traditional local shops.

10.6.3 LAC System—Step 3—Change Indicators

This step is one of the essential parts of the LAC system, when the indicators are selected, that is, the measurable 'indicative' variables of change in the conditions considered adequate and acceptable for each of the aspects analysed (McCool 2013; Takahashi and Cegana 2005). Thus, as indicated in the literature, the most important conditions of the scenario were identified and then the specific indicators that determine changes in these conditions resulting from the development of tourism. One requirement is that these indicators must be easy to measure quantitatively and by observation, so that it is easy to monitor them relatively frequently.

According to the participants of this study, the most important conditions in the tourist development scenario in the historic centre of Porto and the indicators to be observed are (Table 10.3):

10.6.4 LAC System—Step 5—Limits of Acceptable Change

Considering that the indicators refer to the impacts or conditions of an area (Stankey et al. 1985), the limits refer to the level of impact that is acceptable for the different indicators previously mentioned. Like indicators, limits should ideally be quantifiable.

Based on the literature (McCool 2013; Takahashi and Cegana 2005) the central question of this step was 'When thinking about each of the defined indicators, how

Table 10.3 Change indicators

Conditions	Change indicators
Rent costs	Significant increase in the cost of rents
Population historic centre	The decrease in the number of residents in the historic centre
Number of residents versus tourists	Disproportion concerning the two numbers, the number of tourists being much higher
Traditional commerce	The decrease in traditional shops (replacement by major international brands)
Number of short-term rentals in an area	Saturation in the number of short-term rentals in relation to other buildings and residential properties in the area
Number of long-term rent properties vs short-term rent	Significant growth and dominance of short-term rentals compared to residences for long-term rental
Violence rates	The increased crime rate in the area
Trash/dirt on the streets	Observable increase in garbage littered on the streets
Noise levels	Excessive increase in noise decibels in certain areas and times of the day
Parking	Lack of parking spots for the local population
Congestion of public spaces and transport	Intense daily congestion of people in public spaces and transport, as well as vehicles on the streets, causing visible disturbances to the population

Source Own elaboration, based on the results of the focus groups' discussions

much change is the maximum tolerable'? In this way, the defined standards translate into the **limits of acceptable change**. They are the maximum change in conditions that should be allowed.

The participants defined the following limits (Table 10.4), for the indicators previously defined:

As can be seen, one of the most mentioned quantifications in this step of defining the limits was '50%', and one of the terms said was 'balance'. Participants, often faced with the difficulty of establishing specific and quantifiable limits, replied that 'the limit would be a good balance'. This, again, may indicate the perception, by the community, that there is an unbalanced and unfair relationship in the planning and management of tourism, often prioritising the interests and needs of tourists over those of the resident population.

Table 10.4 Limits of acceptable change

Change indicators	Limits of acceptable change
Significant increase in the cost of rents	Proportional to the national minimum wage, with a limit of 50% of this value per person (max. €300 for a one-person studio)
Decrease in the number of residents of the historic centre	No more significant population decrease
Disproportion in the relationship 'number of tourists versus number of residents', the number of tourists being much higher	Do not exceed the limit of eight tourists per resident
Decrease in traditional shops (replaced by major international brands)	A minimum of 50% of the area's traditional commerce must be maintained
(When there is) saturation in the quantity of short-term rentals in relation to other buildings and residential properties in the area	Do not exceed the limit of 50% (for short-term rentals in relation to other buildings and residential properties in the area)
Significant growth and dominance of short-term rentals compared to residences for long-term rental	Limit of 50% for short-term rentals in buildings in the historic centre. Within the same building, at least 50% of the apartments must be reserved for local residents
Increased crime rate in the area	Crime rates do not exceed those that existed before the growth of tourism
Observable increase in garbage littered on the streets	Garbage only inside the containers
Excessive increase in noise decibels in certain areas and times of the day	Limit recommended by health agencies for each space and time
Lack of parking spots for the local population (residents and workers)	Tourist use limit of 30% of existing spots in the historic centre—control and preference for the local residents and workers
Intense daily congestion of people in public spaces and transport, as well as vehicles on the streets, causing visible disturbances to the population	The limit is the quality of life of the local population. Control and balance. Residents should not take more than twice the normal time for commuting

Source Own elaboration, based on the results of the focus groups' discussions

10.6.5 Stakeholder Proposals for More Sustainable Tourism Development

Concluding the focus group sessions and, as a preparation for a possible application of the following steps of the LAC system, proposals for better management and more sustainable tourism development in the historic centre of Porto were defined by the participants. The main proposals, established over the three sessions, based on the problems and indicators previously discussed, can be summarised according to Table 10.5.

Table 10.5 Proposals for more sustainable tourism development

Proposals	References[a]	Groups
Laws and greater control for regulation of the real estate market	13	All
Priority and differentiated prices for residents	12	S/P/R
Decentralisation of tourists' territories	11	S/P/E
Changes in the marketing and segmentation of Porto as a destination/in the tourist segments that the city has been attracting	10	S/P/R/E
Tourist awareness, education and information	8	All
Stricter regulation of short-term rentals in saturated areas	6	S/P/R/E
Strategies to promote longer stays of tourists	6	S/P/E
Greater cooperation between regional destinations	2	S/P

[a]The references refer to the times that each proposal was discussed and identified in the statements, and by which groups (specialists/professionals/residents/environmentalist/merchant)
Source Own elaboration, frequency data taken from the quantitative analysis using webQDA

The most suggested set of proposals, by representatives of all stakeholders' groups, was related to laws and greater control by the government to regulate the real estate market. Some of the proposals, suggested by the participants, were: the establishment of limits for rents, differentiated taxes for properties' owners who rent to tourists or residents, control of the number of apartments for short-term rental accommodation per building, tax incentives for owners to reduce rents and stricter inspection of rental contracts. As previously analysed, the housing issue was also the most mentioned among the negative impacts of tourism, among the indicators of tourism growth and for which more limits were established.

The second most suggested proposal was related to priority access and differentiated prices for the local community, for example, admissions to tourist attractions and public transport, cultural spaces, events, car parking and local shops (cafes, restaurants, local markets). The motivation for this proposal demonstrated, once again, the feeling of inequality and loss in comparison to tourists. As previously seen, the congestion caused by the tourist growth and the increase in the cost of goods and services affected the daily habits of this community, that was forced to stop consuming and attending certain services and places.

The third most mentioned proposal was to decentralise, 'disperse' the tourists from the historic centre of Porto to other areas of the city, as well as to neighbouring municipalities and districts. This proposal can be analysed in conjunction with strategic proposals for longer stays for tourists and greater cooperation between destinations. It was mainly caused by the feeling that there is an excessive and unbalanced concentration of tourism (tourists, attractions, infrastructure and services) in the area of the historic centre.

Due to the type of tourism/tourists that the city has attracted, which, as seen, was one of the main concerns identified, almost all the stakeholders pointed out—as an important step to be taken for a more sustainable tourism growth—changes on the marketing/positioning of Porto as a tourist destination as well as on the tourist segments that have been attracted. They want Porto to change the image of a low-cost short break destination. Residents associated this image with the negative social impacts they experience daily; professionals justified this proposal by saying that this type of tourist that has been attracted spends and consumes very little, which is not good for the economy. The specialists highlighted the fact that this type of tourist is not interested in the culture and heritage of the place, and 'this is the best that Porto has to offer so it should be more consumed' (S1). The environmentalist emphasised that 'the ecological impact that is caused for a two-day trip must be considered'.

Discussing the type of tourists attracted by this 'low cost party destination' image, the residents concluded that 'the major problem is the behaviour of visitors, not the quantity' (P1) and suggested proposals to offer more information and awareness to tourists. The residents discussed, in the focus group sessions, that they feel a lack of interest by the tourists in their true wealth, expressed through their culture, history and traditions. Thus, they consider it important to promote a greater awareness of tourists about the traditions, history and culture of the city and the local community. This measure could contribute to the enrichment of the visitors' experience, as well as to the residents' quality of life. This proposal suggested by the participants is directly related to the creation of the LAC system, which indicates that the visitors' behaviour and attitudes are one of the main factors that define the intensity of the environmental and sociocultural impacts caused by tourism at a destination (McCool 1996).

10.7 Conclusions

Based on LAC, a planning system that proposes to evaluate the limits of acceptable changes from the impacts of tourism on the destination, this work evaluated, in a collaborative process with a variety of twelve stakeholders from the historic centre of Porto (Portugal), the impacts of tourism growth in this area and applied the steps 1, 3 and 5 of the LAC system. These steps consisted, respectively, in making a diagnosis of the area, defining indicators of change and the acceptable limits for each of these indicators, in order to establish the appropriate conditions for more sustainable growth of tourism in the place. Several management problems and current conditions of this development were also identified, which, unfortunately, are far from being considered acceptable, according to the opinion of the majority of the stakeholders interviewed.

The community-oriented, participatory approach of this planning system is one of its great advantages among other planning models. Participatory planning is a response to the challenge of developing more sustainable growth within the tourism sector. The importance that each stakeholder group has in being part of the planning processes is even more evident. The local community is, usually, the most affected by the development of tourism, so, in addition to the knowledge they have about the reality and conditions of the place, they can collaborate in assessing the impacts and defining strategies for the sustainable development of their area.

It was possible to perceive the intensity of some negative impacts that tourism generates in the lives of these people, such as the increase in the costs of rents, goods and services, the loss of local characteristics and traditions and the intense congestion of urban mobility. Various undesirable conditions of this growth have been reported and identified, such as evictions, a saturation of properties available as short-term rentals and excessive noise and garbage on the streets, which highlights the need to seek new plans and take further action. In order to set these conditions within more acceptable limits, some proposals were defined, such as stricter regulation of housing issues and rental conditions for properties, the imposition of a limit on the expansion of short-term rentals and control of noise levels and urban cleanliness.

The maintenance of a scenario of imbalance and growth, without proper planning or measures of regulation and control, like the ones suggested in this chapter, may cause the aggravation of several problems that are beginning to be felt in Porto's historic centre, such as the mischaracterisation and loss of identity, the disappearance of unique traits of culture, traditions and heritage; the increase in gentrification; the worsening of the residents' quality of life, resulting from noise, deprivation and high cost of living (which has not been accompanied by proportional increases in wages), among others.

As discussed by the participants, the study reaffirms that it is necessary to maintain some of the traditional characteristics of this area of the city, to develop conditions to establish residents who are there and to attract new ones, to offer an urban infrastructure that meets the needs of locals and visitors, and to manage the balance between the functions that this area, as the historic centre of a city of great importance at the regional and national level, must offer: housing, employment, commerce, services, leisure and other urban facilities. The lack of more integrated planning to promote a more organised and sustainable growth of tourism may cause this area to lose, once again, its prestige and value, both in the mind of its community and in the tourist market.

Based on the scope of this theme, there are numerous proposals for future research, such as a continuation of the application of the LAC system in the historic centre of Porto. The rapid changes that have occurred in this area suggest that, just as it should occur in the practice of planning, a repetition of a similar study should be carried out, in order to monitor the defined indicators and analyse what changes occurred and what are the principles that should guide future actions. A similar study is also proposed with other groups of stakeholders that were not represented in this investigation, such as tourists and representatives of public bodies responsible for the management and planning of tourism in the historic centre of Porto.

As McCool (2013) concluded his study by proposing that new investigations should test the use of the LAC planning system in different environments, in addition to the wild natural regions, it is proposed that future investigations in this field expand this diversity of applications to other tourist destinations, whether natural or urban areas.

Acknowledgements Third and fourth authors are supported by the project SPLACH—Spatial Planning for Change (POCI-01-145-FEDER-16431), funded by FEDER through COMPETE 2020—Programa Operacional Competitividade e Internacionalização (POCI)—and by national funds through FCT (Fundação para a Ciência e a Tecnologia).

References

Ahn B, Lee B, Shafer C (2002) Operationalising sustainability in regional tourism planning: an application of the limits of acceptable change framework. Tourism Management 23(1):1–15. https://doi.org/10.1016/S0261-5177(01)00059-0

AirDNA (2019) Analyse vacation rentals in Porto | MarketMinder. Accessed on 1 Nov 2019. Retrieved from https://www.airdna.co/vacation-rental-data/app/pt/porto/porto/overview

Almeida García F, Balbuena Vázquez A, Cortés Macías R (2015) Resident's attitudes towards the impacts of tourism. Tourism Management Perspectives 13:33–40. https://doi.org/10.1016/j.tmp.2014.11.002

Andereck K, Valentine K, Knopf R, Vogt C (2005) Residents' perceptions of community tourism impacts. Annals of Tourism Research 32(4):1056–1076. https://doi.org/10.1016/j.annals.2005.03.001

Associação de Turismo do Porto (2019) Porto and North - Visiting. Accessed on 10 November 2019. Retrieved from http://www.visitportoandnorth.travel/Porto-e-Norte/Visitar/Artigos/Porto

Beni MC (2006) Política e Planejamento de Turismo no Brasil. Aleph, São Paulo

Bentz J, Lopes F, Calado H, Dearden P (2016) Sustaining marine wildlife tourism through linking limits of acceptable change and zoning in the Wildlife Tourism Model. Marine Policy 68:100–107

Brida J, Disegna M, Osti L (2011) Residents' perceptions of tourism impacts and attitudes towards tourism policies in a small mountain community. SSRN. https://doi.org/10.2139/ssrn.1839244

Cardoso C, Silva M (2018) Residents' perceptions and attitudes towards future tourism development. Worldwide Hospitality Tourism Themes 10(6):688–697. https://doi.org/10.1108/WHATT-07-2018-0048

CENSOS (2011) Censos - Resultados definitivos. Região Norte - 2011. Accessed on 15 May 2020. Retrieved from https://censos.ine.pt/xportal/xmain?xpgid=censos2011_apresentacao&xpid=CENSOS

Cooper C, Fletcher J, Fyall A, Gilbert D, Wanhill S (2008) Tourism: principles and practice, 4th edn. Pearson Education, Harlow

Cruz R (2009) Turismo, produção do espaço e desenvolvimento desigual: para pensar a realidade brasileira. In: Turismo de base comunitária: diversidade de olhares e experiências brasileiras. Letra e Imagem, Rio de Janeiro

Diario de Notícias (2018) Movimento "O Porto não se vende" promove concentração contra despejos a 09 de junho. Accessed on 20 Nov 2019. Retrieved from: https://www.dn.pt/lusa/movimento-o-porto-nao-se-vende-promove-concentracao-contra-despejos-a-09-de-junho-9381586.html

European Best Destinations (2019) European Best Destinations 2017. Accessed on 14 November 2019. Retrieved from https://www.europeanbestdestinations.com/best-of-europe/european-best-destinations-2017/

Frauman E, Banks S (2011) Gateway community resident perceptions of tourism development: Incorporating importance-performance analysis into a limits of acceptable change framework. Tourism Management 32(1):128–140. https://doi.org/10.1016/j.tourman.2010.01.013

Freire A (2015) Reabilitação urbana no centro histórico do Porto. Universidade Lusíada de Lisboa. Retrieved from http://repositorio.ulusiada.pt/handle/11067/2200

Gonzalez V, Coromina L, Galí N (2018) Overtourism: residents' perceptions of tourism impact as an indicator of resident social carrying capacity: case study of a Spanish heritage town. Tourism Review 73(3):277–296

Gunn CA (1994) Tourism planning, 3rd edn. Taylor & Francis, New York
Hall CM (2004) Planejamento Turístico: políticas, processos e relacionamentos, 2nd edn. Editora Contexto, São Paulo
Horn C, Simmons D (2002) Community adaptation to tourism: comparisons between Rotorua and Kaikoura, New Zealand. Tourism Management 23(2):133–143. https://doi.org/10.1016/S0261-5177(01)00049-8
INE (2019) Estatísticas do Turismo. Accessed on 25 Nov 2019. Retrieved from https://www.ine.pt/xportal/xmain?xpid=INE&xpgid=ine_publicacoes&PUBLICACOESpub_boui=320462327&PUBLICACOESmodo=2
Jornal de Notícias (2018) Porto entre os paraísos que os turistas estão a destruir. Accessed on 2 Dec 2019. Retrieved from: https://www.jn.pt/local/noticias/porto/porto/porto-entre-os-paraisos-que-os-turistas-estao-a-destruir-9750405.html
Kohlbacher F (2006) The use of qualitative content analysis in case study research. Forum Qualitative Sozialforschung 7(1). https://doi.org/10.17169/fqs-7.1.75
Lankford S (2001) A comment concerning "developing and testing a tourism impact scale". Journal of Travel Research 39(3):315–316. https://doi.org/10.1177/004728750103900310
Lima J (2018) Em Miragaia, as ordens de despejo não param de chegar. Público. Accessed on 10 Nov 2019. Retrieved from: https://www.publico.pt/2018/05/21/local/noticia/em-miragaia-as-ordens-de-despejo-nao-param-de-chegar-1830848
McCool S (1994) Planning for sustainable nature dependent tourism development: the limits of acceptable change system. Tourism Recreation Research 19(2):51–55
McCool S (1996) Limits of acceptable change: a framework for managing national protected areas: experiences from the United States. In: Workshop on impact management in marine parks. Kuala Lumpur, Malaysia, pp 1–15
McCool S (2013) Limits of acceptable change and tourism. In: Holden A, Fennel D (eds) The Routledge handbook of tourism and the environment. Routledge, Oxon, UK, pp 285–298
Minayo M (2004) Pesquisa social: teoria, método e criatividade. Vozes, Petrópolis, RJ
Molina, F (2011) A produção do espaço pelo e para o turismo: o caso da praia de Jericoacoara, Ceará, Brasil. Coimbra: Imprensa da Universidade de Coimbra. https://doi.org/10.14195/978-989-26-0244-8_18
Pinto M (2017) O Porto é de todos, não vamos desistir. Público. Accessed on 25 May 2020. Retrieved from: https://www.publico.pt/2017/09/21/p3/noticia/o-porto-e-de-todos-nao-vamos-desistir-1828613
Schetter M, Schetter O (2016) Limits of acceptable change and heritage management on the Island of Mozambique. Governing urban Africa. Palgrave Macmillan UK, London, pp 285–305
Silva A (2017) Turismo e impactos socioculturais: proposta de desenvolvimento sustentável para o centro histórico do Porto. Instituto Politécnico do Porto. Escola Superior de Hotelaria e Turismo
Simpson K (2001) Strategic planning and community involvement as contributors to sustainable tourism development. Current Issues Tourism, 4(1):3–41
Stankey G, Cole D, Lucas R, Petersen M, Frissell S (1985) The limits of acceptable change (LAC) system for wilderness planning. Intermountain Forest and Range Experiment Station. USDA Forest Service. https://doi.org/10.1017/CBO9781107415324.004
Takahashi L, Cegana A (2005) Como monitorar o impacto dos visitantes utilizando o sistema LAC, Limite Aceitável de Câmbio. Revista Turismo em Análise 16(2):206–222. https://doi.org/10.11606/issn.1984-4867.v16i2p206-222
Turismo de Portugal (2020) Visão geral - Turismo em Portugal. Accessed on 5 June 2020. Retrieved from http://www.turismodeportugal.pt/Turismo_Portugal/visao_geral/default.aspx
UNESCO World Heritage Centre (2019). World Heritage List: Historic Centre of Oporto, Luiz I Bridge and Monastery of Serra do Pilar. Accessed on 25 May 2020. Retrieved from: https://whc.unesco.org/en/list/755/
United States Forest Service (2006) T&D Publications. Wilderness and Backcountry Site Restoration Guide. Accessed on 14 October 2019. Retrieved from https://www.fs.fed.us/t-d/pubs/htmlpubs/htm06232815/longdesc/fig2_lac.htm

World Tourism Organization (UNWTO), Centre of Expertise Leisure, Tourism & Hospitality, NHTV Breda University of Applied Sciences, NHL Stenden University of Applied Sciences (2018) 'Overtourism'? Understanding and managing urban tourism growth beyond perceptions, Executive Summary. UNWTO, Madrid. https://doi.org/10.18111/9789284420070

Part III
Experience Design and Management

Chapter 11
Integrating Marketing and Management Planning for Outstanding Visitor Experiences in a Turbulent Era: The Case of Plitvice Lakes National Park

Stephen F. McCool, Paul F. J. Eagles, Ognjen Skunca, Vesna Vukadin, Charles Besancon, and Andjelko Novosel

Abstract Building a viable tourism management plan requires integrative, holistic approaches that provide a foundation for outstanding visitor experience opportunities: diversity of products (the park setting in terms of managerial, biophysical, and social attributes) and programs that connect potential visitors with nature-based products (e.g. marketing). Integrative and adaptive approaches are useful in developing effective responses (such as provide diverse opportunities) to situations where rapidly changing variables (such as visitation) stretch the capacity of managers to respond. These challenges require management to think more holistically than in the past. Plitvice Lakes National Park in Croatia illustrates an example of how management in an era of rapidly growing use can respond in a holistic way to produce higher quality experiences with opportunities to enjoy other nearby offers and continue to contribute to the regional economy. The Park recently experienced significant visitation growth (from about 850,000 annual visitors in 2007 to about 1.75 million in 2018), with peak hourly and daily visitation during the two months of the summer season surpassing the physical capability of the existing visitation system to provide outstanding visitor experiences, resulting in congestion within the Park and consequently degradation of some visitors' experience, increased safety risks as well as some negative impacts on this World Heritage Site outstanding universal value. The

S. F. McCool (✉)
University of Montana, Montana, USA

P. F. J. Eagles
University of Waterloo, Waterloo, Canada

O. Skunca
Granulum Salis, d.o.o., Zagreb, Croatia

V. Vukadin
Park Bureau, d.o.o., Zagreb, Croatia

C. Besancon
Global Park Solutions, Missoula, Montana, USA

A. Novosel
Geocustos, Zagreb, Croatia

A. Mandić and L. Petrić (eds.), *Mediterranean Protected Areas in the Era of Overtourism*, https://doi.org/10.1007/978-3-030-69193-6_11

former General Management Plan (adopted in 2007) provided only relatively vague statements about managing sustainable tourism leaving managers with little formalized direction for promotion and management. The Park revised its General Management Plan from 2016 to 2019 while partially under the scrutiny of the World Heritage Committee due to its designation as a World Heritage Site. The chapter describes how Park managers responded to this situation by engaging its many constituencies to develop an integrated program of promotional changes, capacity building, visitor use modeling, trail construction, development of new opportunities, and contemporary technology to manage visitor use, all occurring within dynamically changing visitor use patterns.

Keywords Marketing and management planning · Visitor experience · Plitvice Lakes National Park

11.1 Introduction

In an era where popular visitation sites, such as cities and destination areas, such as national parks and protected areas, often receive high and growing use, managers are faced with addressing not only peak use challenges, but wicked ones as well. The essential question confronting managers is "How can visitation be managed to provide opportunities for transformative visitor experiences?" This question is always in the manager's mind, either explicitly or implicitly.

Yet, this question exists within a context that is uncertain, changing, and complex. We cannot predict the future with certainty. We do not know what changes in travel and visitation may be in our future. And we are only now coming to grips with the nonlinear character of change: how a small thing can have major consequences. For example, we did not know that a very small thing, a virus (which we can see only with a powerful microscope) that originated in China in December 2019 would have such impacts throughout the world as the novel coronavirus did in the first half of 2020.

In an era of growing wealth, technology, and mobility tourism can be substantially impacted, even for a relatively short time, by disease, economy, disaster, and conflict. More importantly in this era of turbulence, the consequences of tourism have expanded and accelerated as well. From the cities of Europe to the trails of Mt. Sagarmatha (Mt. Everest), Nepal (one climber recently summited the mountain within eleven days of arriving in Tibet because of technological advances in low oxygen environments), some tourist destinations have become congested with not only tourists, but with the results of their negative impacts as well. It also is now clearer than ever before, that global scale tourism is vulnerable to a variety of threats, even if temporarily.

These impacts are partly a matter of the number of tourists, and where visitors temporally reside, partly a function of visitation amounts, partly a function of governance and policy, and partly a function of perceived threats. Negative impacts may result from facilities such as toilets, parking, visits to popular sites and trails; all putting strains on the infrastructure. These impacts may be related to how tourism and infrastructure are managed and the attention to which managers pay to actions occurring on larger scales as well as the number of tourists visiting a destination. However, the positive impacts of substantial economic impact encourage continued growth in volumes which often conflict with the negative impacts.

There is more to the challenge of growing tourism than infrastructure and peak use (see Ferreira and Harmse 2014, for the example of Kruger National Park), however important they may be. Tourist destinations include residents, some benefiting from tourism flows and others seeing only problems. The local political system may hold conflicting attitudes, with some opposed to tourism development and others favoring it. Some see tourism as a way of creating jobs and labor income and others are opposed because of negative impacts on the local quality of life and the biophysical environment. Different government levels may hold divergent views. For example, the local government may see tourism as something to avoid, and the national government may be thinking of the foreign exchange that is produced.

Thus, tourism management recognizes not only the impact of tourism on its immediate area but its context as well. Tourism occurs within a context of multiple interests, where those interests vie and compete for resources, where management and planning are frequently oriented toward reducing challenges on site, but those off-site may be neglected, and where management and marketing may be compartmentalized decisions. And thus, tourism decisions tend to be wicked, that is there is disagreement over goals and cause–effect relationships.

This chapter focuses on one destination, Plitvice Lakes National Park (PLNP) in Croatia, which has successfully addressed the challenges of rapidly accelerating visitation typical of many protected areas in this era of turbulence through the production of a new management plan. To address these challenges and management responses, we begin by describing the wicked nature of the visitor management context that planners face in the twenty-first century, and then use the situation of Plitvice Lake National Park and World Heritage Site to illustrate this situation and how the planning was conducted to address high and growing levels of visitor use. The newest plan, published in 2019, describes the integrated limits on and redistribution of visitation with changing the marketing structure and is compared with the earlier, 2007 plan. We then review how visitor use management was integrated in the last plan. Finally, the lessons learned from this study are described and implications for how they play out in the larger context.

We chose PLNP because the authors worked there in a variety of roles and PLNP is a popular and well-known park and World Heritage Site that experienced fast-growing visitor use as well as social challenges in its management. It is an excellent example of a park in the turbulent world of the twenty-first century.

11.2 Visitor Use Management Is a Wicked Challenge

Historically, visitor use management has been viewed as tame problem, rather than a wicked one. By this, we mean that tame problems tend to be well defined, with end points well described and subject to conventional analysis and by technical solutions that have consequences well described. It is partly that. But the world is complex, it is ever-changing, it is not predictable. Tourism occurs in a turbulent world as demonstrated by the first semester of 2020: all airlines, but particularly internationally dependent lines saw business stopping drops in travel ("Plummeting Demand" in the words of one internet headline). Airlines experienced dramatic declines in travel and shut down thousands of flights. Hotels were hurt economically. Cruise lines were in danger of financially sinking. Popular European destinations, many on or near the Mediterranean, were empty. The European Union closed its borders to outside visitors as did other countries in March. Destinations suffered tremendous economic impact. People were confined to their homes; many people lost their source of employment; panic occurred in hundreds of places. People got sick; other people died.

The world economy sank as did consumer confidence during this global pandemic. All this because of a tiny, tiny virus. None of this was predicted by the travel industry in early January 2020 yet these impacts occurred just weeks later, rolling around the world and throwing the potential tourism season into turmoil. The COVID-19 challenge and its consequences have not been simple problems to solve, partly because of differing objectives and, to put it bluntly, tribal interests were front and center in protection actions. The virus did not know borders, but policy makers acted as if it did.

In a turbulent world, problems are not tame but wicked: wicked problems are linked to other problems, have no stopping rule, are socially tenuous, and are ill-structured (Rittel and Webber 1973; Crowley and Head 2017) and exist within a complex, adaptive, and hierarchal structure (Morris 2020). For example, in the context of visitor use management, use limits are linked to visitor experiences, which are linked to the visitor market (the expectations of those interested in park experiences and are likely to visit the destination), which then are linked to demand for lodging (the amount and type) which then are linked to jobs in the tourism sector. Each of these sectors has both technical and political components. Visitor use limits may be linked to the demand by the central government for foreign exchange which is linked to national politics. Solving one challenge requires addressing another challenge. At each point, a constituency brings its own interest into play and will consider protection of park resources to a greater or lesser degree. And thus, park tourism decisions made at one scale become uncertain, not only because of global trends and patterns, but because of the uncertainty introduced at other scales, linked by constituencies and their interests. However, an integrated and a holistic management plan can address most of these issues within one policy structure.

This state of the world is not new. Seasons (1991) recognized the complexity and change inherent in contemporary social problems, such as park planning: "What seems clear from this discussion is that the degree of uncertainty and increasing complexity with which planners must contend demands new and different modes of thinking and analysis, and roles." Other authors, such as Rittel and Webber (1973) and Allen and Gould (1986) recognized complexity and the notion of wicked problems. Friedmann (1973) applied this notion to planning, Senge (1990) used systems thinking to address organizational complexity, and a more recently variety of scientists have conceived how systems thinking (Strickland-Munro et al. 2010) and a more holistic approach (Kohl and McCool 2016) could be applied to wicked problems in protected areas and tourism. In a tame world, scientists and analysts reduce a holism, in this case a park-tourism system, to a study of parts, first, and then puts them together to develop responses to the challenges to which they are faced. This tame world works on several assumptions: The world is Predictable, Linear, Understandable, and Stable (Kohl and McCool 2016).

Those assumptions apply to how tourism has been conventionally treated. However, reality holds that it is anything but a tame problem. Our day to day activities are easily disrupted: a train is delayed, we experience an unexpected traffic jam, an argument between two people occurs at a park planning meeting, but additional public involvement shows deep divisions within a community. At a larger scale, national-level politicians make commitments of a different kind; a park plan is almost never implemented, although some are. This example is similar to the world over, and the result is park plans left on shelves to gather dust, frustrated park planners and managers, and a bewildered public who thought they were doing something useful.

In a tame world, the dominant planning paradigm, rational-comprehensive planning, works well, its processes are nearly standardized (goals are set, data is collected, alternatives generated, a preferred alternative is presented to decision makers and a plan is approved). But the assumptions are nearly always violated, the context changes (large scale disease interrupts plans), and priorities are re-ordered leading to non-implemented plans, public dissatisfaction, and agency exasperation. Tame solutions tend to focus on numbers of visitors rather than the acceptability or appropriateness of impacts. Planning frameworks that redefine challenges in these terms exist (see for example Stankey et al. 1985).

Some park agency response has been to add more public involvement, and place public involvement earlier in the process to develop trust and build ownership. And this has worked in some places because it recognized that broader involvement creates more ideas to address park tourism planning in situations that deal with some of the consequences of a wicked world. But not all of them. It increases the chance that uncertainty is reduced; it increases the potential resilience of a visitor plan—its flexibility to respond to different situations. It is a move toward managing adaptively.

The COVID-19 virus shows how connected the world is. A virus pops up far away, and tourists stop traveling and don't come to our village, city, or park. Airlines reduce flights, hotels close, local jobs disappear, tax revenues decline, television, for weeks on end, contain bad news and good feelings give way to bad. COVID-19 does not increase linearly but exponentially; it seriously affects older people more than younger people. It cannot be successfully addressed without solving the challenges of youth and older people at the same time but in different ways. And thus, tourism is a wicked problem because one problem cannot be solved in isolation from another.

Because of the difference in assumptions, planning processes made for tame problems do not work for wicked ones. Certainly, some of the activities we do for sub-problems (such as visitor use redistribution) we do regardless of the problem type, but the "whole," the park tourism system, requires we plan differently. Our tourism methodology is integrated with other values and uses, such as marketing, local publics, and national trends and issues, what is happening at the international level, such as the economy, conflict, and disease. Therefore, a park management plan must be holistic in nature and must integrate a wide range of system elements.

11.3 Plitvice Lakes National Park as a Case Example in Working in a Turbulent and Wicked World

Plitvice Lakes National Park and World Heritage Site is situated in the central and inland mountainous (Fig. 11.1) part of Croatia, comprising about 30,000 hectares of land and water. The Park is administered by a Director General and staff, under the supervision of an advisory board and is its own Public Institution. Plans and action taken by the Director General are reviewed by the Ministry of Environment Protection and Energy.

The Park is managed as a National Park under the Croatian Nature Protection Act since 1949 and designated in 1979 as World Heritage Site. Under this designation, Croatia is obligated to protect and present the outstanding universal value that led to the World Heritage listing. This international designation means that UNESCO has an oversight role in insuring that the management of the site does not endanger the value for which the site was designated.

The Park has a unique management obligation, and due to two different legislative structures in place both should be harmonized with each other. One is The Physical Planning Act which defines Physical plans that should be prepared and adopted for the area, which defines what can be constructed/developed within the area, with specified construction zones in which construction is permitted. The second is the Nature Protection Act which stipulates that Park must prepare and adopt a management plan as its key strategic management document. The nature protection sector represented by the sectoral Ministry must approve the proposed physical plan before its adoption. So the physical plan should not permit anything which threatens the Park's nature protection and other management objectives. In the other direction, the management plan should be in line with the Physical plan. There has been conflict and debate over how the two laws are administered.

Figure 11.1 shows the 19 settlements that occur within the park and the physical developments, such as houses, farm structures, and fields occur in or just outside of many of these villages.

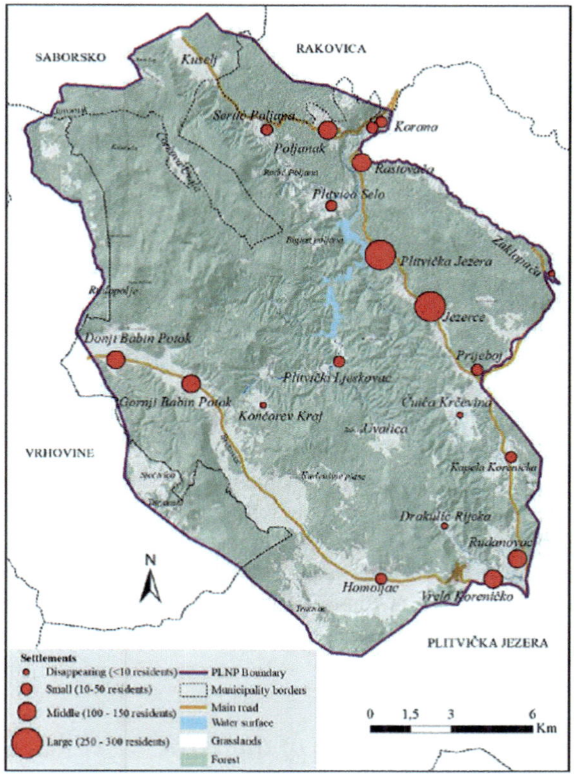

Fig. 11.1 Map of Plitvice Lakes National Park and World Heritage Site. *Source* Plitvice Lakes Management Plan (2019)

PLNP has a well-established, international market profile. The international visitor flows and expenditures to and through the park create a robust national, regional, and local tourism industry. Solid economic benefits occur at all three scales. The park administration also benefits because it is financially self-sufficient based on various tourism fees and charges, giving the park management a strong position in which to undertake planning and to implement any proposed changes. The national and regional markets currently (2020) provide a continuing and growing supply of visitation, and consequent economic benefits and see few problems. Most of the disbenefits occur at the park scale, as high visitor use levels cause negative social and ecological impacts in the park. But these opportunities occur within a context of many challenges (Vurnek et al. 2017); chief among them is the management of visitors.

Political pressures from regional and national constituencies are aimed at tourist growth, sometimes at the expense of negative park-level impacts such as ecological impacts and high tourism peak loads.

The inscription of PLNP as a World Heritage Site occurs when a country agrees to an international convention. World Heritage inscription occurs when a country is a Party to the World Heritage Convention and promotes a national site to become part of the World Heritage Sites list. Adhering to the policies of an international convention is a very high-level legal action, typically superseding national laws and policies. In essence, a country gives up a portion of its sovereignty when it becomes a Party to an international convention.

In 1991, PLNP was placed onto the World Heritage Endangered list by UNESCO because of the military impacts during the Homeland War which gave independence to Croatia. A major concern was the placement of land mines in the park area. Croatia declared its independence from Yugoslavia in 1991. Croatia agreed to Endangered Status in 1992 and the Park was removed from the World Heritage Endangered list in 1997 after the land mine issue was sufficiently dealt with.

The high levels of tourism and the associated development with their social and ecological impacts after 2010 led to UNESCO recommending a reactive mission to the park, which in turn reinforced the need for the management plan, Strategic Environmental Assessment of the Physical Plan and its procedures, which eventually led to harmonizing of the two plans in 2018.

The political ecology of the park demonstrates how the processes and forces of large-scale systems interact with smaller scale park challenges. It shows how important events in a country's history affect park management (park staff was expelled during the Homeland War); it suggests that tourism, once thought as a benign process has entered an era where constituencies at different scales have different expectations for the park.

11.3.1 Description of the Park

The Park has a long history of human use and occupancy going back thousands of years, but most recently, for our purposes, the park was designated in 1949, shortly after WWII when the Yugoslavian government decreed that the area be designated as a National Park. At that time, there were some trails to the waterfalls, lakes, and the canyon so tourists could see them, but there was no unifying management philosophy and protocol. Trails were built and maintained to principal attractions. Hotels, restaurants, and campgrounds were also built and operated by the Park beginning in the 1950s.

The park has had a turbulent recent past, in addition to the growth in tourism. One of the incidents starting the Croatian Homeland War (1990–1995) occurred in the park, some facilities were destroyed or damaged by the war and mines were laid in the park. The newly established country of Croatia recovered from the War. Since 1995, the Park recovered staff, implemented use fees, and developed a well-staffed administration, marketing, research, and management program.

The Park's primary visitor focus is the area of the canyon, lakes, falls, and cascades. Listing as a World Heritage Site came because of the Park's display, one of a few places in the world, of the tufa produced lakes and waterfalls which can easily be reached and viewed. Most of the Park's land involves central European Beech forest and has a limestone base with karst formations in a canyon formed by the White and

Black Rivers. While there are typical features of limestone bedrock (sinks, pits, and caves), in the canyon formed by the rivers, bacteria, and calcium in the water form a tufa which leads to dams and many waterfalls of differing sizes. The resulting tufa creates dams and lakes which then lead to thousands of waterfalls.

These falls, cascades, and lakes produce outstanding beauty that is very attractive and heavily visited. The visitation level in the Park grew steadily (see Figs. 11.2 and 11.3). The intensely visited area is about 1% of the whole park and is where nearly all visitor use and facilities are located. The park plan written in 2007, focused on the biology of the canyon, lakes, and upland forests; only a few pages were devoted to visitation. Statements about visitation were very general. Until the 2000s the park received a sizeable, but somewhat stable visitor use, with management focused on maintaining infrastructure. But beginning about 20 years ago, visitor use began increasing. The 2007 plan indicated 850 thousand visitors in that year; the 2019 plan shows 1.7 million in 2017, a substantial increase over 10 years.

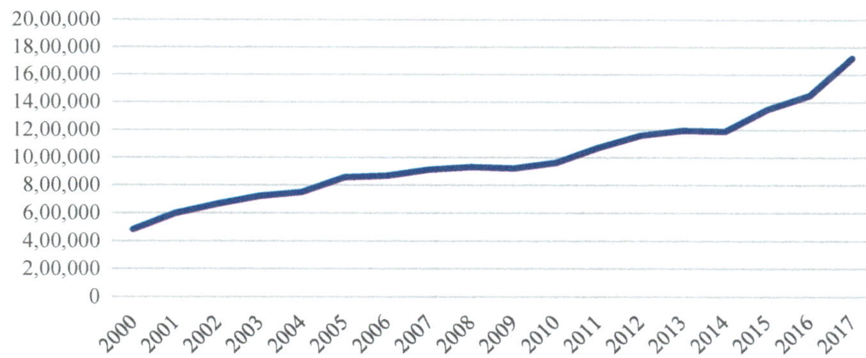

Fig. 11.2 Visitation to Plitvice Lakes, 2000–2017. *Source* The authors based on the official statistics of National Park Plitvice Lakes

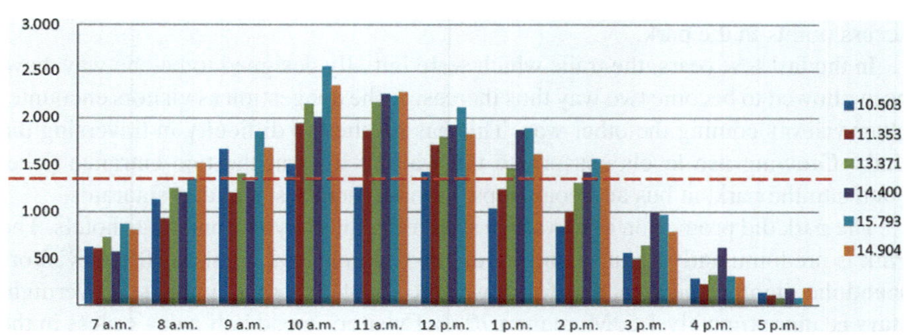

Fig. 11.3 Average hourly visitation distribution, summer 2017 (this includes the two entrances and the average daily visitation for the months of July and August, and for the hours that visitors are allowed. The horizontal dashed line indicated the limit of the system to handle visitation without congestion, about 1300 per hour). *Source* Plitvice Lakes Management Plan (2019)

The surrounding social context is of small farms and communities, and amenity based as well as vacation rental residences near or within the Park boundary on private land. The farms in the Park have not played an important role in economic development of the local area, but do produce traditional products like cheese, jam, and honey. The Park contains the highway which used to be the major travel way from Zagreb (the capital of Croatia) to the famous Adriatic Riviera. There is now a new highway that goes around the Park, but the old highway remains and serves as the main entrances to the park, although some traffic still travels through the highway. The human population of the regional area is declining and getting older.

11.3.2 Visitor Use and Management in the 2007 Plan

As noted above PLNP and WHS is one of the most popular, in terms of visitation, natural sites in central Europe (see Fig. 11.2) and in the Mediterranean region. Its visitation history is one of continued growth numbers. The 2007 plan (PLNP 2007) recognized this as a challenge but did not propose any direct responses to increasing visitation. The 2007 plan did propose more attention should be focused on cultural and historical values, preservation of local and traditional crafts, and several minor adjustments and renovations to the trail system. It proposed that the panoramic buses, which were fueled by diesel be changed. And it proposed that high use densities be examined. But few of these suggestions were acted on as a result of that plan.

Park-owned lodging supply was stable over the period from 2007 to 2018 but local private accommodation supply, comprised mainly of privately owned guest houses, grew significantly in the last few years. While visitor use increased dramatically since 2000, the supply of trails in the park remained about the same. Most of the use occurs between May and September and the hours of 10 AM to 3 PM, indicating that Park visitation is characterized by peak use (Fig. 11.3). Typically, park visitors take 2–6 h to visit the canyon, which consists primarily of walking on trails and boardwalks to the many lakes and waterfalls in the canyon. Buses and boats allow travel between access points in the park.

In the last few years, the trails which were initially designed to be one way, have been allowed to become two way thus increasing the congestion as visitors encounter other visitors coming the other way. This has resulted in difficulty in traversing the trails. Growing use levels exacerbate the waiting lines at the two entrance ticket booths to the park, at bus and boat stops, at toilet facilities, and at restaurants.

The park did promotion of its valued scenery in the canyons, and of its hotels. The Park is predominantly a transit and excursion destination. Less than 30% of visitors spend the night in the area, and for those who do, the average number of overnight stays is approximately 1.5. More than 70% of visitors spend half a day or less in the Park, either in transit or as part of a one-day excursion from another destination (in the coastal area of Croatia or from the capital Zagreb). Nearby attractions, such as caves, large springs, traditional farm foods, and timber milling do not receive many visits.

11.3.3 Management of Visitation Beginning with 2019

Croatian law (Nature Protection Act) requires the preparation of a new management plan every 10 years, so a new planning process was started in 2016 (signed and approved in 2019) in part to address visitation, undoubtedly the largest challenge the Park faced. The plan was designed to not only address these challenges but also the effectiveness of the management called for in the 2007 plan and change the promotion plan to encourage a shift in clientele.

The Park, like most national parks and protected areas, lacked capacity to fully plan for and manage visitation on the trails. The Park staff had no formal training in visitor management concepts or visitor experience management. The Director General of the Park foresaw the need to have contemporary recreation and tourism management concepts applied to the park and put together a team to do so. Authors Skunca and Vukadin were hired to support and facilitate Park staff in development of the new comprehensive management plan. Another consultant (Vladimir Lay) provided information from community residents and park staff. The authors (McCool and Eagles) first visited the Park in 2014 and assessed the current tourism management and development situation. Two capacity building workshops dealing with visitor management (McCool and Bescancon) were carried out, one in 2015 and one in 2017 to prepare Park staff (and other Croatian park staff) to conduct visitor management planning. Other presentations on park management were given by Eagles and McCool during the period 2016–2018. The authors also provided feedback and advice during the planning process on visitor management approaches and tools. These frameworks focus on managing conditions desired by chosen markets rather than numbers of visitors per se. Contemporary tourism marketing concepts, which emphasize quality experiences and regional economic impact, were led with the involvement in the Park by Erika Harms, an international marketing expert. The framework and directions for marketing actions are also integrated in the newly developed management plan.

The visitor plan portion of the 2019 plan contains numerous provisions, but the four themes that are most prominent in terms of managing use and improving the experience include: (1) setting the limits for hourly peak use levels that are defined by acceptable conditions; (2) redistributing these peak levels throughout the day; (3) changing the supply of opportunities offered based on the types of opportunities desired by a new market; and (4) changing the Park promotion to attract a new clientele. Significant visitor management differences between the 2007 and 2019 plans are shown in Table 11.1. Importantly, these actions are integrated so that managing visitation works with marketing to provide a different supply mix. While these are integrated actions, we will present each in turn.

Table 11.1 Item differences about tourism and visitors between 2007 and 2019 plans, Plitvice Lakes National Park Management Plans

Item	2007 plan	2019 plan
Vision	Plitvice Lakes National Park shall remain a UNESCO World Heritage site, and a national leader in the conservation and promotion of unique natural and cultural resources in their valorization by means of sustainable tourism to the benefit of the region and local communities and to the satisfaction of visitors	Plitvice Lakes National Park is a UNESCO World Natural Heritage Site, a place to experience and learn about the outstanding universal value and other natural and cultural values It is an example of good protected area management in cooperation with the local community, where conserved nature is the foundation of sustainable development
Visitor management general objective	Goal 4.1.2.1. Improve the visitor system and visitor services with minimum negative environmental impact	C. Visitation does not undermine Park values, offering visitors unhindered and comprehensive experience, thus presenting conserved world heritage in the best manner possible, while ensuring revenue required for its conservation, building public support for nature conservation and opening possibilities for sustainable local community development
Limits on visitors	No	Yes, 1300 per hour
Ticket sales	Yes, at entrance stations, no limit per hour	Yes, at entrance stations and over internet prior to arrival and by hour
Emphasis on regional visit opportunities	No	Yes
Use of ROS to distribute different opportunities in park	No	Yes

Source The authors

11.3.4 Limits and Redistribution of Daily Use

Encounters on the trail and congested conditions were a principal source of dissatisfaction during use. For example, during the summer use season, a day's visitation can total 15,000–16,000 visitors, which is substantially beyond the physical capability of trails, buses, and boats to transport people without extreme congestion. Two-way trails were very congested during the time when most people visit the

Park. Some people parked cars illegally along highways and roads to gain access to the park. Hotels were full. During peak use periods, good photos of natural features along trails were not really possible. Stops at interpretive signs that communicated important messages could not be made. Board walks crossing lakes and adjacent to streams did not have guard rails, creating a possibility of visitors falling into water. Toilets often had a long line of users. Emergency and law enforcement personnel had difficulty in reaching victims.

Visitor use was not formally limited until the 2019 plan, except that the Park was closed to visitation between certain hours. The 2019 plan limits the hourly visitation to 650 entries per hour per entrance station, resulting in 1300 visitors per hour entering the main valley system. It was felt that at 1300 and below, the existing infrastructure could function effectively. Figure 11.2 shows how the limits of visitation will occur in relation to existing use levels during the July and August busy months. Essentially, this action cuts down the existing peaks of daily use. This peak load limit at mid-day periods resulted in a reduction of visitation at the former highest use periods, with a hope that redistribution in use would occur during lower use time periods, daily and seasonal.

To implement this action in 2020, the Park started selling entrance tickets for hourly times and entrance point through a website. Website implementation of entrance tickets started in 2020 and no evaluation of visitor response is yet available.

The park also, as part of this plan, is studying methods of reinstating the old one-way flow of visitors to reduce the blockage to foot travel that formerly occurred in narrow places in the trail. One-way travel is a method of reducing congestion and other social effects of high visitation.

Both actions will have a significant effect on the flow of visitors through the trails in the lake area and thus reduce congestion. Use can still increase in some periods, such as early morning and late evening, but only to the maximum 650 visitors per hour per entrance.

11.3.5 Change in Marketing for the Plan

Associated with the visitor management plan is a change in how the park and the lodging facilities are marketed (we define marketing in this case as making connections with higher spending and longer staying clientele). First, the supply of visitor opportunities will be greatly expanded. This involves two major thrusts: (1) expansion of the offerings from nearby attractions, such as the Barac Caves a few miles outside the park; and (2) expansion of offerings inside the park, such as traditional farming products produced by villages and settlements which are located primarily in the southeast of the park and other places (see Fig. 11.1 for map showing location of villages and settlements in the Park); and expansion of trail opportunities inside the park designed to facilitate experiences different from the lake trails.

The 2019 plan promotes the park, through placement of advertising by print and electronic media to a clientele that will want to hike, explore, see and test products, and thus adding to the length of stay and visitor spending. In the park, the supply of trails is planned to increase and broaden by adding new opportunities to the trail system. This is done by using an adaptation of the Recreation Opportunity Spectrum (ROS), developed specifically for the park and shown in Figure 11.4 (PLNP 2019).

ROS (Clark and Stankey 1979) has been widely used and adapted around the globe (see for example the Brazilian ROVUC–ICMBIO 2020—and the Latin American ROS–CIPAM, n.d.). ROS has two primary functions in protected areas: (1) protection of specific kinds of visitor opportunities; and (2) limiting the spread of visitor induced and other impacts to specific areas. While ROS was originally designed to address a variety of visitor opportunities, the adaptions mentioned above and PLNP have expanded this concept to include cultural factors. This adaption zones the park for visitation and indicates where new trails may be put and where traditional farming on private land will continue to occur. These new opportunities will encourage visitors to hike in new places and explore villages in the park. This zoning integrates marketing—appealing to potential visitors by adding new visitor experiences to the visitor management plan.

The ROS zoning for the park includes five zones but excludes much of the park where visitation is not allowed (see Vurnek et al. 2019 for another description). This zoning was focused on experiences and was overlaid upon zoning focused on biological aspect as shown in the plan and relevant to visitors.

This marketing will have some impact. New marketing (making connections with higher spending and longer staying clientele) will change the nature of some of the clientele (increase Croatian visits to the park and the region around the Park) and change some of the international visits to those liking to hike, photograph, and explore resulting in longer stays and higher spending.

However, the change in promotion to that proposed in the plan was viewed favorably because it eventually would result in increased spending regionally with the similar number of visitors to what occurs today rather than a continuing increase and the probability of international scrutiny about tourism use. This action would also have the effect of not reducing the quality of the experience. Additional trails would be of little negative environmental impact yet produce longer trips from visitors who have greater interest in seeing other regional attractions and spend more money on their visit.

11.4 Conclusions

Plitvice Lakes National Park and World Heritage Site is an exemplar for protected areas and other tourism destinations receiving growing and high levels of use in the context of a turbulent twenty-first century. It took advantage of a legally required revision for its general management plan in 2019 to address commonly recognized challenges in providing high-quality visitor opportunities, a need for a more economically productive market, and concerns about its recognized World Heritage outstanding universal value. It built capacity to manage visitors, it used contemporary planning

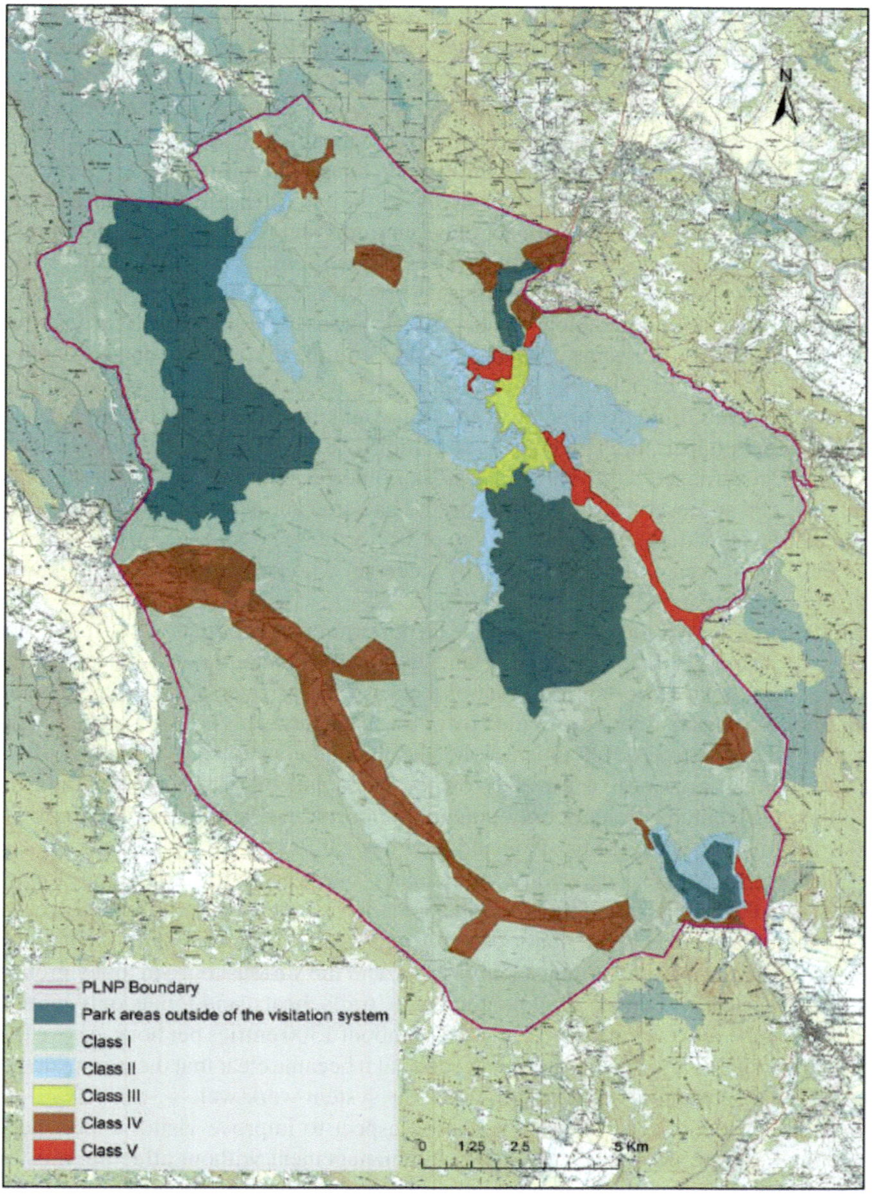

Fig. 11.4 Map of Plitvice Lakes visitor opportunities using Recreation Opportunity Spectrum concept (source 2019 plan). Only the area that receives visitation has been zoned for that person. See text for brief descriptions of visitation zoning

and marketing concepts and techniques to build shifts in its visitor clientele, and it moved to protect its values as a park and as a World Heritage Site. It did this with little change in its internal infrastructure, and instead relied principally on visitor management and marketing changes based on sound science, specified objectives, and marketing principles. The plan emphasized managing for acceptable conditions.

Park administration is implementing these changes in a turbulent environment, where national and regional conflict about visitation, where the resulting foreign exchange and local economic impact plays an important role. And thus, planning is a wicked problem that reflects, in this process, the different views at different scales by the various interests potentially impacted by proposed changes in management. The rise of a highly contagious global disease, caused by the novel coronavirus, demonstrated the uncertainty, turbulence, and complexity of what many planners and scientists have been writing about. Management of tourism is a wicked problem; management has difficulty predicting the future and the relationship between causes and effects is uncertain. The context is complex: different goals held by different scales of governance confront the Park, like other destination. The Park administration faced successfully a dilemma: on the one side are national and international expectations that the Park be managed to protect values and manage for high-quality visitor experiences and on the other by interests who see development as key to raising labor income.

Private farms and settlements provide an opportunity for tourists to see traditional ways of farming, timber processing and living, thus expanding the visitor experience. This is an important observation because the private sector often provides opportunities that cannot be provided by public agencies. This is noted in the special ROS class for a rural development zone and thus demonstrates the utility of ROS. The continued presence of privately owned lands and tourist facilities within the national park emphasizes not only potential opportunities but the need to continue coordinating the Management Plan and the Physical Plan.

The use of physical capability, particularly one-way traffic arguments for the constructed boardwalks and trails within the main park valley were sufficiently robust to be widely accepted by all major power groups, such as the park staff, the national government, the World Heritage Commission, and the various tourism lobby groups. In particular, the system as a whole, the buses, trails, boats, and toilets only worked well when the use level in the park was below about 1300 entries per hour. Above that point, more buses and boats would be needed but it became clear that the infrastructure would need substantial reworking to make the system work well.

The management of visitors is a central aspect to improve visitor experiences. And the Park has done much to change the management without affecting most of its infrastructure, mainly by integrating visitor management (with hourly limits and associated change in ticket purchasing procedures) and marketing (by appealing to a clientele with longer stays and more spending in mind). The Park as an exemplar thus provides several lessons to parks and destinations in similar situations. These are detailed below.

11.4.1 Lessons Learned About Managing High Levels of Visitor Use at a National Park

This paper documents the successful management of a national park created in 1949 and designated as a World Heritage site in 1979. The park survived the war resulting from the breakup of Yugoslavia in 1990–1995. After being designated a site that was World Heritage in Danger because of land mines, the Park returned to full designation in 1995. The first contemporary plan, 2007, contained very little direction for tourism management. The number of visitors doubled from 2007 to 2018, up to 1.7 million entrants, leading to widespread concern that significant social and environmental degradation could occur. The second, more recent management planning process, 2016–2019, was undertaken to prepare a new comprehensive planning document dealing with all aspects of environmental, tourism, and general planning.

This recent planning process and outcome is a successful example of a government institution developing policies to manage impact and tourism volumes in order to maintain tourist service quality and significant environment values. This is the third example of park management planning efforts resulting in the purposive limits on tourist growth. The other two are Pinery Provincial Park, Ontario, Canada in 1971 and Point Pelee National Park, Ontario, Canada in 1989 (Eagles et al. 2020).

Natural World Heritage sites are inscribed because of their outstanding universal value relating to scenery and other superb natural phenomena including, geology, ecosystems, and biodiversity. Contemporary management is physically, financially, politically, and practically capable of ensuring that these values are maintained in perpetuity and utilizes the highest possible standards. The World Heritage Centre of UNESCO is tasked with the responsibility to ensure that management successfully retains the outstanding universal value (UNESCO 2019). Park management and UNESCO forces became allied toward protecting experiences and the outstanding universal value which was in opposition to forces that wished to maintain and increase visitation.

At the same time, the Director General of the park, who had recognized these same issues, took important steps in writing the 2019 plan. He insured that the plan would include tourism management and take action that ensures the continuing quality of the visitor experience and concerns about environmental quality of the Park: he hired a consulting team to write the draft plan, involved internationally recognized consultants to advise and review proposed plan actions, involved the public and park employees in assessing draft actions, sought engagement of the local agricultural community in developing new visitor opportunities, and encouraged a regional approach to marketing. The lesson here is that park planning in this era needs a champion, or an advocate to lead planning, if not by direct involvement then by encouragement and support.

Plitvice Lakes National Park and World Heritage Site had enough management capability to undertake and implement planning. The Park includes a capacity building program to raise expertise in the management of tourism, but it also had the financial and scientific resources to conduct and implement planning. The park management plan planning process proved to be capable of dealing with the natural resource, tourism, and general management issues at the Park. This was an integrated plan, dealing with both resources and an evolving marketing strategy with the objective of increasing economic impact.

The planning process effectively used international expertise, national consultants, park staff members, and national government conservation staff to develop and implement effective management alternatives. This coordinated planning effort was useful and successful. The planning process was sufficiently robust that it survived the loss of the existing park director.

The park's status as a tourism-funded operation ensured sufficient financial resources to undertake inventories, develop useful alternatives, utilize professional planning, and to implement the management plan policies. The park's ownership and operations of most of the hotels, restaurants, souvenir shops, transit operations, and all site attractions meant that tourism policies could be fully considered, and changes implemented quickly and effectively, without complicated and expensive negotiations with concessionaires or contractors.

The problems with tourism development on private lands within the park illustrate the issue of legal conflict around development planning in a park, or any tourism destination. In essence, for tourism planning to be fully effective the governance powers must have the authority over all tourism development in the regional area, or risk planning failure.

Many park plans are not implemented. The plans may have been written by consultants who are from someplace else with the staff and local community not holding much ownership in the planning process, nor understanding the actions proposed in the plan. In this case, consultants were both local and international, staff members were involved and highly supportive of policies that reduced peak tourist loads to levels that were consistent with the current infrastructure limits. The park staff members were well aware of tourism problems and were happy to see a planning process that enabled their concerns to be assuaged. The professional staff complement, most of whom were local citizens were generally open to change and management improvement. Their support was critical in getting park-level and local community political acceptance. Because of this ownership of the plan, it was eventually approved by the government and by UNESCO.

The park plan reduced the disbenefits—social and ecological in nature—accruing to former visitor use while maintaining national foreign exchange benefits and enhancing regional economic benefits through a change in the visitor market and expansion of opportunities available. This shift is huge and shows what can be done to demonstrate and maintain both ecological values of protected areas and World Heritage Sites and enhance regionally the economic value of them.

Finally, the context of the plan implementation demonstrated its wicked, complex, and uncertain nature. The turbulent character of this context cannot be ignored and is a great influence on plan implementation. The Park did not use a simplistic definition of the challenges it faced, such as defining them as "over tourism" but realized it had many different challenges that were linked and that they existed within a turbulent and uncertain context. Importantly, not only did a global disease have an impact on implementation of park planning actions, it demonstrated that the park operated in an uncertain environment where actions and activities occurring distant from the park had an effect on it. The integrated character of park planning made the park more resilient and adaptable in the face of this uncertainty.

References

Allen GM, Gould EM Jr (1986) Complexity, wickedness, and public forests. J Forest 84(4):20–23

CIPAM (Consortium on International Protected Area Management (n.d.) ROVAP: El Rango de Oportunidades para Visitantes en Áreas Protegidas. Colorado State University, Fort Collins, CO, 27 p

Clark RN, Stankey GH (1979) The recreation opportunity spectrum: a framework for planning, management, and research. Department of Agriculture, Forest Service, Pacific Northwest Forest and Range Experiment Station. Gen. Tech. Rep. PNW-GTR-098. Portland, OR, 32 p

Crowley K, Head BW (2017) The enduring challenge of 'wicked problems': revisiting Rittel and Webber. Policy Sci 50(4):539–547

Eagles PFJ, Novosel A, Skunca O, Vukadin V (2020) Developing targets for visitation with sustainable tourism. Chapter 3 In: Spenceley A (ed.) Handbook of sustainable tourism. Edward Elgar Publishing

Ferreira S, Harmse A (2014) Kruger National Park: tourism development and issues around the management of large numbers of tourists. J Ecotourism. https://www.researchgate.net/deref/http%3A%2F%2Fdx.doi.org%2F10.1080%2F14724049.2014.925907

Friedmann J (1973) Retracking America; a theory of transactive planning. Anchor Press, New Jersey, 289 p

ICMBIO (Instituto Chico Mendes de Conservacao da Biodiversidade) (2020) ROVUC: Rol de oportunitidades de visitacao em unidades da conservacao. Government of Brazil, Ministry of Environment, Brasilia, 2 p

Kohl J, McCool SF (2016) The future has other plans: planning holistically to conserve natural and cultural heritage. Fulcrum Publishing, Golden, CO, p 320

Morris WC (2020) Recreation as a social-ecological complex adaptive system. Sustainability 12:752–768

PLNP (2007) National Park Plitvice Lakes Management Plan. Ministry of Culture Republic of Croatia, Zagreb, Croatia, 99 p + appendices

PLNP (2019) Plitvice Lakes National Park Management Plan 2019–2028. Plitvice Lakes National Park Public Institution. Plitvice Lakes, Croatia, 354 p

Rittel H, Webber M (1973) Dilemmas in a general theory of planning. Policy Sci 4:155–169

Seasons M (1991) Planning for uncertain futures: thoughts for practice. Plan Canada 31(6):31–37

Senge PM (1990 [2006]). The fifth discipline: the art & practice of the learning organization. Random House, New York

Stankey GH, Cole DN, Lucas RC, Petersen ME, Frissell SS (1985) The limits of acceptable change (LAC) system for wilderness planning. USDA Forest Service, Intermountain Forest Experiment Station (INT-176), Ogden, UT, USA, 37 p

Strickland-Munro JK, Allison HE, Moore SA (2010) Using resilience concepts to investigate the impacts of protected area tourism on communities. Ann Tourism 37(2):499–519

UNESCO (2019) Operational guidelines for the implementation of the world heritage convention. World Heritage Center, Paris, France, p 177

Vurnek M, Brozinčević A, Rendulić Ž, Miculinić K, Vukadin V, Škunca O (2019) Management zonation and its implementation at a UNESCO World Heritage Site: a case study for the Plitvice Lakes National Park, Croatia. Eur J Environ Sci 9(2):87–96

Vurnek M, Brozinčević A, Čulinovic K, Novosel A (2017) Challenges in the management of Plitvice Lakes National Park. Republic of Croatia, National Parks Manage Conserv. https://doi.org/10.5772/intechopen.72375

Chapter 12
The Lack of Policy, Planning, and Governance: The Mismanagement of Visitor Pressure in Cumalıkızık, Bursa—A World Heritage Site

Sina Kuzuoglu and Burcin Kalabay Hatipoglu

Abstract Using a rural World Heritage Site from Turkey, this qualitative study questions tourism governance from the perspective of social-ecological systems (SES) theory. For exploring our research aim, we collected data from multiple primary sources in addition to secondary resources between the years of 2018–2020. We first describe the planning phases within the historical context of tourism development in Cumalıkızık village and explore the governance mechanisms, interactions, and tensions between the stakeholders. Our findings suggest that the co-management processes that are not effectively functioning prevent the community from taking corrective action and responding to visitor pressures. Failing to establish an adaptive and resilient system, the village is losing its identity. Through these results, we recommend public authorities to take responsibility and be more accountable for tourism development. Our results indicate that the community would benefit from product and service innovation that would sustain authenticity and foster culture-based tourism.

Keywords Governance · Sustainable tourism · Overtourism · World heritage site · Turkey · Adaptive co-management

12.1 Introduction

This chapter examines the governance processes for managing community resources at a rural World Heritage Site (WHS) in Turkey. As cultural heritage sites have started to attract an increasing number of visitors, governance mechanisms in these areas have become entangled with tourism management issues (Islam et al. 2018). For many nations, the inscription of cultural heritage assets as a WHS is now more of a political nature, representing both international recognition and a potential tool to

S. Kuzuoglu (✉)
Department of Recreation and Leisure Studies, University of Waterloo, Waterloo, ON, Canada
e-mail: skuzuogl@uwaterloo.ca

B. K. Hatipoglu
School of Business, Canberra and IIRG, The University of New South of Wales, Sydney, Australia
e-mail: b.hatipoglu@unsw.edu.au

A. Mandić and L. Petrić (eds.), *Mediterranean Protected Areas in the Era of Overtourism*, https://doi.org/10.1007/978-3-030-69193-6_12

attract foreign tourists (Huang et al. 2012; Human 2015; Jones et al. 2020). However, WHS's aim, that is, to preserve cultural and natural assets, may also be overshadowed by political and economic priorities.

While the introduction of tourism activity into heritage sites can provide the necessary financial resources to conserve cultural and natural assets, developers, businesses, and residents' economic expectations may not always be compatible with WHS's conservation framework. Without developing the social capacities of tourism, growth places the sustainable development of tourism at risk (Higgins-Desbiolles et al. 2019). This risk may be amplified in those sites where a core community resides and works within the WHS's perimeters. Understanding the relationship between the core community and their connection with their heritage and natural environment is, in such contexts, paramount since they predominantly rely on resources in their immediate surroundings to maintain their livelihoods (Poulios 2014). The study of these concerns through the social-ecological systems (SES) lens can help us in understanding how communities in rural WHSs can develop and utilize resources to cultivate their capabilities to "cope with" and "adapt to" change and, at the same time, maintain their unique character and functions (Bui et al. 2020; Ruiz-Ballesteros 2011).

This chapter focuses on Cumalıkızık village, which, together with the Khans area and Sultan Kulliyes of Bursa, Turkey, was inscribed as a WHS in 2014. Six core WHS components demonstrate the urban and rural socioeconomic system established by the Ottoman Empire in the early fourteenth century. Cumalıkızık village is, in this context, a one-of-a-kind example for showcasing an Ottoman village as it has preserved its historical texture and traditional lifestyle for the last seven centuries (Tas et al. 2009). The village has received increasing public attention in the last two decades, causing an overwhelming number of domestic and foreign visitors wandering the village's narrow streets. The case site is potentially a critical demonstration of co-management structures at a rural heritage site, including managing visitor pressures and linking the core community with the living heritage site through resource governance. Thus, the question arises whether co-management structures in tourism governance facilitate adaptive communities that can manage change while maintaining the cultural uniqueness of living heritage sites?

To address the research question, the empirical analyses presented in this chapter are carried out during 2018–2020 based on the theoretical background and research methodology outlined in Sects. 12.2 and 12.3, respectively. We then describe the planning phases within the historical context of tourism development in Cumalıkızık and then explore the governance mechanisms, interactions, and tensions between the stakeholders. The last section presents the discussion and conclusions.

12.2 Theoretical Background

12.2.1 Living Heritage Sites and Governance

Effective policies, plans, and governance systems ensure that tourism does not negatively affect the cultural heritage sites and their surrounding areas (Bramwell and Lane 2011; Heslinga et al. 2019). However, suggesting an effective governance system adapted to all heritage sites is unrealistic since each heritage site is unique in character and managerial, economic, social, environmental, and cultural context.

The coordination and cooperation among multiple stakeholders involved in WHSs often entail tensions (Çahantimur and Öztürk 2020). Particularly in developing nations, governance arrangements do not always stimulate collaboration among stakeholders due to lack of structure and civil society involvement, illiteracy, and public dissatisfaction (Islam et al. 2018). Despite these potential setbacks, community participation, and a move toward effective governance processes are increasingly viewed as vital for managing tourism at heritage sites (Plummer and Fennell 2009). A resource manual by the United Nations Educational, Scientific and Cultural Organization (UNESCO) (2013, p. 15), for example, recommends a "shift" in heritage management toward inclusive approaches in which local communities are more involved in the governance of the cultural assets in their surroundings. As described in the manual, this new governance model depends on maintaining a balance between individual and community goals, efficient use of resources, and accountability.

Once the collaboration is established between the community members and other actors to manage heritage sites, the potential for informed decision-making, a larger pool of resources, and better management of conflicts among these actors increases (Plummer and Fennell 2009). Despite this possibility of significant benefits, community involvement in decision-making often remains limited, not going beyond positioning residents merely as informants in heritage management (Jones et al. 2020; Li et al. 2020). Participation of communities is reported to international authorities by "tokenistic consultation" as exemplified in the management plans of the UK heritage sites (Landorf 2009, p. 506) and by "distribut[ing] power horizontally but to local politicians" as experienced in Turkey (Human 2015, p. 171).

Another scholarly perspective underlines a self-governance model administered by relevant tourism stakeholders (e.g., Folke et al. 2005; Brondizio et al. 2009; Ruiz-Ballesteros 2011). Even though self-governance is successfully applied in some sites, it has risks for unbalanced tourism growth. These risks are more recently exemplified by overtourism in popular cultural tourism destinations (e.g., Venice and Barcelona) due to market-oriented governance approaches (Koens et al. 2018). More than one-third of European residents regard overtourism as a threat to cultural sites (Adie et al. 2020). As its most prominent symptom, overcrowding affects both the tourist experience and the residents' quality of life (Mihalic and Kušcer 2018). It creates negative externalities on the social, natural, and physical environment (Koens et al. 2018). Furthermore, a significant risk for WHSs is that they can lose their internationally acclaimed status (Seraphin et al. 2018).

In this perspective, introducing a change from overtourism to no-tourism is one option (Seraphin et al. 2018). At the same time, scholars also discuss employing degrowth strategies before these sites experience irreversible impacts of tourism (Cheung and Li 2019; Higgins-Desbiolles et al. 2019). However, in destinations with exceptionally high dependence on income through tourism activity, not all stakeholders would be willing to accept these changes. Before taking any action, the risks of overtourism can be diagnosed through a sustainability lens to assess various alternative strategies and outcomes (Mihalic 2020). Rather than risking the degradation of natural and cultural resources in a site, a coherent tourism strategy that balances economic goals with the strengthening of the community resilience should be a potential target (Bui et al. 2020; Cheer et al. 2019).

12.2.2 Adaptive Co-management Strategies

For a long time, heritage management took an expert-driven approach that emphasized preserving tangible assets under scientific knowledge guidance (Poulios 2014). In this approach, future developments and resource use were predictable and controllable in a linear manner (Plummer and Fennell 2009). However, we recognize that heritage sites and their surroundings are complex and dynamic (Islam et al. 2018), which require flexible approaches to their governance (Cochrane 2010). In the living heritage approach, the tangible and intangible heritage is connected to the communities through their everyday practicing of local traditions. In this view, heritage is an "essential part of human life" (Orbaşli 2000, p. 18). Indeed, change and uncertainty are part of any system (Ruiz-Ballesteros 2011). Similarly, heritage is also "constantly changing, re-valuated, interpreted in various ways by different actors" (Tengberg et al. 2012, p. 17).

In living heritage sites, conservation frameworks that prioritize the continuity of the community's linkages to heritage and intangible assets over preserving the material fabric can be more promising for the sites' future (Poulios 2014). Inherently, community members' contemporary lifestyles will affect the WHS, and similarly, the site will also change the community's social aspects (Chandani et al. 2019). When the community's linkages with their heritage are sustained, the members could then safeguard the continuance of both the site's historical characteristics and the intangible aspects of their heritage (Poulios 2014).

The bilateral connection between the core community and the site can create and maintain a resilient SES. How the core community functions within the environment determines its pathway toward sustainable development (Ruiz-Ballesteros 2011). The community's potential for change, adaptability, and capacity for resilience are major indicators for defining this pathway (Larsen et al. 2011; Ruiz-Ballesteros 2011). Frequently, factors like increased visitor pressure and infrastructure deficiencies put multiple strains on WHS and the community (Mihalic 2020; Mihalic and Kušcer 2018). Furthermore, the twin forces of economic development through tourism and the fulfillment of the WHS requirements for heritage preservation compete for community resources.

From the SES theory perspective, tourism governance suggests creating synergies across multiple actors to increase tourism's benefits for stakeholders (Heslinga et al. 2019). The system's functional interdependencies require a multilevel approach for addressing governance, extending from local to global (Brondizio et al. 2009). Forming a co-management structure for resource governance is one way to organize the interactions between the multiple actors of a system for sustainable tourism (Folke et al. 2005; Islam et al. 2018; Plummer and Fennell 2009).

Co-management of a heritage site would ideally demonstrate formal decision-makers, (e.g., the central and local governmental agencies), and expert groups' commitment to an alliance with the core community members and other relevant stakeholders (Brondizio et al. 2009). Such a governance structure would, thus, signify distributed participation in a complex system (Larsen et al. 2011) and a "joint management of the commons" in heritage sites (Islam et al. 2018, p. 1893).

These governance arrangements would benefit from "continuous learning, implementation, and adaptation by all actors" to respond to the uncertainties in the system's environment (Plummer and Fennell 2009, p. 154). Mechanisms like mediation and negotiation can be used to decide between "competing visions of recovery and risk" across levels (Brondizio et al. 2009; Larsen et al. 2011 p. 486). In an adaptive co-management arrangement, the involved actors come together to interpret problems through communication and negotiation; they make decisions jointly; they are autonomous; and they develop their capacities through learning and adaptation (Plummer and Fennell 2009). Active engagement of the core community in decisions concerning the management of heritage assets and how to blend experts' scientific knowledge with traditional knowledge is targeted for good governance (Landorf 2009; Li et al. 2020; Tengberg et al. 2012; UNESCO 2013). Feedback and continuous learning are predecessors of adaptation and social capacity improvements needed to develop sustainable tourism over time (Cochrane 2010; Higgins-Desbiolles et al. 2019; Mandić 2019).

In summary, the literature provides examples for both success and failure in self-organization, collaboration, and governance at sites for achieving the dual goals of economic development through tourism and preservation of cultural heritage. What is promising for the living heritage sites is that if the core communities' resilience is improved, they can potentially cope with disturbances and adapt to the changing environment. Thus, in this chapter, we examine how the selected case WHS manages change over time and whether it maintains its unique character by engaging in co-management structures in tourism governance.

12.3 Research Methodology

We use the SES perspective to offer insights that would contribute to a deeper understanding of how tourism–stakeholder–visitor interactions in WHSs can be better managed. Other studies have also used SES for examining community resilience and governance arrangements at tourism sites (e.g., Bui et al. 2020; Ruiz-Ballesteros 2011; Larsen et al. 2011). In the literature, there are ongoing arguments about whether qualitative or quantitative methods would be most suitable in adopting the SES perspective, with advocates for each side (Bui et al. 2020). In our single case study, we found qualitative research more appropriate to provide a more in-depth description of the phenomenon under examination (Eisenhardt and Graebner 2007). More specifically, to examine how the core community started engaging in tourism activity and how they responded to the changes in their physical and social environment, it was necessary to gain a historical perspective, obtain information on the tourism offering of the WHS, and attain empirical data from multiple stakeholders. The interpretive analysis adopted in the chapter allowed researchers to understand how the community behaved during change based on the participants' lived experiences (Creswell 2014).

12.3.1 Research Setting

Cumalıkızık is one of five Kızık villages that remained in Bursa, in the northwestern part of Turkey (Fig. 12.1). During the early Ottoman times, each of the

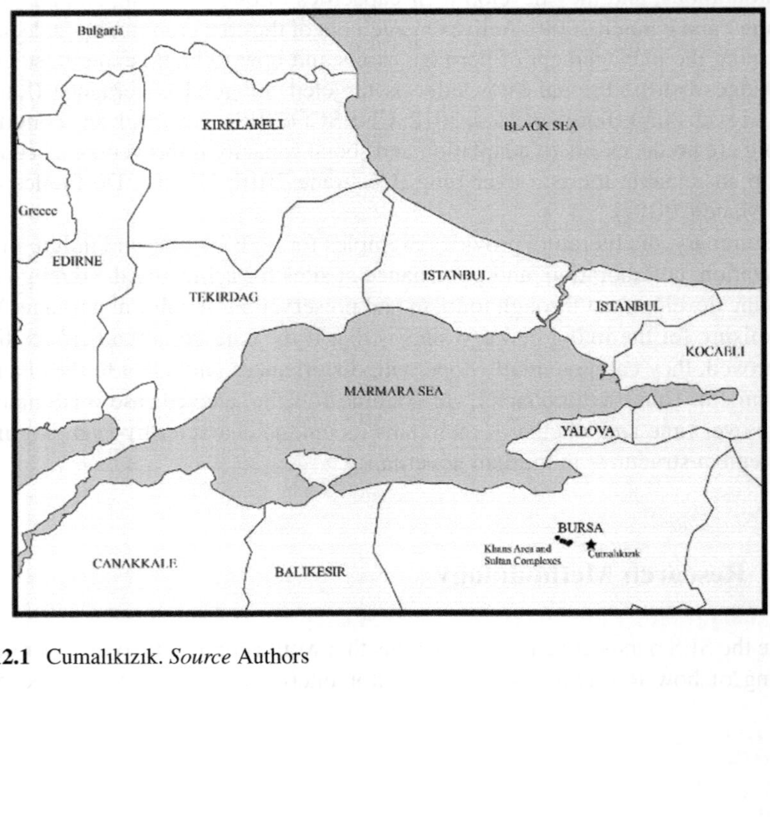

Fig. 12.1 Cumalıkızık. *Source* Authors

Kızık villages had a unique purpose in the communal life. Cumalıkızık's status as a waqf village, which gave an assurance that its entire agricultural production was purchased, suggested that the villagers contribute to maintaining the certain structural elements. The socioeconomic model that resulted from this is believed to have been instrumental in the preservation of Cumalıkızık village in its original state for more extended periods (Akan 2013; Pekerşen et al. 2019).

The village of Cumalıkızık remained an unknown gem to the outsiders for the better part of its modern history. In its current status, the village of Cumalıkızık is first and foremost recognized for being home to exemplary rural spatial planning and vernacular structures of the Ottoman Empire's early times. The village "demonstrates a continuity of architecture with natural setting" (Ozorhon and Ozorhon 2014, p. 185). Natural stone-paved narrow streets, small public spaces at the intersection of these streets, and the two- or three-story-high houses with inner courtyards represent the lifestyle during the Ottoman Empire years (Tas et al. 2009). Recent research suggests that 180 of the original 270 historic houses remain in Cumalıkızık, and villagers occupy 150 of them (Pekerşen et al. 2019).

Cumalıkızık was designated as a culturally protected site in 1981, followed by the completion of the architectural conservation plans in 1992 (approved in 1994). The integration of the conservation framework into the community members' livelihood was perceived to be essential. This perspective was also central in the works of Cumalıkızık Conservation and Revitalization Project 1998, coordinated by the Bursa Local Agenda 21 group (Oren et al. 2002). The nationally acknowledged status of Cumalıkızık was solidified in the international arena when it was introduced to the tentative list of WHSs in 2000. A multi-stakeholder project, governed by central and local government institutions and Bursa Chamber of Architects, was implemented during 2007–2008 (Tas et al. 2009). The project aimed to establish conservation standards and offer proposals for socioeconomic development in Cumalıkızık with the community's contribution.

In 2014, Cumalıkızık was inscribed as a WHS under the heading "Bursa and Cumalıkızık: The Birth of the Ottoman Empire." Cumalıkızık's listing was part of a serial nomination that included the Khans district and the Sultan complexes (Hüdavendigar, Yıldırım, Yeşil, and Muradiye) in central Bursa. While the inhabited part of Cumalıkızık, which is enlisted as the core zone, sits on roughly nine hectares, an expansive buffer zone around the village (about 191 hectares) is a natural protected site (Akan 2013). The buffer zone does not have any residents and instead is allocated to the continuation of agricultural production (Akan 2013).

12.3.2 Data Collection

We aimed for trustworthiness (credibility, transferability, dependability, and confirmability) in our data collection and qualitative data analysis through the following several steps (Lincoln and Guba 1985). First, we triangulated data from multiple primary sources in addition to secondary resources. We used diverse data collection methods (e.g., interviews, participative observations, and social media analysis), thus satisfying the credibility criterion for qualitative rigor. The second author's long-term engagement with the site management team and a longitudinal approach to data collection also contributed to establishing the confirmability of the chapter.

We visited the village numerous times during 2018 and 2019 together and separately (for a detailed description of the data collection procedure, see Table 12.1). Our first research trip was introductory, during which we met with the president of the Cumalıkızık Village Women's Education, Solidarity and Development Association, attended a meeting with the site manager of the WHS, and interviewed the president of the Cumalıkızık Agricultural Development Cooperative. This first visit was intentionally scheduled on a Sunday, which allowed us to get a sense of the scale and scope of the tourism activity in Cumalıkızık on the busiest day of the week. The remaining visits to the village were content-specific, in which the first and second ones focused on the documentation of the food and souvenir items in Cumalıkızık (weekdays), respectively. In the subsequent visits, we gathered tourism development perceptions in Cumalıkızık using semi-structured interviews. The second author also contributed as a panelist to the annual event of the WHS in July 2019 and gathered comments from the public regarding cultural heritage preservation and tourism in Bursa.

We selected our interviewees based on a purposive sampling approach (Bryman 2012). The interviewees worked at the Site Management Unit of the Municipality, experts who have previously worked at the site, a local non-profit organization, and in tourism within the research site. Additionally, we also chose a group of residents that were not directly affiliated with the tourism sector.

All research visits to Cumalıkızık have included an observational component in the form of observer-as-participant for "gaining knowledge of total situations" (Pearsall 1970, p. 342). Since many of the residents did not feel comfortable with video and/or voice recording, detailed notes were taken throughout the interviews and field observations. Additionally, pictures of the streets, menus, and souvenirs were taken as evidence of the tourism offering. Between the years 2019 and 2020, we collected data for the visitor numbers and followed the changes in the village through communication with the Bursa Metropolitan and Yıldırım Municipalities.

Furthermore, to continue triangulating data sources for visitor satisfaction, we examined the customer ratings and comments for Cumalıkızık and local establishments through the TripAdvisor site. The secondary data sources included the site management plan, academic sources, and print and social media.

12.3.3 Data Analysis

The analysis consisted of multiple stages. We have first conducted a historical analysis on Cumalıkızık going back to the 1980s, emphasizing the interventions by various stakeholder groups and the responses to those programs by the core community.

Table 12.1 Data collection (2018–2020)

Source	Type	Emphasis	#
WHS site manager	Group briefing	History of the WHS inscription process, challenges	1
WHS site manager	Informal meetings	Challenges	2
Cumalıkızık Village Women's Education, Solidarity and Development Association	Semi-structured interviews, face-to-face	History of tourism development, challenges in governance	4
Cumalıkızık Agricultural Development Cooperative	Semi-structured interview, face-to-face	Changing livelihood of the village from agriculture to tourism	1
Academic expert	Unstructured interview	The social aspects of the village in the 1980s, resident attitudes toward tourism	1
Restaurant Owners	Semi-structured interviews, observation	Product and service offering, the origin of the ingredients	26
Shop owners	Semi-structured interviews, observation	Product offering, the origin of the products	14
Visitors	Unstructured interview	Experiences, visitation to the museum	15
Residents	Group Interview (6–8 people each)	Attitudes toward conservation, expectations from tourism	2
Bursa Metropolitan Municipality (a resident of Cumalıkızık)	Semi-structured interview, face-to-face	Changes in the village	1
Bursa Metropolitan Municipality	Semi-structured interview, telephone	Visitor numbers, challenges in governance, managerial intervention	3
Yıldırım Municipality	Semi-structured interview, telephone	Negotiation processes with the shop owners	1
TripAdvisor (2015–2020)	Examination of the ratings and comments	Customer satisfaction	1553
Cumalıkızık Village	Printed and Social Media	Visitor experiences	Various

How the community members addressed and interpreted change was investigated. Second, the environmental and social effects of tourism on the lives of community members and the physical characteristics of the village were recorded.

Later, we systematically analyzed the village establishments' product and service offerings, considering their connection to cultural heritage and localism. We compared these findings with visitor evaluations through publicly available commentaries on TripAdvisor. Our participant observations were used to either accept or reject the emerging patterns. We identified the gaps between what is offered by the community and what is expected by the visitors. We examined the co-management structure that is established between the public agencies and the community. Based on the theoretical background, we identified the cultural heritage conservation and development approach used by the site management office, its effects on the community, and strategies for conflict resolution. Finally, our research process was accompanied by periodical updates with the key stakeholders of the research (i.e., municipalities and WHS site management and residents). Through these steps, we attempted to display how external pressures shape the socio-ecological systems in the living heritage site of Cumalıkızık. The draft version of the analysis was shared and discussed with site management to ensure the credibility of our findings.

12.4 Findings

12.4.1 Tourism Development in Cumalıkızık Village

In the early 1980s, the village's new status as a culturally protected site placed severe restrictions on what was and was not permissible to do in private properties. The residents remember those days with negative connotations because the new status meant that they could not renew or renovate their run-down houses to fit their changing needs. The renovations were costly for a small community whose income rested on the little agricultural income they gained from their land in and around the village (Ahunbay et al. 2014). Even more, the community had developed negative feelings toward the experts that became frequent visitors to the village.

There were only two small coffee shops in the village that served as social gathering spaces. The community was highly traditional and would not let outsiders enter their houses or interview their families. Culturally interested individuals, international and domestic, visited the village in small groups. However, at this point in history, the community was very conservative. In such a restrictive setting, the site manager shared that most of the early academic research was conducted without entering the houses and focused on analyzing the architectural assets rather than the village's social aspects.

In the village, women were confined mostly to household chores while the men were more involved in agricultural practices. The women we interviewed revealed that

> [they] were not allowed to walk through the road between the two coffee-houses.

In addition to the income generated from agricultural production and household-based animal husbandry, villagers also sold home-made products to the few visitors that wandered their streets. Women interacted with visitors through small openings in their house doors and were hesitant to engage with male visitors.

The village population gradually dropped from 1900 to less than 900 in the 2000s (Akan 2013). Consequently, the community needed support for slowing the out-migration, maintaining their livelihoods, and preserving the cultural heritage assets in the village. Stakeholders, including the municipality, non-profit organizations, and volunteers, gathered and put together the Cumalıkızık Conservation and Revitalization Project in 1998 (Tas et al. 2009). They presented tourism as the most suitable tool for improvements in the villagers' livelihoods. The sale of locally produced materials alongside hosting local events, such as raspberry festivals, would complement the community's expectations while support preserving the tangible heritage assets (Schneider and Esin 2000).

During the Revitalization Project implementation, two streets, public buildings, and four houses were renovated and repurposed as an accommodation facility, a restaurant, and a souvenir shop (Oren et al. 2002). Courses targeting youth and women in handcrafts (e.g., traditional embroidery) and tourism services (e.g., accommodation and foreign language) were provided in the village. Simultaneously, the Bursa Metropolitan Municipality initiated a food-based project to assist the livelihood of Cumalıkızık and some other surrounding villages.

Many women were reluctant to engage in tourism activity for fear of their families and neighbors. With the municipality's encouragement, a group of women founded the Cumalıkızık Village Women's Education, Solidarity and Development Association to organize their efforts in a more structured way. The Association's founder shares their first experience as

> We were 32 women, making gözleme (flatbread with various fillings) and selling at the shopping center on the weekends. Other women from the village came to visit, but they ridiculed us for working here.

Later, the Association members first sold their products at the village center on Sundays, and then with the municipality's encouragement, converted their houses into small food-based businesses. Gradually, other women also joined in, and at the time of writing, the Association has more than 130 members.

The preparation of the Cumalıkızık conservation plans and its introduction to the WHS tentative list made the village well-known among local and national administrators and scholars and professionals. Nevertheless, it did not promptly increase the visitation numbers as the residents or the municipality expected. According to the Küpeli House manager, a heritage house museum organized by UNESCO, approximately 20–30 visitors arrived daily in the village during 2000 (Çetin 2017). The supply and demand side of tourism grew significantly after the production of a popular television series set in the village in 2003 (Çetin 2010). The success of the first food service establishments, combined with the increased visitor activity, prompted other residents to convert parts of their houses into food outlets.

Since the designation of Cumalıkızık as a culturally protected site, much work has been done to conserve and revitalize the village's cultural assets, including historic houses, a mosque, a hammam, a fountain, a cemetery, and three trees (UNESCO

2014). The municipality has upgraded the infrastructure to fit with the community's changing needs (e.g., sewage and water systems). Since 2012, the Governate of Bursa has financed the buildings' conservation costs (Ahunbay et al. 2014).

The inscription of Cumalıkızık as a WHS in 2014 has further increased the spotlight on the village, and the site management unit observed a steady increase in visitor numbers. There is no accurate way of reporting the total number of visitors because statistical data is not systematically collected. However, the rising number of visitors to the Ethnography Museum and the Küpeli House (3000 on the weekends and 500 during the weekdays) indicate the change in tourism activity.

12.4.2 *Tourism-Induced Social and Environmental Changes*

The increased tourism activity in Cumalıkızık transformed the source of the community's livelihood from an agriculture-dependent into a tourism-dependent one. Gradually, as the number of visitors increased, the community decreased their agricultural activity in their gardens and the buffer zone and used less local produce for food production. Animal herding also decreased as the sight of the cows going through the village disturbed the visitors. In the absence of awareness on the importance of biodiversity preservation and its linkages to cultural heritage, local fruit species have started disappearing from the village.

Arguably, the community misses out on an opportunity to offer a unique product to visitors by ignoring local produce's importance in rural tourism. Many visitors to the village complain about the mediocre quality of the food items. Unaware of their nature-based assets, villagers continued to invest in their built capital to increase their tourism capacity.

Just like the changes in their natural environment, the social aspects of the village have also altered. The community's rigid and conservative social structure has transformed into a more open society due to being exposed to outsiders. We observed women as managers of their businesses, working together with their family members, and engaging with diverse visitors while serving at their establishments.

Besides economic empowerment, tourism had several adverse effects on the social structure of the community. According to the residents, rivalry among the business owners is fierce. In our encounters with the food businesses, we also experienced badmouthing and contradictory answers about the origins of food items. Many replied to our inquiries as all the items were produced in-house using local agricultural proceeds in their establishment, although most "other" businesses procured them through wholesalers. A similar narrative regarding the souvenir items is also present in Cumalıkızık. Thus, we understand that the concept of local production is deeply engrained in the minds of the residents who have a material benefit from tourism activity. However, the pressure to keep up with the demand often deviates from this ideal of local production but keeps up the appearance of local authenticity. This widening gap between the expected and the actual in terms of both food and souvenir items contributes to the deterioration of tourism activity itself. The loss of authentic and moral behavior is reflecting negatively on the visitors' experience. Some of the comments from the visitors are as follows:

> Wherever you talk to somebody, they try to sell you breakfast or some other item. (TripAdvisor, International Visitor, February 2020)
>
> We got bored of hearing the same promotional conversations. (TripAdvisor, Domestic Visitor, December 2019)
>
> On every doorstep, there is someone that is shouting for attracting visitors. (TripAdvisor, Domestic Visitor, October 2019)

The greed has overtaken the businesses to the extent that, as reported in local news one of the establishments that hosted 3000 patrons over an extended weekend in June 2019 (overcapacity) caused the death of a person and 50 others to have food poisoning (Tıkır and İnan 2019).

12.4.3 Tourism Offering and Visitor Experiences

Since the 2000s, we observe that tourism development in Cumalıkızık has gradually distanced itself from the village's architectural assets, cultural value, and uniqueness, and become highly commercialized. Many community members have converted the first floors of their houses and inner courts into food businesses, souvenir shops, and a few accommodation facilities. In addition to selling their produce in front of their doorsteps, they had started utilizing the municipality's self-standing stalls. We also observed many unregistered stalls between the streets and the house owners displaying their products on their heritage houses' walls. Specifically, over the weekends, the crowds and the stalls visually impede the houses, making it difficult to view the architectural elements and experience life in the village (see Fig. 12.2).

The visitor profile is diverse, consisting of mostly domestic visitors from Bursa and other surrounding cities. The visitor data for the Ethnography Museum suggest a range of 5–12% foreign visitors for the years 2015–2020 (Bursa Site Management Unit 2020). Based on our field observations that are later confirmed by the Women's Association and the WHS management unit, like the country's changing tourist profile, the more recent international visitors are from the Middle East, replacing Western culture-based tourists. At its current state, the village streets are overcrowded with visitors on the weekends. There are no local craft shops or sales of traditional local products (e.g., Bursa towels or handicrafts made from silk). In the absence of add-on services like guided tours and authentic experiences linked to the local culture's learning, the visitors stay for a short time, and the destination records a low-multiplier effect. We also find that the visitors complain about hygiene issues at food establishments and low value for money.

> It is a pity. Cumalıkızık used to be an authentic village. Now it is all commercialized. (TripAdvisor, Domestic Visitor, February 2020)

Fig. 12.2 Stalls in Cumalıkızık. *Source* Authors

> This place is still in good enough condition to become a wonderful historic village in the right hands where the stories of the past could be celebrated, but instead, it is a giant opportunity overlooked. (TripAdvisor, International Visitor, August 2019)
>
> I came here mainly because it is a UNESCO site and to look at the Ottoman style architecture. This place is over-run by vendors selling the same stuff and restaurants serving the same food. (TripAdvisor, International Visitor, July 2015)

Examples for the loss of authenticity are too many to provide here. However, we observed how the community was surrendering to tourism's commercialization on every trip that we conducted to the village. Whether visitors value these experiences is not clear, but, overall, many culture-based visitors are disappointed. There is no authentic story told about Cumalıkızık's importance in history or the value of its cultural heritage through its tourism offering and its brand (e.g., tours, food, and products) (Taheri et al. 2018). Consequently, television series that are filmed in and around the village become the sole promotional elements emphasized by tour guides in Cumalıkızık.

12.4.4 Governance Arrangements

WHS's planning and management involve multiple stakeholders, including local authorities and governmental entities at the regional and national levels. In Turkey, the Ministry of Culture and Tourism is the highest governmental entity that makes regulations on the conservation of historical, cultural, and natural assets, and the General Directorate of Monuments and Museums maintains, develops, promotes, and evaluates these assets (Akan 2013). The Ministry appoints the WHS managers; however, the site management is often funded by the metropolitan municipalities. In Cumalıkızık's case, the Yildirim Municipality and the Bursa Metropolitan Municipality are the regional administrative entities, respectively. In this context, the site management unit becomes entangled with the political paradigm reigning in the municipalities.

UNESCO representatives are more of an advisory board to the municipal forces and the site management office that undertake the decision-making processes. In addition to the public stakeholders, Cumalıkızık Village Women's Education, Solidarity and Development Association, and Agricultural Development Cooperative are the leading non-profit organizations in the village. A critical framework guiding the conservation and development efforts at Cumalıkızık is the WHS Management Plan (2013–2018), which is to be revised and resubmitted in 2020.

We observe that the stakeholders' efforts before and during the WHS inscription processes have resulted in the Cumalıkızık's community to be aware of their cultural heritage and be engaged in conserving these assets (Fig. 12.3). In each of the heritage preservation projects that we examined for the years between 1998 and 2014, the community was considered an important stakeholder group. The community contributed to the conservation efforts under the supervision of experts and the municipality. Through formal awareness-raising meetings, collaborative workshops, and summer schools, key values and expectations of the community were established by the site management (Dostoğlu et al. 2016; Tas et al. 2009). These consultations

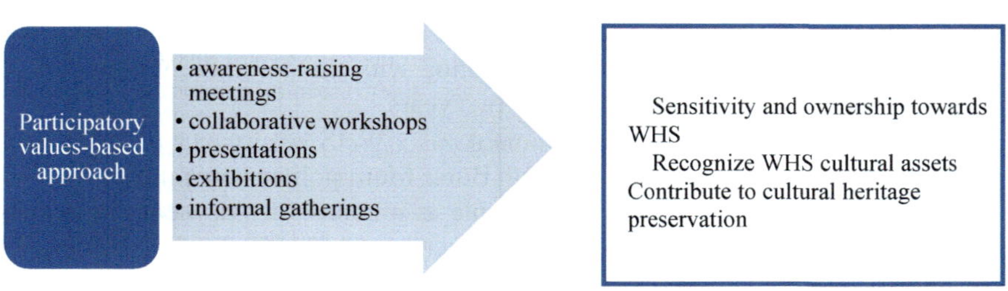

Fig. 12.3 Cultural heritage conservation efforts for community participation *Source* Authors

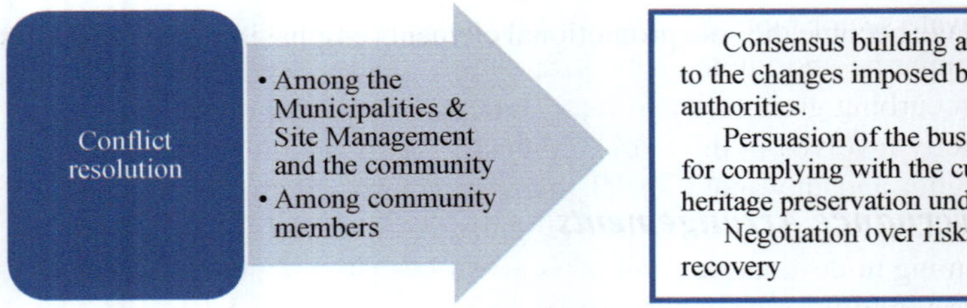

Fig. 12.4 Strategies for conflict resolution in Cumalıkızık. *Source* Authors

with the local community and other stakeholders, constituted the basis for the values-based approach to heritage conservation (Tengberg et al. 2012) in Cumalıkızık. This approach resulted in the management's approval of community members' active use of the heritage structures in Cumalıkızık subsequent to their renovation as long as they did not damage the heritage preservation framework and they were only allowed to make minimal interventions on the structures (Poulios 2014).

Several ways of addressing issues regarding the management of shared resources existed in the community (Fig. 12.4). For instance, when the public authorities converted the Municipality Culture Center to a restaurant, the community objected strongly. They demanded change, and the establishment was converted back to a public space. The Women's Association has taken an active role in the WHS development and implementation phases. The Association acts as a liaison between the community and the public authorities and uses persuasion with the business owners for compliance with shared values.

Our observations suggest that a consensus on how tourism should be developed among the stakeholders is non-existent, and the conflicts are not resolved quickly. Even though the community objects outsiders to own stalls or establishments in the village, some investors have acquired the property in the village for operating food businesses and employ a workforce from other parts of Bursa. The disagreements with these groups are even more, and the Association needs to remind them repeatedly of some common values.

The Association frequently consults with site management over the issues that raise concern. On the other hand, the site management notes that the community is not always receptive to their suggestions for improvement. We find that the site management uses meetings and negotiation techniques to convince the members to improve the tourism offering. For example, to address the inappropriate souvenir sales on the streets, they have been negotiating with the community members to move the stalls to a shared space for over two years.

The key stakeholders involved in Cumalıkızık's WHS governance push for their interests and priorities (Fig. 12.5). For both Bursa Metropolitan and Yıldırım Municipalities, increased visitor activity is desirable as it contributes to their public visibility. Therefore, their recommendations and actions involve increasing the number of visitations rather than supporting an integrated strategy for tourism development.

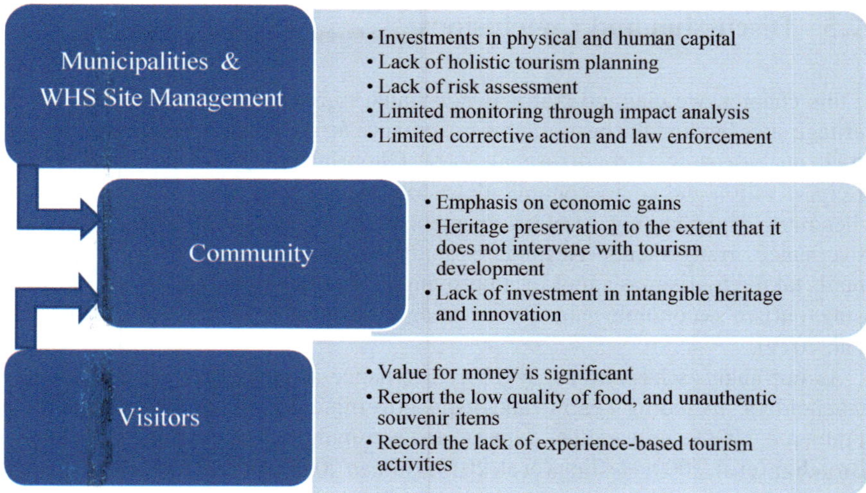

Fig. 12.5 Visitor pressures in Cumalıkızık. *Source* Authors

Furthermore, the municipality is reluctant to enforce regulations on the community since they would not jeopardize their chances to win the elections.

The interests of the site management and the experts rely on the continued protection and display of the physical heritage assets. The WHS Management Plan (2013–2018) does not foresee adverse externalities of tourism development in Cumalıkızık. Although the plan notes certain supportive elements like maintaining the agricultural fields in the buffer zone or food production in hygienic conditions, these aspects are not integrated into a holistic tourism development and management plan for cultural heritage sites.

The community members display a typical example of high economic expectations from tourism development that is also observed in other emerging economies (Bui et al. 2020). Even though the Cumalıkızık community complains about the adverse impacts of visitation and how it lowers their quality of life, they have become dependent and tolerant of visitors. They cannot build consensus on an action plan to regulate tourism activity, and they lack the human and social capital resources for leading change on their own. Unfortunately, the village is led by the community's aspirations at its current state, and the market-based approach has allowed the visitors to have control.

12.5 Discussion and Conclusions

In this chapter, we claimed that it is not viable to recommend one system for all heritage sites but find the best suitable governance arrangement for the rural community in our case study (Brondizio et al. 2009). Our primary argument is that dual goals of conservation and socioeconomic development can be achieved through tourism when integrated planning, local involvement, and education and training support the governance arrangements (Heslinga et al. 2019; Jones et al. 2020). This approach entails taking tourism as a tool for improving the communities' quality of life while using tourism's economic gains for investing in sustainable development (Hatipoglu et al. 2019).

As our analysis has shown, WHS's governance arrangements involve multiple stakeholders, including central and local governmental authorities, chambers of commerce, NGOs, universities, experts, and communities. Confirming prior studies (Brondizio et al. 2009; Heslinga et al. 2019; Human 2015), there are overlaps between central and local agencies, and gray areas exist in governance. Conflicts are inevitable; however, both formal and informal conflict resolution mechanisms are necessary for taking timely action.

Even though all key stakeholders are instrumental in the conservation of cultural heritage, the core community at the living heritage site of Cumalıkızık should continue to stay at the center of governance arrangements. As opposed to the government agencies, the community will be more flexible in dealing with the changing socioeconomic challenges of tourism and therefore respond in more effective ways to the community–heritage–environment interactions (Heslinga et al. 2019). Indeed, the community, which already links cultural heritage conservation to the continuity of their livelihoods, will look for ways of protecting it. However, for a complete transition to a "living heritage approach," where experts take a secondary role, the community members need to be fully empowered (Poulios 2014). In Cumalıkızık, empowerment can be enhanced through investments into the community members' human and social capital by specifically targeting culture-based tourism development. Furthermore, the community should be committed to preserving and connecting to their intangible heritage, which has been missing until now. In this vein, it is in the residents' best interest if public authorities can create opportunities for uncovering and preserving the village's intangible heritage.

The co-management processes in Cumalıkızık shared between the public authorities (Municipality and WHS Site Management), and the community is not functioning effectively. Overcrowding and social and environmental disturbances in the village are symptoms of a failing governance system (Higgins-Desbiolles et al. 2019). Ideally, this governance arrangement should enable both partners to learn collectively about the impacts of change (e.g., visitor pressures) and take corrective action by trial and error (Plummer and Fennell 2009). Learning to live with change, developing adaptability and resistance would help this SES function and survive (Ruiz-Ballesteros 2011). However, in its current trajectory, the community fails to establish an adaptive system and thus, gradually loses its identity.

Our findings suggest that the community is slow in accepting change, but once a change occurs, they want to hang on to the new status rather than develop strategies to deal with the ongoing change and uncertainty. Opposition to experts' suggestions for improvement is not because the community can self-organize and find better solutions, but due to fear of losing what they think they have. The community cannot develop internal control mechanisms that are essential for monitoring collective resources (Ruiz-Ballesteros 2011). Thus, the lack of consensus on how tourism should be developed and monitored slows down the community for taking corrective action and display timely resistance to visitor pressures. The co-management tools that the partners are utilizing like consensus building, persuasion, and negotiation are promising and should be further supplemented with increased sharing of knowledge and decision-making, and collective learning (Brondizio et al. 2009; Islam et al. 2018; Plummer and Fennell 2009).

As opposed to other academic studies, we will not be supporting a self-governance arrangement for Cumalıkızık at this stage, especially since the absence of structured monitoring mechanisms and intervention policies may have an adverse impact on the integrity and authenticity of the cultural heritage assets. We acknowledge that the absence of corrective action risks WHS's future and other possible inscriptions in the country. We agree with Koens et al. (2018) that public authorities should take responsibility and be more accountable for developing tourism activity.

Over time, stakeholders have presented tourism as a panacea for socioeconomic development in the village, without supporting the community members with tools to assess risks and build mechanisms for coping with change. As such, members have considered increased tourism activity as the final goal and have not planned for investing in community resources such as collective learning, culture preservation, new products, or services. Without product and service innovation that balances authenticity and interpretation, sustaining culture-based tourism will be at risk (Heslinga et al. 2019).

Tourism has a low-multiplier effect in Cumalıkızık, and it is spiraling down under the pressures of the market. Thus, considering the socioeconomic challenges of the WHS, various suggestions can be made for improvement. An agricultural revitalization strategy to support food-based tourism is essential, and when combined with localism and product differentiation, it can be a good starting point for the community. We understand that agriculture is not economically viable for the community anymore, and they will not readily engage. At this stage, the public authorities need to intervene, persuade the community by developing an understanding of holistic tourism development, and provide technical and financial support framework by working together with experts from universities and non-profit organizations.

Our analysis suggests a need to regain the village's authentic tourism experience to attract and satisfy culture-based visitors. Brand heritage can be strengthened via product and service innovations reflecting the cultural heritage of Cumalıkızık and providing tour guides with historical information (Taheri et al. 2018). Most importantly, the Cumalıkızık brand needs to be communicated so that the visitors' experience at the site complements its purported values (Taheri et al. 2018, p. 68). Evaluation of the benefits accruing to the stakeholders through a systematic impact assessment will guide the co-management partners for effective interventions (Hatipoglu et al. 2019). Smart technologies can follow visitor activity in all six WHS components and obtain reliable visitation data (Mandić 2019). When these efforts are combined with visitor segmentation analysis, public administrators can better control demand and supply at the WHS (Mandić 2019).

This chapter makes several theoretical contributions. We have brought together social-ecological systems and living cultural heritage sites and contributed to our understanding of governance arrangements at tourism destinations. Empirical evidence from our case study challenges the emerging literature on the advised self-governance models for rural communities by revealing the entangled social and political relationships that are often observed during the community–heritage–visitor interactions in living heritage destinations. We suggest that these complex set of relationships and the barriers for adaptive co-management structures can be better understood through a historical analysis of the community, including their core values, changing lifestyles, and involvement in conservation and tourism development. Our analysis confirms that adaptive co-management structures will be necessary for the community to tackle social and physical disruptions. Overall, the site's long-term success will rely on how the system controls these disruptions and determines its fate (Brondizio et al. 2009; Ruiz-Ballesteros 2011). Consequently, co-management structures need to be underpinned by holistic and integrated tourism planning, risk assessment, and law enforcement rather than leaving the WHS to be shaped by the market. Regarding the living heritage approach, evidence from our research emphasized empowerment, capacity improvement, and full commitment to tangible and intangible heritage for the community.

We acknowledge several limitations in this chapter. We have chosen to use a single case study, which is problematic for representativeness. A multiple case research design would allow us to draw parallels with the governance challenges that other rural communities are experiencing (Eisenhardt and Graebner 2007). Comparisons with urban heritage sites and cross-country studies can further broaden our perspective, confirm or deny the emergent findings, and improve our research's transferability. Nevertheless, the chapter presents empirical insights into the WHS governance processes for managing shared resources and offers ways of improving them with stakeholder participation. The chapter opens the avenue for future research examining the linkages between the WHS management plans, risk assessment, and tourism planning and development. Furthermore, the inclusion of tourism development experts in the WHS inscription processes, including the preparation of management plans, should be examined. Finally, we still need more research on how tourism's economic gains can be used to conserve cultural heritage, improve the quality of life of the communities, and take steps toward sustainable development simultaneously.

Acknowledgements An earlier version of this chapter was presented at the International Conferences on Tourism (ICOT) 2018 in Kavala, Greece. We would like to acknowledge the support of Bursa WHS Management Office and Cumalıkızık village community members in our data collection.

References

Adie BA, Falk M, Savioli M (2020) Overtourism as a perceived threat to cultural heritage in Europe. Curr Issues Tourism 23(14):1737–1741

Ahunbay Z, Ayrancılar T, Polat A, Uray A (2014) Conservation of the vernacular heritage in the villages of Bursa, Turkey. In: Vernacular architecture: towards a sustainable future. CRC Press, pp 59–64

Akan TV (2013) Bursa (Khans Area—Sultan Complexes) and Cumalıkızık Management Plan (2013–2018). Bursa Metropolitan Municipality, Bursa

Bramwell B, Lane B (2011) Critical research on the governance of tourism and sustainability. J Sustain Tourism 19(4–5):411–421

Brondizio ES, Ostrom E, Young OR (2009) Connectivity and the governance of multilevel social-ecological systems: the role of social capital. Ann Rev Environ Resour 34:253–278. https://doi.org/10.1146/annurev.environ.020708.100707

Bryman A (2012) Social research methods. Oxford University Press, New York, NY

Bui HT, Jones TE, Weaver DB, Le A (2020) The adaptive resilience of living cultural heritage in a tourism destination. J Sustain Tourism 28(7):1022–1040

Bursa Site Management Unit (2020) Ethnography museum visitor data. Personal communication

Çahantimur AI, Öztürk RB (2020) A participatory approach to the sustainable development of an historic landscape; management plan proposal for Hisar district, Bursa/Turkey. Hist Environ: Policy Pract, 1–34

Çetin T (2010) Cumalıkızık köyünde kültürel miras ve turizm algısı (perception of cultural heritage and tourism in Cumalıkızık village). Milli Folklor 22(87):181–190

Çetin O (2017) Cumalıkızık'ın gözdesi 'Küpeli Ev' (The favorite 'Küpeli House' in Cumalıkızık). Retrieved from http://www.gazetebursa.com.tr/magazin/cumalikizikin-gozdesi-kupeli-ev-h17274.html

Chandani K, Karuppannan S, Sivam A (2019) Assessing the values of living heritage sites in Kathmandu Valley: a community perspective. J Cult Heritage Manage Sustain Dev 9(1):93–110

Cheer JM, Milano C, Novelli M (2019) Tourism and community resilience in the Anthropocene: accentuating temporal overtourism. J Sustain Tourism 27(4):554–572

Cheung KS, Li L-H (2019) Understanding visitor–resident relations in overtourism: developing resilience for sustainable tourism. J Sustain Tourism 27(8):1197–1216

Cochrane J (2010) The sphere of tourism resilience. Tourism Recreation Res 35(2):173–185

Creswell JW (2014) Research design: qualitative, quantitative, and mixed methods approaches. Sage Publications Ltd, Thousand Oaks, CA

Dostoğlu N, Başdemir A, Durmaçalış B, Çalıkuşu E (2016) Awareness raising process in World Heritage Site Bursa. TÜBA-KED Türkiye Bilimler Akademisi Kültür Envanteri Dergisi 14:209–221

Eisenhardt KM, Graebner ME (2007) Theory building from cases: opportunities and challenges. Acad Manage J 50(1):25–32

Folke C, Hahn T, Olsson P, Norberg J (2005) Adaptive governance of social-ecological systems. Ann Rev Environ Resour 30(8):441–473

Hatipoglu B, Ertuna B, Salman D (2019) Corporate social responsibility in tourism as a tool for sustainable development. Int J Contemp Hospitality Manage

Heslinga J, Groote P, Vanclay F (2019) Strengthening governance processes to improve benefit-sharing from tourism in protected areas by using stakeholder analysis. J Sustain Tourism 27(6):773–787

Higgins-Desbiolles F, Carnicelli S, Krolikowski C, Wijesinghe G, Boluk K (2019) Degrowing tourism: rethinking tourism. J Sustain Tourism 27(12):1926–1944

Huang C-H, Tsaur J-R, Yang C-H (2012) Does world heritage list really induce more tourists? Evidence from Macau. Tourism Manage 33(6):1450–1457

Human H (2015) Democratising world heritage: the policies and practices of community involvement in Turkey. J Soc Archaeol 15(2):160–183

Islam MW, Ruhanen L, Ritchie BW (2018) Tourism governance in protected areas: investigating the application of the adaptive co-management approach. J Sustain Tourism 26(11):1890–1908
Jones TE, Bui HT, Ando K (2020) Zoning for world heritage sites: dual dilemmas in development and demographics. Tourism Geographies, 1–23. https://doi.org/10.1080/14616688.2020.1780631
Koens K, Postma A, Papp B (2018) Is overtourism overused? Understanding the impact of tourism in a city context. Sustainability 10(12):4384
Landorf C (2009) A framework for sustainable heritage management: a study of UK industrial heritage sites. Int J Heritage Stud 15(6):494–510
Larsen RK, Calgaro E, Thomalla F (2011) Governing resilience building in Thailand's tourism-dependent coastal communities: conceptualising stakeholder agency in social–ecological systems. Glob Environ Change 21(2):481–491
Li J, Krishnamurthy S, Roders AP, van Wesemael P (2020) Community participation in cultural heritage management: a systematic literature review comparing Chinese and international practices. Cities 96:102476
Lincoln YS, Guba EG (1985) Naturalistic inquiry. Sage, Beverly Hills, CA
Mandić A (2019) Nature-based solutions for sustainable tourism development in protected natural areas: a review. Environ Syst Decisions 39:249–268
Mihalic T (2020) Conceptualising overtourism: a sustainability approach. Ann Tourism Res 84:103025
Mihalič T, Kuščer K (2018) Impacts of overtourism on satisfaction with life in a tourism destination. Paper presented at the Submitted to Travel and Tourism Research Association's (TTRA 2019) European Chapter Conference, Bournemouth, Ljubljana
Orbaşli A (2000) Is tourism governing conservation in historic towns? J Architectural Conserv 6(3):7–19
Oren U, Woodcock DG, Var T (2002) Sustainable tourism development: a case of Cumalikizik, Turkey. Tourism Anal 6(3–4):253–257
Ozorhon G, Ozorhon I (2014) Learning from Mardin and Cumalıkızık: Turkish vernacular architecture in the context of sustainability. Arts 3(1):175–189. https://doi.org/10.3390/arts3010175
Pearsall M (1970) Participant observation as role and method in behavioral research. In: Filstead WJ (ed) Qualitative methodology: firsthand involvement with the social world. Markham, Chicago, pp 340–352
Pekerşen Y, Güneş E, Selçuk B (2019) Kültürel miras turizmi değerlerinin korunması ve sürdürülebilirliği kapsamında yerel halkın tutumu: Cumalıkızık örneği (The attitude of local people in the scope of the protection and sustainability of cultural heritage tourism values: the case of Cumalıkızık). Türk Turizm Araştırmaları Dergisi 3(3):350–368
Plummer R, Fennell DA (2009) Managing protected areas for sustainable tourism: prospects for adaptive co-management. J Sustain Tourism 17(2):149–168
Poulios I (2014) Discussing strategy in heritage conservation. J Cult Heritage Manage Sustain Dev 4(1):16–34
Ruiz-Ballesteros E (2011) Social-ecological resilience and community-based tourism: an approach from Agua Blanca, Ecuador. Tourism Manage 32(3):655–666
Schneider E, Esin Ş (2000) Turkey "Bursa Local Agenda 21 Cumalikizik Conservation and Revitalization 98" project. Bursa Metropolitan Municipality, Bursa
Seraphin H, Sheeran P, Pilato M (2018) Over-tourism and the fall of Venice as a destination. J Destination Market Manage 9:374–376
Taheri B, Farrington T, Curran R, O'Gorman K (2018) Sustainability and the authentic experience. Harnessing brand heritage—a study from Japan. J Sustain Tourism 26(1):49–67
Tas M, Tas N, Cahantimur A (2009) A participatory governance model for the sustainable development of Cumalıkızık, a heritage site in Turkey. Environ Urbanization 21(1):161–184
Tengberg A, Fredholm S, Eliasson I, Knez I, Saltzman K, Wetterberg O (2012) Cultural ecosystem services provided by landscapes: assessment of heritage values and identity. Ecosyst Serv 2:14–26

Tıkır EF, İnan M (2019) Köy kahvaltısından zehirlendikleri öne sürülen 3 kişiden 1'i öldü (1 person dead among 3 suspected of poisoning at village breakfast). Retrieved from https://www.dha.com.tr/yurt/koy-kahvaltisindan-zehirlendikleri-one-surulen-3-kisiden-1i-oldu/haber-1659588

UNESCO (2013). Managing cultural world heritage: resource manual. Retrieved from http://whc.unesco.org/en/managing-cultural-world-heritage/

UNESCO (2014). Bursa and Cumalıkızık: the birth of the Ottoman Empire. Retrieved from http://whc.unesco.org/en/list/1452

Chapter 13
Aspects of Intensive Short Day Trips on the Remote Pelagic Location: Incorporating Visitor Survey, Radar Tracking and Environmental Risk Analysis in NATURA 2000 Monitoring of the Blue Cave, the Island of Biševo

Hrvoje Carić and Ivo Beroš

Abstract The NATURA 2000 sites in the Adriatic Sea are large and complex areas with impaired and limited conservation implementation. Seasonally they are exposed to the intensive maritime traffic with significant day trip pressures visiting most attractive and sensitive locations. Inadequate management and planning of tourism and conservation often results in both diminished attractiveness for visitors and increased risks to the protected ecosystems. The location of case study presented is a remote attraction on a pelagic island. Blue Cave's scenic beauty attracts day trips that have increased by almost seven folds in the last 15 years. This sudden boost caught the local economy in an overtourism spiral with risks being more evident as the visitor numbers grow. Risks are spatio-temporal, taking place in a very limited area and in narrow time frames. This chapter investigates aspects of intensive short-term visits (a couple of hours of sightseeing, field trips, excursions) to the sea cave as a significant regional and international attraction. In order to disclose a potential framework, the authors have used tools and methods across the following disciplines: maritime traffic monitoring, environmental risks analysis, visitor crowding perception. Authors are confident that described framework allows for more informed management and decision-making and urges for protected area management to be open to interdisciplinary research practices.

Keywords Maritime traffic monitoring · Environmental risks analysis · Visitor crowding perception · Sea cave · NATURA 2000 · Heat mapping

H. Carić (✉) · I. Beroš
Institute for Tourism, Zagreb, Croatia
e-mail: Hrvoje.Caric@iztzg.hr

I. Beroš
e-mail: ivo.beros@iztzg.hr

A. Mandić and L. Petrić (eds.), *Mediterranean Protected Areas in the Era of Overtourism*, https://doi.org/10.1007/978-3-030-69193-6_13

13.1 Introduction

How many visitors and what level of recreational use can a protected area accept without undermining its biodiversity as the core value, and maintain its long-term preservation? The question was initiated more than 90 years ago in the USA's nature conservation and management (Sumner 1942, according to Manning 2007) and have evolved significantly through policy and scientific literature since.

Trends of tourism growth are marked with the increasing demand of visits to the protected areas thus resulting with pressures to develop various visitor facilities and infrastructure that in return often both increase risks of damage to the natural values and diminish visitors' experiences (Eagles et al. 2002; Gössling and Peeters 2015; IVUMC 2016). Decades of everlasting tourism growth paradigm in the Mediterranean have generally been ignoring issues of carrying capacity and overtourism (Manning 2007; Mrđa et al. 2014; Peeters et al. 2018) that are not adequately monitored and therefore hard to manage (Carić and Jakelić 2018). The relationship between perceived crowding and visitor satisfaction relates directly to visitor satisfaction (Buckley 2009; Saveriades 2000). The influence of perceived crowding on visitor satisfaction is widely researched in the USA, and Europe (Arnberger and Mann 2008; Kalisch and Klaphake 2007; Luque-Gil et al. 2018; Manning et al. 2009; Shelby et al. 1989; Vaske and Shelby 2008; Zehrer and Raich 2016) and the results of those studies reveal that crowding has a negative impact on satisfaction. To manage the disturbance of visitors due to crowding, managers need to know how the use levels affect perceived crowding. The visual approach, founded on image-based questionnaires, is the suggested method to determine crowding norms (Manning et al. 1996). The method is often used for visitor management in protected areas (Kalisch and Klaphake 2007; Needham et al. 2004). Another approach is to predict the level of disturbance by use of data gathered from measured use levels and time-stamped questionnaires (Klanjšček et al. 2018).

When it comes to the management of main attractions in the Mediterranean islands, the priority is optimising visitor flows. However, what happens when the protected area is an attraction integral to a destination, as often is a case in such a setting. This chapter is looking into a sensitive marine area that attracts short day trips with potential overcrowding effects. It is based on the recent experience of establishing an action plan for the EU funded infrastructure on Biševo island with a focus on the Blue Cave visitor management issues. During this process, the authors have worked in collaboration with a public company,[1] the regional NATURA management authority,[2] and the local government.[3] The process led to an innovative use of the following analytical tools aimed to perceive potential risks of overcrowding more clearly:

- maritime traffic monitoring (the radar tracking),

[1] Nautički centar Komiža http://www.komiza.hr/nauticki-centar/.

[2] Javna ustanova More i Krš http://www.dalmatian-nature.hr/.

[3] City of Komiža http://www.komiza.hr/.

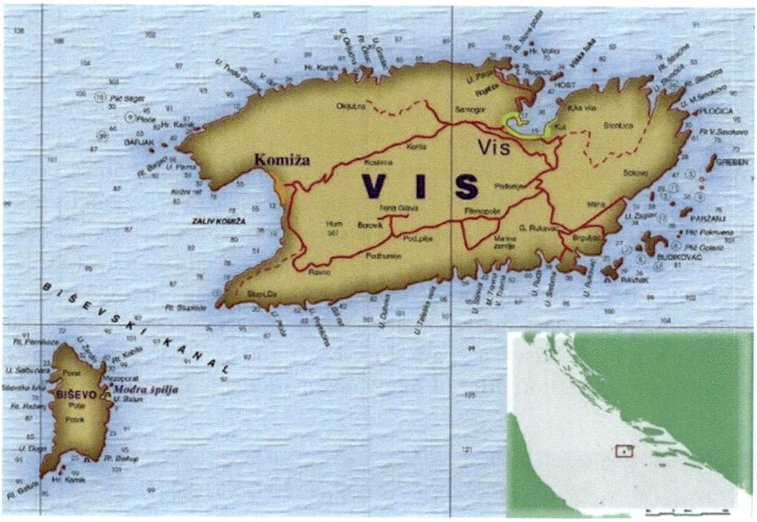

Fig. 13.1 Island Biševo and Vis *Source* Author. Fieldwork 2018

- environmental risks analysis (standardised NATURA impact analysis),
- visitor crowding perception (the survey).

13.2 The Case Study

Biševo is a small (5,9 km2) remote island with many abandoned settlements and a long history. It is located south east-off the island of Vis and is considered to be one of the most remote islands in Croatia (Fig. 13.1.). It is rich in biodiversity and is a part of the ecological network with four NATURA 2000 sites[4] with two geomorphological phenomena: Blue Cave, and the Sea monk cave.

The NATURA 2000 sites are managed by a public institution that lacks resources and, at the moment, does not have a formal management plan. On the national and systemic level, problems of non-conformity were cited by European Commission in 2016, and as a result, the Croatian authorities have improved legislative framework, but the overall state of NATURA management is still substandard (EC 2019). Hence, big and complex areas with substandard conservation implementation pose a risk for ecosystems when faced with overtourism and lack of proper tourism management and planning. Mediterranean climate, the attractiveness of the Blue cave, and remote access limit Biševo's tourism offer to short visits and day trips or excursions leaving limited economic impacts with noticeable environmental risks.

[4]Natura codes HR2001097, HR3000098, HR3000469, HR1000039.

The Blue Cave is located on the eastern side of the island of Biševo, near a small bay Mezzoporat that has docking and anchoring area. The cave has one smaller opening used for entering point for a visit with a small boat. The second opening is submerged and located on the south side and is responsible for the phenomena. The sunlight passes through the sea and then reflects from the bottom of the cave creating attractive blue illumination of the sea and silver illumination of the submerged objects (Fig. 13.2). This enchanting play of light has made the cave internationally attractive, especially in the era of fast travel, internet and social media.

Visits to the cave are controlled and organised by a licensed company (NC Komiža) and provide significant income for the small town of Komiža and the island of Vis through the ticket sales and the local site-seeing offer. However, due to the rapid growth of tourism in the last ten years (Fig. 13.3), the excursions to the Blue cave become part of the excursion offer of many destinations in the wider region.

Fig. 13.2 Interior of the Blue Cave (*Photo credit: Zoran Jelača*)

Fig. 13.3 The number of tickets sold for the entrance to the Blue Cave for the Period 2004–2019 *Source* NCK 2019

Due to differences in the home ports, the maritime transfer of visitors to the Blue Cave can vary from 20 min to 3 h. The waiting time at the dock in the Mezoporat before the entering cave depends on the crowding and seasonality and can vary 10 min to 90 min. The average stay in the Blue cave is 5 to 10 min but can be longer if circumstances allow (Figs. 13.4, 13.5 and 13.6).

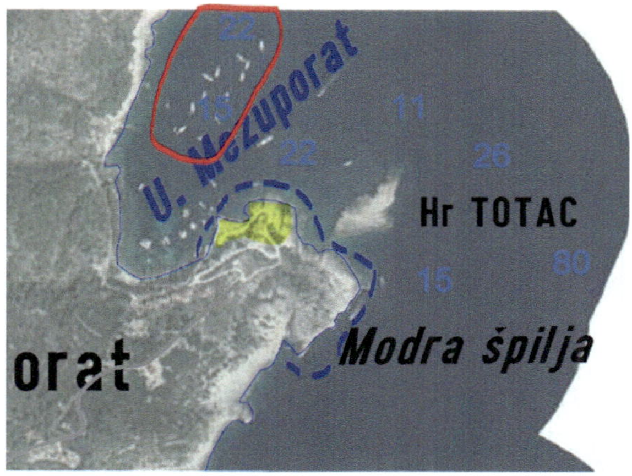

Fig. 13.4 Description of the Blue Cave visitation (The yellow—the dock and the waiting area for the transfer to and from the Blue Cave. The blue line—Blue Cave touring boat route. Red—the anchorage area for visitors arriving by sailboats and yachts.) *Source* Author. Fieldwork 2018

Fig. 13.5 Tour Boat Exiting the Cave (*Photocredit: Hrvoje Carić*)

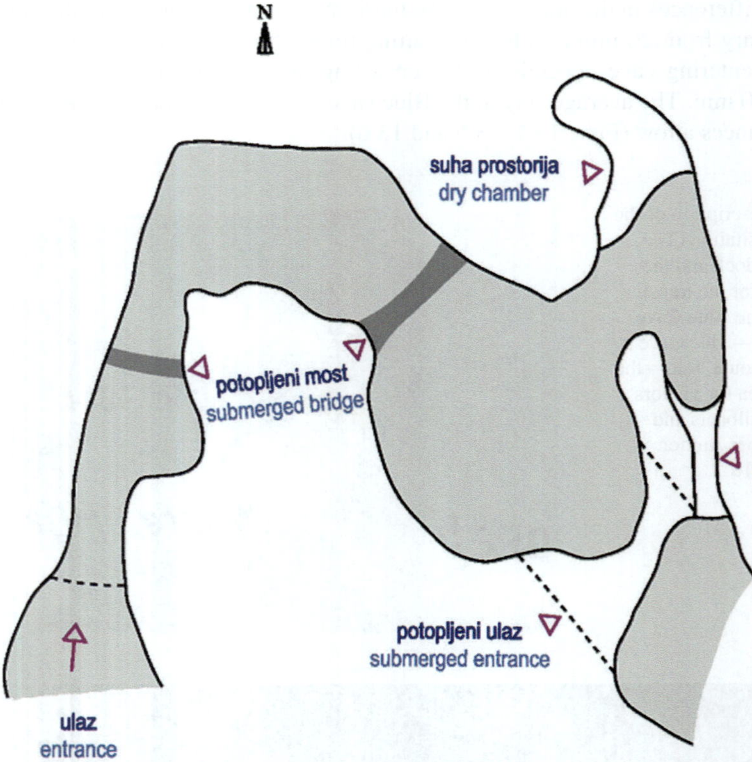

Fig. 13.6 Approximate plan of the Blue Cave (Overall dimensions 40 m x 40 m. Diameter of the main chamber 15 m.) *Source* H. Carić Fieldwork 2018

13.2.1 Maritime Traffic Monitoring

Insights into the spatial and temporal distribution of vessels for part of the Biševo island surrounding area was made with VTMIS (Vessel Traffic Monitoring and Information System) and CIMIS-a (Croatian Integrated Maritime Information System[5]) that enables precision monitoring of vessels in the VTS (Vessel Traffic Service). Croatian Ministry of the Sea, Transport and Infrastructure—Maritime Safety Directorate is the managing authority of named systems and have kindly provided data for Maritime traffic density and Passage line counting presented below.

Maritime traffic density heat map was produced for the given area and a time interval, based on the data from the Radar tracking and AIS (Automatic Identification System[6]). Each AIS and radar signal is represented by a pixel that is imprinted on the

[5]CIMIS is based on Directive 2010/65/EU and related establishment of a Maritime Single Window Interface.

[6]Official system used by authorities to track the traffic in the real time.

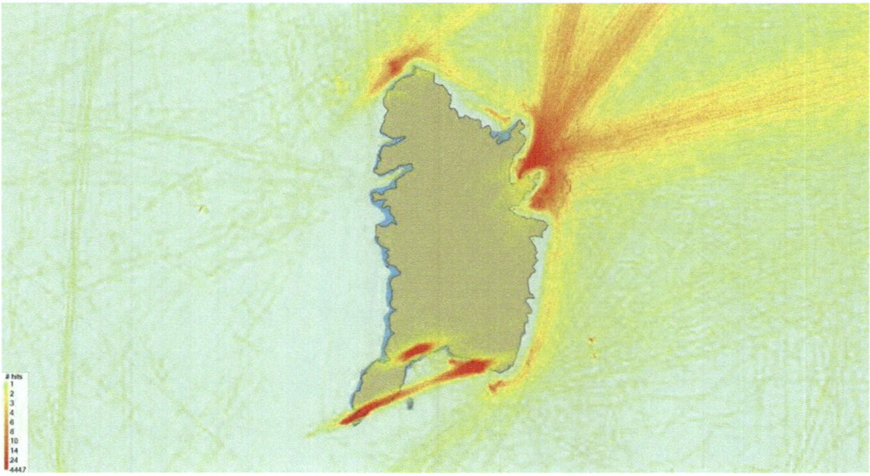

Fig. 13.7 Heat Map of Maritime Traffic Density for One Month (August 2018) in Biševo Waters (Longitude: 43°00,11'N—42°56,85'N; Latitude: 15°56,52'E—16°04,43'E) *Source* Croatian Ministry of the Sea, Transport and Infrastructure—Maritime Safety Directorate, 2018

associated location on the map (Fig. 13.7). In this way, pixel multiplication produces a gradation of colour from yellow to red, as it is visible on the scale in the corner.

The above heat mapping allows a better understanding of corridors and related environmental impacts that are taking place in the NATURA 2000 ecological network. Since the red areas and lines intensify as traffic intensifies, it can be said that emissions and risks from the boat traffic intensify in the same manner. Engine exhausts, underwater noise, marine litter, biocides, bilge and oil discharges are making impacts that are not currently monitored, but with the above heat map, it is possible to assume areas exposed to the highest risks.

Passage line counting was produced by the VTS service. Based on pre-defined lines, the system counts crossings and directions are presented in Fig. 13.8.

During the period analysed above (August 2018), the total number of crossings was 7.232, and the number of sold tickets was 50.257—most of the arrivals were from the north (direction Split). This data can be produced for day to day management and provide valuable information on traffic directions and intensities across different time units (hours, days, weeks).

13.2.2 Environmental Risks Analysis

Tourism benefits are extensively recognised in the literature as providing income for nature conservation, income to the local community, and well-being benefits to the visitors (Dolnicar et al. 2013). The Millennium Ecosystem Assessment (MEA

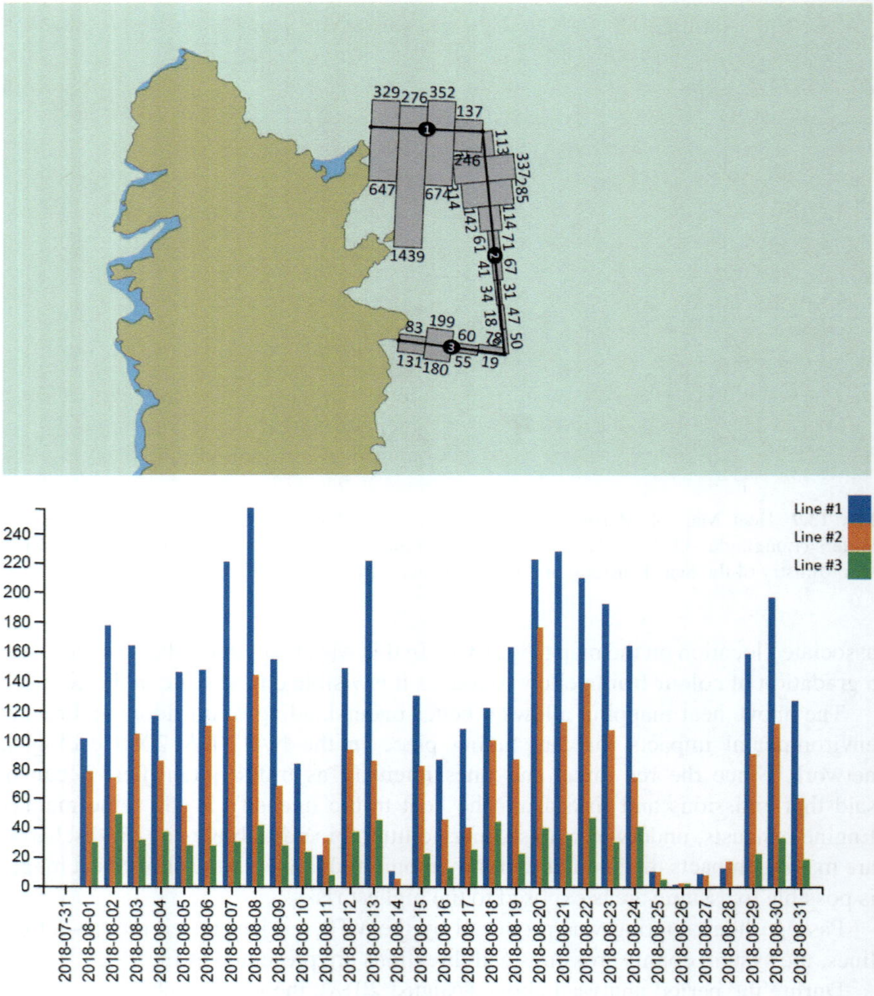

Fig. 13.8 Map and Graph: Number of passengers and directions of vessels for three lines during 1 to 31 August 2018 in the waters surrounding the Blue Cave *Source* Croatian Ministry of the Sea, Transport and Infrastructure—Maritime Safety Directorate, 2018

2003) promoted the idea of non-material benefits of nature" or cultural ecosystem services, emphasising the link between environmental health and human health that is measured in tourism through visitor satisfaction. Environmental risk induced by human activities is a threat of effects on living organisms and the environment by, among others, emissions, wastes, and resource depletion. Therefore, the analysis bellow is offering an approach in recognising human-induced pressures in the context of nature as the tourism resource and attraction.

The four Natura 2000 sites have been assessed in relation to existing threats in accordance with the formal Reference List of Threats, Pressures, and Activities.[7] There are 22 threats and pressures disclosed for the sites organised in 9 thematic areas: A-Agriculture, D-Transportation, E-Urbanisation, F-Biological resource use, G-Human intrusions & disturbances, H-Pollution, I-Invasive taxa, J-Natural System modifications, M-Climate change. The Table 13.1 represents the overview of formally recognised negative impacts and those that were disclosed as emerging in the recent period (marked with shaded orange box and red-letter P). The last official update of the state of the threats and pressures was conducted in 2015, and the Table 13.1 was amended with potentially emerging issues as the result of the 2018 and 2019 fieldwork of the first author.

Since Blue cave is part of the more comprehensive marine and island geological and biological system, it is crucial to consider the ecological footprint that day trips generate. Therefore, the Table 13.1 was amended with P (shaded orange box) as potential and emerging threats, pressures and impacts. The most significant environmental threat and pressure is the travel to the Blue cave that can last from 10 min. up to 3 hours. Modes of travel are speed boats of various sizes (carrying 10 to 30 passengers), former fishing boats and similar slower boats (carrying 20 to 100 passengers). Most significant impact related to this mode of travel is underwater noise and engine air emissions, but other relevant impacts are marine litter, bilge and oil/gas spills, uncontrolled anchoring in *posidona oceanica* and other sensitive habitats.

Threats and pressures the Blue Cave visitations relate to the cave micro-environment and visitor health and are:

- Damage to the walls and sea bottom of the cave
- Changes in the microclimate of the cave and air content
- Changes in microbiology (invasive species)
- Stability of the cave due to limestone wear
- Climate change
- Littering

Initially, the threats and pressures must be attempted to be formed into quantifiable parameters and indicators of potential impacts. Therefore, the framework of Blue Cave monitoring methodology should include the following:

1. Physical parameters—exploring all available cave space and channels (record damage to the rocks, breaking the rocks, vandalism, waste, etc.).
2. Habitat parameters—developing checklist of previously recorded organisms and sampling; sampling cave walls (*Bacteria, Cyanobacteria, Protozoa, Fungi, Algae*, etc.).
3. Microclimate and air content parameters: water and air temperatures, relative humidity, airflow, carbon dioxide, particles, VOC/engine exhaust gases.
4. Monitoring biodiversity of the sea with particular attention to the endangered and allochthon invasive species. Monitoring of the seawater parameters.

---

[7] Eionet: Central Data Repository http://cdr.eionet.europa.eu/help/natura2000.

Table 13.1 Recognised and Emerging Negative Impacts of four NATURA Sites related to Biševo Island with Description and Ranking (Ranking: H = high, M = medium, L = low, P = potential, emerging)

Threats and pressures code	Negative impacts	Biševo mainland HR2001097	Biševo sea HR3000098	Vis Archipelago HR3000469	Remote islands HR1000039
	A. Agriculture				
A02	modification of cultivation practices				L
A04.03	abandonment of pastoral systems, lack of grazing				M
	D. Transportation and service corridors				
D01.01	paths, tracks, cycling tracks	L			
D03.02	shipping lanes		P	L1	P
	E. Urbanisation, residential and commercial development				
E01.04	other patterns of habitation				L
E03	discharges		P	L	P
E03.01	disposal of household / recreational facility waste		L		
E03.04	other discharges		L		
	F. Biological resource use other than agriculture & forestry				
F02	fishing and harvesting aquatic resources		M	M	L
F04.02.02	hand collection	H			
F05	illegal taking/ removal of marine fauna		M		
	G. Human intrusions and disturbances				
G01	outdoor sports and leisure activities, recreational activities	M	P	P	L
G01.01	nautical sports		M	P	
G01.01.01	motorised nautical sports		P	M	
G01.07	scuba diving, snorkelling		M		
G02	sport and leisure structures	M	P	P	
	H. Pollution				
H03.03	marine macro-pollution (i.e. plastic bags, styrofoam)		P	M	
H05.01	garbage and solid waste	L	L	M	
H06.01	noise nuisance, noise pollution		P	M	
	I. Invasive, other problematic species and genes				
I01	invasive non-native species		H		
	J. Natural System modifications				
J01.01	burning down	M			
	M. Climate change				
M01.01	temperature changes (e.g. rise of temperature & extremes)	P	L	P	P

Source Natura Standard Data Form, http://natura2000.eea.europa.eu/

5. Development of guidelines and recommendations based on monitoring that deal with the protection of cave and visitor safety.

13.2.3 Crowding Perception of the Visitors (the Visitor Survey)

Perceived crowding is a value judgement on the number of people an individual encounters (Shelby et al. 1989) that relates directly to customer satisfaction (Buckley 2009; Saveriades 2000). The research on recreation conflicts and crowding in national parks, forests, and other outdoor settings is widely spread in the USA (Arnberger and Mann 2008; Vaske and Shelby 2008; Shelby et al. 1989; Manning et al. 2009). The relationship between perceived crowding and visitor satisfaction is also researched in Spanish protected area (Luque-Gil et al. 2018), for a winter sports outdoor setting (Zehrer and Raich 2016) and German and New Zealand National Parks (Kalisch and Klaphake 2007; Ryan and Cessford 2003).

The visitor survey was presented here was conducted during the period from April to October 2019. A total of 778 questionnaires were completed. A personal interview with visitors of Island Biševo and Blue Cave is used as the data collection method. The survey was initially developed in Croatian and then translated into English, German and Italian versions as well. The survey questionnaire consisted of 49 questions encompassing: the socio-demographic characteristics of visitors, characteristics of visits, visitors' knowledge of special areas on Biševo, their perception of crowding, their perception of key dissatisfaction factors, and satisfaction with elements of tourism supply. Authors intended to confirm that perceived crowding has a negative impact on visitors' satisfaction as well as to determine socio-demographic and visitation characteristics that cause a negative perception of visitors.

The study applied a random sampling of visitors who arrived as a part of an organised excursion from nearby Komiža and Vis (short trip, 33.1%) or other ports (long-trip, 52.5%), and others, who arrived independently (14.4%); of these, 53% were female and 47% were male. Most of them belonged to the age groups 26–35 (26.1%) and 18–25 (24.7%). One-quarter came from Croatia, followed by the United Kingdom (12.1%) and United States (8.9%). The others were distributed among 23 other countries. Most questionnaires were collected during the high-season (July and August, 53.6%), 26.9% of questionnaires were collected during the shoulder season (June and September), and the rest were collected during low-season (April, May and October).

In the study, perception of crowding was measured on the well-known nine-point Likert scale (Heberlein and Vaske 1977) which is used in much research in the United States (Manning 2007; Shelby et al. 1989), Canada (Needham et al. 2004) and Europe (Arnberger and Mann 2008). Furthermore, responses 1 and 2 are gathered as "Not at all Crowded", responses 3 and 4 as "Slightly Crowded", responses 5 to 7 as "Moderately Crowded" and responses 8 and 9 as "Extremely Crowded". Most visitors (46.8%) perceived crowding as "Moderately Crowded", followed by those who perceived crowding as "Extremely Crowded" (21.7%). The other visitors perceived crowding as "Slightly Crowded" (14.4%) and "Not at all Crowded" (17.7%). All data were processed using the statistical programme SAS. To find factors influencing perceived crowding Chi-squared test of independence for categorical variables (Agresti 2007) and Cramer's V measure for the effect size of the Chi-squared test (Cohen 1988) were used. The results revealed that there was a small to medium level of association between age and the perception of crowding ($p < 0.0001$, Cramer's V = 0.14, d.f. = 3). The variables associated with the perception of crowding were season (medium to high, $p < 0.0001$, Cramer's V = 0.27, d.f. = 2), and duration of stay (small to medium, $p < 0.0001$, Cramer's V = 0.14, d.f. = 3) but the visitor type was not associated with the perception of crowding.

The critical factors of dissatisfaction examined in this survey were "Crowding at the docking area" (KF1), "Long wait to enter the Blue Cave" (KF-2), "Nothing much to do while waiting to board the tour boat" (KF-3) and "Exposure to heat and sun while waiting" (KF-4). The perception of dissatisfaction was measured on the four-level scale ranging from 1—"Did not experience" to 4—"Had a strong negative influence". Moreover, the scale was collapsed into two categories: 0—"Did not have negative influence" and 1—"Had a negative influence" so that further calculations were made with the collapsed scale. 65.8% of visitors did not experience negative influences in any of the factors, and for 10.4% visitors, all the factors were perceived as having a negative influence. Each of the given factors negatively influenced about 21% of visitors.

The results obtained via Chi-squared test of independence (Table 13.2) show that the critical factors of dissatisfaction were highly associated with the perceived crowding and visitor type, and, to a smaller degree with a duration of stay and season. The key factors of dissatisfaction were not associated with age.

The results presented in Table 13.2 show that visitor type had a significant impact on the key factors of dissatisfaction and, implicitly, on perceived crowding. Within the group of the dissatisfied visitors, those that travel longest to the cave have the most significant share. It can be explained as most of those trips are boat trips to multiple

Table 13.2 Chi-squared test of independence results between the key factors of dissatisfaction and the selected variables (For each test variable, the value of Chi-square statistics, p-value and Cramer's V are presented)

	KF1	KF2	KF3	KF4
Age	4.58/0.33/-	9.32/0.054/-	4.86/0.30/-	7.07/0.13/-
Perceived crowding	110.1/< 0,0001/0.39	125.4/< 0,0001/0.42	82.4/< 0,0001/0.34	65.6/< 0,0001/0.30
Season	8.91/0.011/0.11	13.6/0.001/0.13	15.3/0.0005/0.14	9.75/0.0076/0.11
Duration of stay	29.4/< 0,0001/0.20	44.1/< 0,0001/0.25	35.6/< 0,0001/0.22	25.2/< 0,0001/0.19
Visitor type	23.8/< 0,0001/0.18	66.8/< 0,0001/0.30	53.9/< 0,0001/0.27	16.7/0.0002/0.15

Source The authors

islands and top attractions in a single day (island hopping). The visit to the Blue Cave is just a stop in a busy schedule of an ambitious itinerary, decreasing the comfort of visitors (some of them cove a distance of 100 nautical miles). Consequently, the stress levels are increased, resulting in perceived quality of experience as the related literature suggests (Buckley 2009; Neuts and Nijkamp 2012; Saveriades 2000). Since visitors of island-hopping day trips are mainly younger, the correlation between age and the crowding perception is more significant.

In the questionnaire, there were 16 questions about visitor satisfaction with elements of tourism supply. Their satisfaction was measured on a seven-point Likert scale ranging from 1—"Very bad" to 7—"Excellent". In Table 13.3, the mean scores for visitors who were (column 1) and were not (column 0) negatively influenced by particular key factor of dissatisfaction are presented. Two sample t-test for the

Table 13.3 Mean score for satisfaction with elements of tourism supply for visitors who had a negative influence (Column 1) and who didn't have a negative influence (Column 0) on key factors of dissatisfaction

Elements of tourism supply	KF1		KF2		KF3		KF4	
	0	1	0	1	0	1	0	1
1. Scenic beauty and attractiveness of the Blue Cave	6.84	6.55	6.84	6.57	6.83	6.57	6.84	6.55
2. Cleanliness of the Blue Cave	6.75	6.50	6.76	6.51	6.74	6.57	6.78	6.45
3. Cleanliness of the sea in the bay (Mezoporat)	6.62	6.34	6.63	6.32	6.62	6.35	6.63	6.26
4. Cleanliness of the coast in the bay (Mezoporat)	6.54	6.18	6.56	6.14	6.56	6.09	6.55	6.08
5. Information about the Blue Cave in the bay area	6.07	5.39	6.06	5.49	6.08	5.37	6.04	5.46
6. Quality of the guided small boat tour (dock-cave-dock)	6.45	6.19	6.47	6.13	6.45	6.13	6.45	6.20
7. Hospitality of the Nautical Centre Komiža staff	6.52	6.15	6.53	6.14	6.51	6.14	6.52	6.16
8. Availability of toilets	5.70	5.02	5.73	4.96	5.75	4.84	5.71	4.90
9. Cleanliness of toilets	6.04	5.27	6.04	5.33	6.05	5.27	5.98	5.35
10. Waste disposal	6.19	5.54	6.24	5.46	6.25	5.33	6.21	5.33
11. Refreshment offerings	6.07	5.51	6.16	5.31	6.12	5.28	6.09	5.35
12. Souvenir offerings	6.01	5.52	6.05	5.43	6.03	5.34	6.02	5.49
13. Ticket price value for money	5.97	5.09	5.98	5.09	5.95	5.09	5.97	5.08
14. Information about the Blue Cave prior to the excursion	5.72	4.76	5.75	4.65	5.71	4.68	5.68	4.81
15. Transfer from the starting harbour to docking on Biševo (Mezoporat)	6.40	5.80	6.39	5.83	6.39	5.80	6.33	5.99
16. Service provided by the excursion organiser (travel agency)	6.48	5.96	6.48	5.95	6.49	5.85	6.45	6.03

Source The authors

difference of means was performed for each pair (key factor, the element of tourism supply) and results revealed that there is a significant difference between visitors who were negatively influenced and those who were not. These findings lead to the conclusion that perceived crowding (via key factors of dissatisfaction) has a negative impact on visitor satisfaction.

13.3 Discussion

Nature conservation policy implementation is lagging behind and, in the context of the immense tourism demand, concerns are mounting. Informed management and decision-making need reliable information, and this chapter describes the possible sources that can be correlated. First of all, surveying results make it evident that peak crowding pressures cause visitor dissatisfaction. Especially those that travel long distances and subsequently have the most immense ecological and environmental impact (air and noise emissions predominately). Hence, arrivals from the distant ports of departure that stay less than 2 hours, should be discouraged. Furthermore, environmental impact monitoring should be based on traffic heat maps. Density and intensity of speed boats and sailboats give away optimal positions for locations for monitoring: underwater noise, antifouling contamination, marine litter, hydrocarbon spills, engine emissions, etc.

Shipping lanes and seasonal maritime traffic influence marine Natura 2000 sites and for the area here in question the population of Bottle-nosed Dolphin (Tursiops truncates) is of interest. Besides the formal requirement to protect and research the population, the dolphins represent the opportunity to develop tourism along the lines of dolphin watching, citizen science, volunteering, education, biology research, etc. There are numerous opportunities to develop tourism content to utilise the attractive heritage that was recently recognised by UNESCO and organised under the title of the Geopark Vis Archipelago.

13.4 Conclusion

The work presented here aims to bridge nature conservation and tourism interests and to allow for a framework that would help management challenges through interdisciplinary research practices. Correlating overtourism and risks to biodiversity and environmental protection is a daunting task that is rarely seen in the literature. Reasons are primarily difficulties in acquiring bio-indicators related to visitation monitoring, especially for sea caves. Also, visitation standards and limits (physical and psychological) that could chart more precise irritation indexing of visitors are scarce. Having this in mind, the authors believe that by correlating cross-disciplinary practices, a framework could be established that would yield improvements in management and in practical sciences. Simultaneously covering areas of maritime traffic monitoring, environmental risks analysis, and visitor crowding perception, have provided the following findings:

1. The radar traffic heat mapping allows a better understanding of maritime corridors and related potential environmental impacts in the ecological network.
2. Environmental risks are much easier to name and comprehend once the heat mapping of maritime traffic and visitor monitoring is introduced and correlated.
3. The visitor survey gives insight into the factors influencing the negative perception in relation to the characteristics of visit, travel/transfer to the cave and other elements of the experience.

Future effort should be directed in better understanding the tourism and biodiversity risks as the basis for a comprehensive monitoring and management model. The model's negative externalities from the business perspective are the willingness of visitors' not to return nor to recommend the experience. In contrast, the negative environmental externalities are increased emission levels (for example, engine exhausts, underwater noise, marine litter, biocides, bilge and oil discharges noise, waste) and habitat disturbance (for example, uncontrolled anchoring, trampling, erosion).

Acknowledgements Croatian Ministry of the Sea, Transport and Infrastructure—Maritime Safety Directorate, Nautički centar Komiža, Blue World, Javna ustanova More i Krš, ADIPA Society for Research and Conservation of Croatian Natural Diversity.

References

Agresti A (2007) An introduction to categorical data analysis. Wiley

Arnberger A, Mann C (2008) Crowding in European forests: a review of recent research and implications for forest management and policy. Forestry 81(4):559–571

Buckley R (2009) Ecotourism: principles and practices. CABI

Carić H, Jakelić M (2018) Informirano upravljanje posjećivanjem u zaštićenim područjima. Prostor znanstveni časopis za arhitekturu i urbanizam 26(1 (55)):146–155

Cohen J (1988) Statistical power analysis for the behavioral sciences, 2nd edn. Erlbaum, Hillsdale, NJ

Dolnicar S, Lazarevski K, Yanamandram V (2013) Quality of life and tourism: a conceptual framework and novel segmentation base. J Bus Res 66:724–729

Eagles, Paul FJ, McCool, Stephen F, Haynes, Christopher DA (2002) Sustainable tourism in protected areas: guidelines for planning and management. IUCN Gland, Switzerland and Cambridge, UK. xv + 183 pp

EC (2019) The EU environmental implementation review 2019—country report Croatia. EC Directorate General Environment, Brussels

Gössling S, Peeters P (2015) Assessing tourism's global environmental impact 1900–2050. J Sustain Tour 23(5):639–659

Heberlein TA, Vaske JJ (1977) Crowding and visitor conflict on the Bois Brule River. Water Resources Center, University of Wisconsin-Madison

IVUMC (2016) Visitor use management framework: a guide to providing sustainable outdoor recreation, Edition One, Interagency Visitor Use Management Council

Kalisch D, Klaphake A (2007) Visitors' satisfaction and perception of crowding in a German National Park: a case study on the island of Hallig Hooge. For Snow Landsc Res 81(1/2):109–122

Klanjšček J, Geček S, Marn N, Legović T, Klanjšček T (2018) Predicting perceived level of disturbance of visitors due to crowding in protected areas. PloS one 13(6)
Luque-Gil AM, Gómez-Moreno ML, Peláez-Fernández MA (2018) Starting to enjoy nature in Mediterranean mountains: crowding perception and satisfaction. Tour Manag Perspect 25:93–103
Manning R (2007) Parks and carrying capacity: commons without tragedy. Island Press, Washington, USA
Manning RE, Freimund WA, Lime DW, Pitt DG (1996) Crowding norms at frontcountry sites: a visual approach to setting standards of quality. Leis Sci 18(1):39–59
Manning R, Lawson S, Valliere W (2009) Multiple manifestations of crowding in outdoor recreation: a study of the relative importance of crowding-related indicators using indifference curves. Leisure/Loisir 33(2):637–658
MEA (2003) Ecosystems and human well-being: a framework for assessment. Island Press, Washington, DC
Mrđa A, Carić H, Bojanić Obad Šćitaroci B (2014) Značaj koncepta turističke nosivosti za prostorno planiranje - Dosadašnja istraživanja, razvoj koncepta i metodološki pristupi. Prostor: znanstveni časopis za arhitekturu i urbanizam 22(2):212–227
Needham MD, Rollins RB, Wood CJ (2004) Site-specific encounters, norms and crowding of summer visitors at alpine ski areas. Int J Tour Res 6(6):421–437
Neuts B, Nijkamp P (2012) Tourist crowding perception and acceptability in cities: An applied modelling study on Bruges. Ann Tour Res 39(4):2133–2153
Peeters P, Gössling S, Klijs J, Milano C, Novelli M, Dijkmans C, Eijgelaar E, Hartman S, Heslinga J, Isaac R, Mitas O, Moretti S, Nawijn J, Papp B, Postma A (2018) Research for TRAN committee overtourism: impact and possible policy responses. European Parliament, Policy Department for Structural and Cohesion Policies, Brussels
Ryan C, Cessford G (2003) Developing a visitor satisfaction monitoring methodology: quality gaps, crowding and some results. Curr Issues Tour 6(6):457–507
Saveriades A (2000) Establishing the social tourism carrying capacity for the tourist resorts of the east coast of the Republic of Cyprus. Tour Manag 21(2):147–156
Shelby B, Vaske JJ, Heberlein TA (1989) Comparative analysis of crowding in multiple locations: results from fifteen years of research. Leis Sci 11:269–291
Sumner EL (1942) The biology of wilderness protection. Sierra Club Bull 27:14–22
Vaske JJ, Shelby LB (2008) Crowding as a descriptive indicator and an evaluative standard: results from 30 years of research. Leis Sci 30(2):111–126
Zehrer A, Raich F (2016) The impact of perceived crowding on customer satisfaction. J Hosp Tour Manag 29:88–98

Chapter 14
Research, Education and Tourism as Place Marketing Tools: The Case of the Jonian Dolphin Conservation in Taranto, Italy

Nicolaia Iaffaldano and Sonia Ferrari

Abstract In fragile environments that are already seriously threatened, such as the city of Taranto and its Gulf, special attention and the involvement of all stakeholders is required (Selby and Morgan in *Tour Manag* 17(4): 287–294, 1996). In these destinations, tourist experiences and resources have the potential to become instruments to increase visitors' environmental consciousness, increasing the awareness about the value of biodiversity and the importance of its conservation. Today Taranto is struggling to overcome the negative image associate with an Ilva steel plant by enhancing its natural resources. The Jonian Dolphin Conservation, an association that carries on activities of research and protection of the sea and cetaceans in the Gulf of Taranto, operates in these directions and represents an essential example of blue economy enterprise and a sustainable way to exploit sea resources. This paper aims to study how an environmentally sustainable activity can favour tourism development and urban repositing in a place affected by a negative image.

Keywords Place marketing · Sustainable tourism · Experience · Dolphin observation · Environmental awareness · Urban image

14.1 Introduction

The city of Taranto is considered a place affected by severe pollution and deterioration of environmental resources which reflects negatively on its urban image. For this reason, it seems necessary to invest in urban image renovation through an effective place marketing plan, to increase local community's awareness about environmental sustainability and to favour the conservation of natural resources and biodiversity. The Jonian Dolphin Conservation (https://www.joniandolphin.it), an

N. Iaffaldano
University of Bari, Bari, Italy
e-mail: nicolaia.iaffaldano@uniba.it

S. Ferrari (✉)
University of Calabria, Rende, Italy
e-mail: sonia.ferrari@unical.it

A. Mandić and L. Petrić (eds.), *Mediterranean Protected Areas in the Era of Overtourism*, https://doi.org/10.1007/978-3-030-69193-6_14

association, established 11 years ago to carry on activities of research and protection of the sea and cetaceans in the Gulf of Taranto, operates in these directions and represents an essential example of blue economy enterprise and a sustainable way to exploit sea resources.

The activity of the Jonian Dolphin Conservation (JDC) is based on cetaceans watching with daily boat trips and volunteering camps, and various educational activities mainly intended for schools. Additionally, they are engaged in several research programmes. Within Taranto, as a destination, the excursions offered by JDC from March to December, are becoming a tool to reduce tourist seasonality. They are usually sold out all around the year and attract numerous domestic and international tourists, overcoming the prejudice that Taranto is just a polluted city that has nothing to offer to visitors. Visitors are about 12,000–14,000 every year, with 35–40% of foreign presence according to the report of tickets sold by the JDC. The association aims to collect funds for its research activities and increase visitors' awareness of environmental sustainability by shaping and providing exceptional visitor experiences. During the daily tours, they sensitise guests by education, for example, about the damage caused to the marine ecosystem by plastic. The encounter with dolphins and sperm whales is an inspiring and frequently emotional moment for visitors. Dolphins are very loved and charismatic animals and those who see them usually share their experience through social media by posting photos and videos, thus increasing the overall involvement and promoting the excursion and site. The reviews on the official website, which refers to the Tripadvisor rating, suggest that the guests' levels of satisfaction are very high.

Linking tourism, culture and environment is essential in place marketing to become competitive at an international level (Boisen et al. 2018; McCool 1995). This integrative approach to tourism development through the promotion of naturalistic itineraries, local fish and mussel farming products might stimulate the improvement of Taranto's urban image (Brosius et al. 2005). It is essential to underline that dolphins, which have always lived in the gulf, are strongly linked with the city's history, as they are also the protagonists of numerous legends about the establishment of Taranto. Today the city is struggling to overcome the negative image associate with an Ilva steel plant by enhancing its natural resources. Recently, local municipality, the JDC, Taranto University and several research institutions have supported the establishment of a Blue Oasis. The oasis is a fundamental step for the establishment of a Cetacean Sanctuary in the Gulf of Taranto. The recent establishment of the Mar Piccolo Regional Park in the area in front of Taranto and the Blue Oasis represents a significant phase in reviving and promoting the city through the exploitation of its peculiarities and potential.

This chapter aims to show that a subject that works for environmental conservation purposes can also offer engaging educational activities and unique experiences with a tourist value. The study applies qualitative research design and structured interviews with relevant stakeholders to discuss how JDC contributes to raising awareness on environmental issues through sustainable activities related to sustainable tourism and urban image repositioning.

14.2 Towards Environmental Sensitiveness and Tourism Sustainability

In fragile destinations in terms of natural and cultural heritage, substantial tourism development and a growing number of visitors can generate different types of effects, with negative and positive implications. The negative impact of tourism, and especially of overtourism, on the physical and natural environment has been broadly studied in the last decades (Butler 1980; Clarke 1997; Hernandez-Maskivker et al. 2019; Mowforth and Munt 2003; Swarbrooke 1999). The damages of tourism to the natural environment are caused not only by the large number of visitors, who in some cases stimulate the overtourism but also by the inappropriate behaviour of many visitors (Capocchi et al. 2019; Cheer et al. 2019; Peeters et al. 2018). In fact, tourists are not often respectful of nature, consequently creating temporary or permanent damages to the delicate equilibrium of ecosystems and affecting natural environment elements such as flora, fauna, habitats and landscapes. The leading cause of such tourist behaviour, which can represent a danger for the protected ecosystem, is the low level of visitors' preparation and knowledge (Heslinga et al. 2019). All that often led to residents' unfavourable attitudes towards tourism, with consequent frictions and tensions among hosts and guests and social sustainability problems.

The most effective tools to address these problems and prevent their negative consequences are monitoring, environmental education and teaching, to show visitors how to behave respectfully. Fortunately, today the attention to the sustainability of visitors' flows and tourism impacts on the natural environment, biodiversity and ecosystems conservation is very high. Tourism is recognised as an industry that can help support environmental-friendly initiatives, increasing the awareness about the value of natural resources and the importance of their conservation (Cole et al. 2008; Ferrari and Gilli 2016; Ferrari and Pratesi 2012; Haukeland et al. 2011).

Investing in environmental education programmes (i.e. guided tours, laboratories, conferences, training courses), teaching wilderness ethic, informing visitors about natural areas appropriate fruition and promoting forms of responsible tourism can influence tourists' behaviour and attitude. These are some of the instruments to create among residents and visitors a higher awareness about the importance of local natural heritage, its role and the advantages that tourism can bring to the community. These activities are integrated part of programmes and initiatives of many public and private sector stakeholders in Taranto today. All of them are designated to improve tourism sustainability and to invest in strengthening and promotion of urban identity.

Postmodern society appears dominated by the experience economy (Pine and Gilmore 1998, 1999). The overlapping and sometimes merging of different spheres of activities such as art, education, leisure, tourism, television, business and entertainment is one of the most significant phenomena of this era. It led to the disappearance of the border between art and everyday life, and the transformation of art, culture, education and history into consumer goods (Harvey 1990; Urry 1990). Today entertainment, emotions and playfulness are the dominant elements in an increasingly flexible, ambiguous, multidimensional social environment that is interested above all in living the present (Fabris 2003; Ferrari 2006; Rifkin 2001).

Consumers pursue well-being. They seek playful experiences, escaping all that is monotonous and boring. Their choices appear to be increasingly dominated by fashions, emotions and the search for the pleasure of the senses. Consumption becomes a complex and holistic experience, since the client purchases not only the good itself but also service and information components, experiences and culture (Rifkin 2001).

Experiences could be defined (Schmitt 1999) as the result of situations that have happened, that trigger stimuli in the sensory system, the heart or the mind. Experiences have effects on the sensory, affective and cognitive systems and influence the individuals' behaviour over time (i.e. consumption patterns and lifestyles). The reaction to experience can be a thought, an emotion or a feeling (LaSalle and Britton 2003).

As a consequence, more and more often, people are looking for emotions and unique and unforgettable experiences. The experience is often an instrument of transformation and inner change, a way to achieve moments of intense satisfaction; these are linked above all to experiences called extraordinary, optimal or flow (Arnould and Price 1993). Optimal experiences are connected with moments of joy that occur when the experience allows achieving something unexpected, which goes beyond expectations, needs and desires and which is, even, unimaginable before the experience itself (Csikszentmihalyi 1990). This could be the case of cetacean observation, the object of this study, which, as will be demonstrated, stimulates emotions and feeling of great amazement.

Environmental education can benefit from this continuous search for memorable experiences, thus giving rise to more useful and engaging learning instruments and communication tools. Frequently these tools seem essential to increase community awareness about environmental issues and biodiversity conservation. This is especially true in fragile environments that are already seriously threatened, such as the Gulf of Taranto, and require special attention and the involvement of all stakeholders. In these destinations, tourist resources have the potential to become instruments to increase visitors' environmental consciousness. In fact, many scholars affirm that a more excellent environmental knowledge by tourists can reinforce their environmental sensitivity (Amyx et al. 1994; Huang and Shih 2009; Peterson 1982; Townsend 2000).

Studies show that recently the attention of environmental education has shifted from the purely cognitive aspects to the focus on increasing environmental awareness, sustainability and biodiversity conservation (Nazir and Pedretti 2016). In this sense, field learning through experiences, especially in natural environments, potentially increases the effectiveness of environmental education activities by reducing the gap between theoretical knowledge and real life (Brody 2005; Bögeholz 2006; Jeronen et al. 2017; Lavie Alon and Tal 2015; Palmer and Suggate 1996). This is also consistent with the great success that "edutainment" initiatives have had in recent years. These are activities which aim to educate young people by entertaining and offering them highly involving experiences. The term edutainment combines entertainment and education activities (Aksakal 2015; Colace et al. 2006), intending to learn through playful and recreational experiences (Wilson 2018). These experiences are often linked to the enjoyment of activities through innovative technologies, such as virtual realities, which increase the involvement of the participant by stimulating all the senses (Okan 2003).

14.3 The Case of the Jonian Dolphin Conservation

14.3.1 The Place

The complex and varied geomorphology of the Gulf of Taranto, with its low waters alternate with steep underwater canyons that sink not far from the coast, is an ideal habitat for dolphins. The canyons are the real biodiversity treasure of the Gulf seabed. According to the JDC, the dolphin species that could be found there are mainly Stenella Striata (Striped dolphin), Tursiops (Bottlenose dolphin) and Grampo (Risso's dolphin).

The presence of dolphins in the Gulf since ancient times led to the development of myths regarding their role in the foundation of the Taranto. The dolphin depicted on the coins that were minted in Taranto since the sixth century B.C. is associated with the legend of Taras, son of Poseidon, and the nymph Satyrion. Taras was the founding hero of the colony where he arrived safe and well on the back of a dolphin after being shipwrecked. Beyond mythology, this suggests the positive attitude of the inhabitants of Taranto towards the dolphins, which have been chosen to be a symbol of the city.

Due to its geological conformation, the seabed of the Gulf of Taranto could hide deposits of hydrocarbons. For this reason, the Ministry of Economic Development in Italy is considering drilling in the area 12 miles away from the coast. The idea of drilling in the Gulf of Taranto has generated severe concerns from both the Apulian Region authority and some ecological movements, mainly because of the use of offshore research methods[1] which require the release of shock waves with compressed air (air-gun) into the sea. However, at the moment, no activity has been taken. The creation of the Blue Oasis in the Gulf of Taranto, aimed at being included in the list of the Specially Protected Areas of Mediterranean Importance (SPAMI) for the protection of cetaceans, has led to the adoption of various measures within limits established by the Marine Environmental Strategy (Caffio 2019, p. 63; ACCOBAMS, *Agreement on the Conservation of Cetaceans of the*

[1]Offshore activities include the exploration of crude oil and natural gas deposits by drilling exploratory wells (exploration phase). Then start up the production phase with the drilling of production wells. These wells are used to extract the oil and/or natural gas (extraction phase). Within this process, air-gun is used. This is a technique of hydrocarbon exploration in the sea that employes a strong pulse of compressed air "shot" into the water and that sends shots to the seabed. The echo emitted from the subsoil provides information about the underground structures. An amendment to the M5S political party aims to ban the airgun technique, with penalties of up to 120,000 euros and the revocation of permits (Retrieved from: Il sole 24 Ore, 15 January 2019).

Black Sea, Mediterranean Sea and Contiguous Atlantic Area, www.accobams.org). Furthermore, the European Directive on the Marine Strategy implemented in Italy with the Legislative Decree No. 190/2010 (Gazzetta Ufficiale Serie Generale n. 270 del 18-11-2010) states that, in order to achieve good environmental status, human activities of energy production in the sea must not be a source of pollution, and noise must be compatible with the proper functioning of ecosystems. For the first time, marine noise has become a quality parameter of the marine environment. Vessel traffic, seismic investigations, underwater sonar emissions, seabed drilling for hydrocarbon deposits and marine wind turbines can cause behavioural changes and sometimes irreversible damage to cetaceans (Caffio 2019, p. 61). Therefore, conservation actions must necessarily include the right measures to reduce the harmful impacts of human-made noise, as well as pollution from waste, and above all the vast amounts of plastics in the sea. Additionally, Taranto has recently become one of the ports on cruise travellers' routes, and next year, 20 ships are expected to disembark approximately 30.000 daily visitors, which is a considerable number, that could potentially create concerns about the cetaceans in the area.

14.3.2 Methodology

The research aimed to investigate the impact of the activities of JDC on raising awareness regarding environmental issues. Along with that, we aimed to analyse how JDC activities support the sustainability of tourism development and urban image regeneration. Particular attention was paid to the effectiveness of activities as providing information to visitors about the appropriate use of natural resources, and the promotion of responsible tourism via experience design and education activities.

To carry out this study, we apply a series of structured interviews and a qualitative research design. This approach to the investigation was the most appropriate considering both, the limitations and the focus of the study, concentrated not on the outcomes but the processes and the sociocultural context (Merriam and Tisdell 2009). Also, the qualitative method provides flexibility and adaptability to the research aims and is more interactive, in-depth and sensitive when studying human behaviours, ways of thinking, attitudes and motivations (Carson et al. 2001; Silverman 1998). Twenty-five in-depth and semi-structured interviews with stakeholders were conducted through Skype between September and October 2020, each lasting approximately 70–90 min. A convenience sampling approach—the snowball method—was applied. This method is useful for contacting the right type of respondents for the specific aims of the research (Biernacki and Waldorf 1981). The respondents (Table 14.1) were selected based on the author's expert opinion and profound knowledge of the phenomenon and site under investigation. Table 14.1 describes in which companies or institutions the interviewees are employed.

The content of the interviews changed significantly, depending on the professional characteristics of the respondents. The respondent permitted to record and following to transcribe the interviews. The number of interviews was decided according to the

Table 14.1 The sample of interviewees

Institution or Association	No. of interviewees
A. Municipal Department and other local authorities	3
B. Jonian Dolphin Conservation's volunteers	3
C. Environmental and sociocultural associations	7
D. Trade and economic promotion subjects	2
E. High schools and University	4
F. Guests	6
Total	25

principle of saturation, which states that data collection should end when no new categories emerge from the data (Silverman 1998).

A subsequent step was to structure meanings using narrative, leaving the stories of participants as they were told. The narrative provides more in-depth information about the participants' experiences. Data analysis followed a deductive coding approach. The data were managed by hand through a thematic analysis. The thematic content analysis helped to identify four main themes, including *the dolphins' observation as a memorable experience; the experience as an education and awareness-raising activity; the urban renovation process of Taranto; the contribution of the environmentally responsible activities to the tourism development and the image repositioning of Taranto.* In the next paragraph, the results of the interviews are presented according to the above themes.

14.3.3 Findings

14.3.3.1 The Dolphin Observation as a Memorable Experience

We wanted to understand which type of experience the JDC deliver and its guests' reactions. The findings regarding this theme show that the impressions and reactions of all participants are similar to each other. One of the guests of JDC, for example, said:

> The perception of the moment is very subjective, but only a few guests are not strongly emotionally involved during the experience.

Another resident explained:

> Many cry when they see the animals free in their environment. In the vast majority of cases, seeing animals stimulates powerful emotions.

The visitors have different reactions, as a volunteer from JDC (group B) explained:

> Children are entranced: they often expect the evolutions that can be seen in a dolphinarium, but that seldom occur in nature. On the contrary, the jumps that dolphins do are spontaneous, and we subsequently explain their behaviour. It is wonderful to work with children: they are much more sensitive than adults; they are the first to be surprised and what they learn is reported in the family afterwards.

Another volunteer (group B) added:

> We help visitors imagine what is going to happen during the crossing. It is our concern to teach them all the differences between free and captive animals. We try to explain to them what our activity of study and protection of cetaceans is, and why animals must be observed in freedom and their natural environment. …Participants pay for the ticket, thus financing the research. Therefore, the boat trip is not only a playful observation; it is also a part of our research.

One of the visitors (group F) said:

> When you are on board, you are strongly involved as a participant, because you are a protagonist: you are given a pair of binoculars and participate in the sighting. Then, when a dolphin is sighted, it is tracked and monitored, and those are real scientific tracking activities. The guest is 100% involved in the cataloguing process of this scientific study. So, it is also thanks to us that the monitoring campaign for that specific day is done.

Concerning the sharing on social media by guests during the sea crossing on board, a volunteer of JDC (group B) explained:

> Guests are keen to show their experience live, to share it with friends on social media like Facebook and Instagram, and that is a form of advertising for us! At a certain point, we often recommend them to forget about cameras, camcorders and other devices. We suggest them to enjoy dolphins and observe them because if they stare at the monitors or displays of their devices, they miss many moments. Participants usually post images, so every boat trip becomes viral.

A tourist (group F) added:

> During the day in the boat we tend to take photos and videos to share them on social media. Our behaviour onboard is a sort of frenzy and excitement because we want to take and shoot every moment.

An experience is often an instrument of transformation and inner change, a way to achieve moments of intense satisfaction and moments of joy that occur when experience allows achieving something unexpected, which goes beyond expectations and is even unimaginable before the experience itself. This is the case of cetacean observation that represents an extraordinary experience (Arnould and Price 1993; Csikszentmihalyi 1990) fostering the feelings of great amazement for visitors. This is evident from the statements of many of the interviewees.

14.3.3.2 The Experience as an Education and Awareness-Raising Activity

Regarding this second theme, we were interested in understanding if and at what level the studied experience represents a useful environmental education tool and a way to increase awareness about sustainable tourism and biodiversity conservation's issues.

A representative of the public administration (group A) stated:

> Surely, the activity of the JDC has a strong impact in terms of awareness and attention to the city of Taranto, thanks to a storytelling that differs from what has been done until a few years ago.

A tourist (group F) additionally supported these conclusions:

> There are educational moments on board focused on the problems caused by plastic and on the ecosystem in general. All that is kept in tourists' memory, because everything is explained by making examples with something concrete …When we, fortunately, saw dolphins and turtles in the sea, they explain to us that plastic can harm them.

The same opinion about the importance of learning first-hand experiences in the field was shared by a representative of high schools (group E), suggesting that:

> We do inform on the catamaran, for example about pollutants and the cleaning of the sea. Students pick up trash with a net. They learn that even a plastic cap, which seems insignificant, can damage the environment very much, as if animals swallow it, for example, can die. Students are highly sensitised by these educational initiatives because observation is more engaging than just listening.

Another interviewee of the same group (E) suggested that after the experience, he noted that there were changes in students' behaviour and lifestyle:

> Before, when we were only talking about microplastics, they were not so involved but now, thanks to the practical activities carried out at sea, they are much more aware and sensitive and have adopted more careful attitudes even in their daily life.

This might be a crucial lesson for further development of a recently inaugurated project in which three schools are involved in a network called T.U.R.S.I.O.P.E.S. (an acronym on Italian standing for "*Taranto unisce in rete le scuole ioniche che operano nell'educazione allo sviluppo sostenibile*" that is: "*Taranto joins in a network of the Ionic schools that operate in the education for sustainable development*"). In general, the interviewers support the conclusions concerning the importance of environmental awareness and pro-environmental education and behaviour of young generations for the future of sustainability of interregional development.

A representative of the public administration (group A) supported these claims and further explained the importance of infrastructural improvement for the achievement of this goal

> The awareness on environmental issues will be more effective and involving thanks to the project of a 'green' aquarium which was authorised by the Government last October, as today a traditional aquarium is no longer judged very positively, especially if you have the opportunity to see free dolphins in the gulf. For this reason, now it has been designed as a 'green' aquarium. Of course, it will not be an aquarium with fish in the tanks and things like that, but something absolutely innovative and environmentally sustainable. Probably, thanks to some cameras, we will be able to observe the free dolphins living in Mar Grande.

A representative of an environmental association (group C) invoked a more holistic approach to this initiative by involving other environmental education initiatives as creating an experiential multimedia museum in Taranto, which could be an environmental education centre about the protected marine areas of the Mediterranean. The environmentalists (group C) emphasised the trend at the European level suggesting the closing of dolphinariums. These opinions are diverging from the expectations of the educational institutions (group E) seeing them as the places where students and younger populations can sensitise with wild animals. All of this suggests that soon, tight cooperation and communication between all relevant stakeholders will be needed to achieve consensus.

The interviews showed us that offering environmental education experiences in a natural environment where it is possible to observe free animals in their habitat is a unique and extraordinary way to communicate. It conveys values of attention to the natural environment and biodiversity that would hardly be so strongly received in any other way. The direct experience and personal involvement in the observation and research activities are incredibly useful in raising awareness and represent an unforgettable and unique moment for the guest.

14.3.3.3 The Urban Renovation Process of Taranto

In the last few years, the local institutions tried to implement a broad diversification of the local economy in Taranto. As a consequence, they are favouring the development of a comprehensive series of productive activities and projects related to the sea and the blue economy, far from the great Ilva steel plant. Regarding the urban renovation process of Taranto, it is not clear if the local community is understanding and supporting this path in all its aspects and aims (Camarda et al. 2015; Greco and Di Fabbio 2014; Le Xuan and Tricari 2014).

A representative of an environmental association (group C) explained:

> I think that Taranto inhabitants know that the change process in progress can be a real and positive development path… They should understand that the change is necessary, even though it is not a simple one.
>
> Probably there will be resistance because a large part of the population, although supporting the change, is still working in the Ilva and in the linked activities and industries… In Taranto, there is no other way but sustainable development.

From the opposite point of view, another interviewee (group D) explained that:

> The local community in Taranto is sceptical towards this project of urban redevelopment that seems to be underway. In many cases, residents seem to be unaware: for example, mussel farmers have not noticed anything. What saddens me the most is to see that now Taranto inhabitants are disillusioned and very biased to believe in the possibility of social redemption.

14.3.3.4 The Contribution of the Environmentally Responsible Activities to the Tourism Development and Image Repositioning of Taranto

Today, in Taranto, some private and public subjects are carrying on programmes and initiatives that go in the direction of the repositioning of Taranto's urban image, in order to improve tourism sustainability, to develop tourism and to invest in place marketing tools to strengthen urban identity and image.

Regarding these issues, an interviewee (group D) said:

> Tourism in a city like Taranto was a utopia because there was no awareness of being able to finance an activity structured like that of the JDC through tourism. Now the course has been a bit reversed. At the beginning of their activity, they were considered downright crazy, because they were thinking of carrying out a scientific research activity funded by tourism in Taranto. All that seemed utopian.

The representative of the JDC further explained (group B):

> We can now perceive the process of change that is being realised in Taranto. It involves an increasing interest for natural resources and new positioning, that is different from the previous image of a city polluted by Ilva. We are the advocates of this change. The new brand, "Taranto capital of the sea", shows the tail of a sperm whale. This brand was created thanks to the development of our researches on the territory; it reveals that there is a close collaboration with the current municipal administration, also because it became clear that the cetaceans living in the Gulf of Taranto are a vital strength point of urban offer. What we always repeat on board of our boat is that Taranto is not only Ilva but also something else: Ilva is there with its various problems, but there is much more to see; there is plenty of different beauties.

A trade and economic promotion subject (group D) emphasised the role of the JDC in the attraction of tourists and temporal redistribution of tourist flows from May to September:

> The association has been working for 10 years, and I do not believe that it is necessary to increase the number of boats for the observation of dolphins, as they might become a disturbing element. He added: The tour takes place on a traditional small boat, i.e. a catamaran, with a limited load capacity, 35 seats, with good stability so that you can enjoy a more authentic experience.

The findings of the interviews pointed out that the eco-sustainable activities carried out by the JDC are tools to create among residents and visitors a greater awareness about the importance of natural heritage, its role in the sustainable tourism development, and the advantage that tourism can bring to the community (Sanna and De Bernardo 2015; Tricarico and Zandonai 2018).

14.4 Discussion and Conclusions

The interviews show that the experiences offered by JDC have a substantial impact on the participants, provoking significant inner changes, new attitudes and greater environmental awareness. In this sense, they can be considered useful, as they contribute to achieving multiple objectives of the organisers, including entertaining, educating, raising awareness, fundraising, promoting city's image and fostering tourism development. The experiences offered are unique, highly involving and exciting. Environmental education can benefit from the continuous search for such experiences by consumers and tourists, thus giving rise to more useful and engaging learning instruments and place marketing communication tools. These instruments are often essential to increase community awareness about environmental problems and biodiversity conservation. They might be recognised as a new way of enabling social protagonism in the management of collective goods and the development of innovative services.

The findings of this research emphasise the importance of the initiatives as JDC, for the sustainability of natural ecosystems like the Gulf of Taranto. The city of Taranto has initiated the transformation of its image by focusing on the blue economy, aware of the importance of increasingly sustainable forms of development. The observation of dolphins seems a small piece of this mosaic, but in reality, it has played and is still playing an essential role. In fact, it was the first time that a development initiative involved, unique and memorable experience, a scientific research activity, a tourist attraction, environmental awareness and education and tools for urban reposition. The success of this project initiated the change of mentality and attitudes that could lead to significant results in terms of place marketing for Taranto.

The involvement of stakeholders in sustainable development projects and exchange of knowledge is a challenge for researchers, practitioners and policymakers, aiming to establish evaluation parameters in the granting of space assets and financial resources. Among the evaluation parameters, the exchange of knowledge and experience in open and collaborative spaces should be indeed a primary objective to ensure that best practices and solutions play a guiding and inspiring role within destination ecosystem.

References

Aksakal N (2015) Theoretical view to the approach of the edutainment. Procedia—Soc Behav Sci 186:1232–1239. https://doi.org/10.1016/j.sbspro.2015.04.081

Amyx DA, DeJong PF, Lin XL, Chakraborty G, Wiener JL (1994) Influencers of purchase intentions for ecologically safe products: an exploratory study. In: Park CW, Smith DL (eds) AMA winter educators' conference proceedings. American Marketing Association, Chicago, IL, pp 341–347

Arnould E, Price L (1993) River magic: extraordinary experience and the extended service encounter. J Consum Res 20(1):24–45

Biernacki P, Waldorf D (1981) Snowball sampling: problems and techniques of chain referral sampling. Sociol Methods Res 10(2):141–163. https://doi.org/10.1177/004912418101000205

Bögeholz S (2006) Nature experience and its importance for environmental knowledge, values and action: recent German empirical contributions. Environ Educ Res 12:65–84
Boisen M, Terlouw K, Groote P, Couwenberg O (2018) Reframing place promotion, place marketing, and place branding-moving beyond conceptual confusion. Cities (London, England) 80:4–11. https://doi.org/10.1016/j.cities.2017.08.021
Brody M (2005) Learning in nature. Environ Educ Res 11:603–621
Brosius JP, Tsing A, Zerner C (eds) (2005) Communities and Conservation: histories and politics of community-based natural resource management. Altamira Press, Walnut Creek, CA
Butler RW (1980) The concept of a tourist area cycle evolution: implications for the management of resources. In Butler R (ed) The tourism area life cycle, vol 1. Chanel View Publications, Bristol, pp 5–12
Caffio F (2019) *Mari di Taranto: Golfo, Mar Grande, Mar Piccolo tra Puglia, Lucania, Calabria.* Scorpione Editrice, Taranto
Camarda D, Rotondo F, Selicato F (2015) Strategies for dealing with urban shrinkage: issues and scenarios in Taranto. Eur Plan Stud 23(1):126–146
Capocchi A, Vallone C, Pierotti M, Amaduzzi A (2019) Overtourism: a literature review to assess implications and future perspectives. sustainability 11(12):1–18
Carson D, Gilmore A, Perry C, Gronhaug K (2001) Qualitative marketing research. Sage, Thousand Oaks
Cheer JM, Milano C, Novelli M (2019) Tourism and community resilience in the anthropocene: accentuating temporal overtourism. J Sustain Tour 27(4):554–572
Clarke J (1997) A framework of approaches to sustainable tourism. J Sustain Tour 5(3):224–233
Colace F, De Santo M, Pietrosanto A (2006) Work in progress: Bayesian networks for edutainment, 36th ASEE/IEEE Frontiers, in Education Conference, https://doi.org/10.1109/fie.2006.322573
Cole DN, Yung L, Zavaleta ES, Aplet GH, Chapin FS, Graber DM., … Parsons DJ (2008) Naturalness and beyond: protected area stewardship in an era of global environmental change [George Wright Society]. Georg Wright Forum 25(1):36–56
CsikszentmihalyI M (1990) Flow: the psychology of optimal experience. Harper & Row, New York
Fabris GP (2003) Il nuovo consumatore: verso il post-moderno. Franco Angeli, Milano
Ferrari S (2006) Modelli gestionali per il turismo come esperienza: emozioni e polisensorialità nel marketing delle imprese turistiche. Cedam, Padova
Ferrari S, Gilli M (2016) Protected natural areas as innovative health tourism destinations. In: Smith M, Puczkó L (eds) The Routledge handbook of health tourism, pp 419–429
Ferrari S, Pratesi CA (2012) National parks in Italy—sustainable tourism marketing strategies [Finnish Journal of Tourism Research]. Matkailututkimus 8(1):7–23
Greco L, Di Fabbio M (2014) Path-dependence and change in an old industrial area: the case of Taranto, Italy. Camb J RegNs, Econ Soc 7(3):413–431
Harvey D (1990) The condition of postmodernity. Blackwell, Malden
Haukeland JV, Grue B, Veisten K (2011) Turning national parks into tourist attractions: nature orientation and quest for facilities. In: Fredman P, Tyrväinen L (eds) Frontiers in nature-based tourism. Routledge, New York, pp 76–99
Hernandez-Maskivker G, Ferrari S, Cruyt ANJ (2019) Exploring community stakeholders' perceptions of mass tourism: the case of Bruges. Tour: Int Multidiscip J Tour 14(1):77–94
Heslinga JH, Hartman S, Wielenga B (2019) Irresponsible responsible tourism; observations from nature areas in Norway. J Tour Futur
Huang PS, Shih LH (2009) Effective environmental management through environmental knowledge management. Int J Environ Sci & Technol 6(1):35–50
Iaffaldano N, Ferrari S (in press) Analysis and segmentation of the touristy demand in a marine protected area on the basis of sustainability indicators: the case of Tremiti Islands. Italy. In Albrecht J (ed) Visitor experience management in nature-based tourism. CABI Publishing, Wallingford

Jeronen E, Palmberg I, Yli-Panula E (2017) Teaching methods in biology education and sustainability education including outdoor education for promoting sustainability—A literature review. Educ Sci 7(1):1. https://doi.org/10.3390/educsci7010001
Lasalle D, Britton TA (2003) Priceless: turning ordinary products into extraordinary experiences. Harvard Business School Press, Boston
Lavie Alon N, Tal T (2015) Student self-reported learning outcomes of field trips: the pedagogical impact. Int J Sci Educ 37(8):1279–1298. https://doi.org/10.1080/09500693.2015.1034797
Le Xuan S, Tricari L (2014) Imprese Comuni: rigerezione urbana e community enterprises nel Regno Unito. Maggioli Editore, Santarcangelo di Romagna
McCool SF (1995) Linking tourism, the environment and sustainability. CABI Publishing, New York
Merriam SB, Tisdell EJ (2009) Qualitative research: a guide to design and implementation. Jossey-Bass, San Francisco
Mowforth M, Munt I (2003) Tourism and sustainability, 2nd edn. Routledge, London
Nazir J, Pedretti E (2016) Educators' perceptions of bringing students to environmental consciousness through engaging outdoor experiences. Environ Educ Res 22(2):288–304. https://doi.org/10.1080/13504622.2014.996208
Okan Z (2003) Edutainment: is learning at risk? Br J Educ Technol 34(3):255–264
Palmer JA, Suggate J (1996) Influences and experiences affecting the pro-environmental behavior of educators. Environ Educ Res 2:109–122
Peeters P, Gössling S, Klijs J, Milano C, Novelli M, Dijkmans C, Eijgelaar E, Hartman S, Heslinga J, Isaac R, Mitas O, Moretti S, Nawijn J, Papp B, Postma A (2018) Research for TRAN committee—overtourism: impact and possible policy responses. European Parliament, Policy Department for Structural and Cohesion Policies, Brussels
Peterson N (1982) Developmental variables affecting environmental sensitivity in professional environmental educators (Unpublished thesis). Southern Illinois University, Carbondale, IL
Pine B, Gilmore J (1998) Welcome to the experience economy. Harvard Business Review, July–August: 97–105
Pine B, Gilmore J (1999) The experience economy: work is theatre & every business a stage. Harvard Business School Press, Boston, Harvard
Rifkin J (2001) The age of access: the new culture of hypercapitalism. Penguin, New York, NY
Sanna F, De Bernardo V (eds) (2015) Sviluppo locale e cooperazione sociale. Beni comuni, territorio, risorse e potenzialità da connettere e rilanciare. Ecra, Roma
Schmitt B (1999) Experiential Marketing. The Free Press, New York
Selby M, Morgan NJ (1996) Reconstruing place image. A case of study of its role in destination market research. Tour Manag, 17(4):287–294. https://doi.org/10.1016/0261-5177(96)00020-9
Silverman D (1998) Qualitative research: meanings or practices. Inf Syst J 8(1):3–20. https://doi.org/10.1046/j.1365-2575.1998.00002.x
Swarbrooke J (1999) Sustainable tourism management. CABI, Wallingford, UK
Townsend C (2000) The effects of environmental education on the behavior of divers: a case study from the British Virgin Islands (Unpublished master's thesis). The University of Greenwich, London
Tricarico L, Zandonai F (2018) Local Italy. I domini del "settore comunità" in Italia. Milano: Fondazione Giangiacomo Feltrinelli
Urry J (1990) The tourist gaze: leisure and travel in contemporary society. Sage, London
Wilson R (2018) Nature and young children: encouraging creative play and learning in natural environments. Routledge, Abingdon, Oxon

Part IV
Intelligent Nature-Based Tourism Development

Chapter 15
Nature, Tourism, Growth, Resilience and Sustainable Development

João Romão

Abstract This perspective article takes into account several previous studies focusing on the relations between territorial resources, tourism dynamics, economic performance (both in terms of growth and resilience), sustainable development and smart specialization in a large number of European regions. Supported by different methodologies, the results of different international comparative analysis and an individual case-study are used in order to define and to investigate the particular relations between natural features of the territories and their utilization for tourism purposes, along with different types of impacts. Revealing that Southern European regions are mostly following strategies of cost-leadership based on the provision of mass tourism products and services with low value-added, the results of these works also show that the long-term economic benefits of this type of path-dependence evolution are limited, both in terms of economic growth and also in terms of the socio-economic resilience of these regions when facing negative external shocks. In particular, the high importance of the tourism sector within regional processes of specialization makes them especially vulnerable both to the impacts of overtourism and no-tourism. Possible strategies for diversification of the regional economic structures are discussed, taking into account the current conditions of regional development.

Keywords Endogenous resources · Tourism-led growth · Sustainability · Europe · Resilience · Smart specialization

15.1 Introduction

The utilization of natural resources for tourism development has been broadly analyzed, both in terms of their positive aspects (the creation of unique and differentiated tourism products and services based on the specific features of the destinations) and also focusing on the possible negative implications (potential degradation or destruction of sensitive ecological resources due to overuse). In most of the cases,

J. Romão (✉)
Yasuda Women's University, Hiroshima, Japan
e-mail: joao_romao@me.com

A. Mandić and L. Petrić (eds.), *Mediterranean Protected Areas in the Era of Overtourism*, https://doi.org/10.1007/978-3-030-69193-6_15

these analyses focus on specific case-studies, creating difficulties for the comparison between different areas and destinations (Weaver and Lawton 2007; Lu and Nepal 2009). In fact, comparative analyses of the impacts of tourism dynamics in environmentally sensitive areas on regional economic growth and sustainable development are scarce.

Recent extensive studies on European regions addressed these problems in an international context: first by analyzing the relation between natural resources and tourism demand (Romão 2015) and with tourism competitiveness (Romão et al. 2017); and then, in a broader sense, investigating the impacts of nature and tourism on economic growth (Romão and Nijkamp 2018) or on regional sustainable development (Romão and Neuts 2017). These analyses cover an extensive set of European NUTS-2 regions, and they are often supported by techniques for spatial analysis, allowing for the identification and discussion of specific aspects of the Mediterranean region. More recently—and only focused on regions where tourism and hospitality services are assumed as priority sectors within smart specialization strategies (mostly located in the Mediterranean area) the relations between tourism dynamics, regional growth and socio-economic resilience are scrutinized (Romão 2020). These international comparative studies are complemented by a specific case-study focused on aspects of resilience in a tourism-dependent South European region (Romão et al. 2016).

This article frames the previous analyses within the scarce related literature and uses the results obtained as a starting point for the discussion of overall policy and managerial implications related to destination differentiation, environmental protection, economic growth and regional development. In particular, aspects related to overtourism (Dodds and Butler 2019) and no-tourism (the severe reduction or complete suspension of tourism activities due to the COVID-19 pandemics) [Gössling et al. 2020] are discussed. The existence of a large number of areas with abundant classified natural resources coexisting with massified forms of tourism, creating low value-added for the regional economies and limited impacts on growth and sustainable development appear as a major problem identified for most of the Southern European regions. Strategies of differentiation aiming at increasing the value added by the tourism sector, while limiting the number of tourists emerge as essential strategic options.

The above-mentioned studies are presented in detail in Sect. 15.2, starting with a discussion of the relation between nature, productivity and tourism (2.1), followed by an analysis of its implications on regional growth and sustainable development (2.2) and concluded by an assessment framed by the concept of resilience and integrated into the context of the definition of smart specialization strategies in European regions. Section 15.3 discusses all these results, suggesting that, as a result of their path-dependent processes of tourism development, Southern European regions are particularly vulnerable both to the problems related to overtourism and no-tourism. Section 15.4 concludes the chapter.

15.2 Over-Specialization in Tourism

15.2.1 Nature, Productivity and Tourism

The problematic relation between natural resources and tourism dynamics in the Mediterranean area is introduced in this Section, by considering two studies analyzing the same large sample of 237 NUTS-2 European regions—from Austria, Belgium, Bulgaria, the Czech Republic, Denmark, Estonia, Finland, France, Germany, Greece, Hungary, Italy, Latvia, Lithuania, Luxemburg, the Netherlands, Poland, Portugal, Romania, Slovakia, Slovenia, Spain and the United Kingdom (excluding small islands). The data cover the period 2006–2012 but the characteristics and relationships under analysis to correspond to structural aspects of the territories, being subject to relatively slow processes of transformation, in particular those related to the ecological features of territories, including their classification, regulation and utilization. These two studies are complemented by a third one, comprising the same set of regions, but focused on the impacts of immaterial aspects of the territories (like the qualifications of the work-force or regional innovation capabilities) on the productivity and competitiveness of the tourism sector. Due to unavailability of relevant data, some Mediterranean regions (mostly from the Balkan region) are not considered in the studies. However, some general tendencies identified in the econometric models computed may be of interest for all the Southern European regions.

During the period under analysis, a significant development in urban tourism could be globally observed, with the related rising importance of the contribution of several metropolitan areas in different locations of the continent for the growth of tourism in Europe. However, the Mediterranean "macro-region" was still a major tourist destination within the European context. As documented in Romão (2015) [Table 1], Île de France (Paris), Cataluña (Barcelona) and Inner London were among the 10 regions with higher volumes of tourism demand in 2012, while Outer London, Merseyside (Liverpool) and Berlin were among the 10 regions with higher growth rates of tourism demand between 2006 and 2012.

Moreover, the Southern part of Europe also revealed a very high concentration of valuable natural resources, as expressed by the share of the regional territories classified and protected within the Natura 2000 network (following harmonized criteria for biodiversity in the European Union). Figure 15.1 shows the spatial concentration in the Southern part of Europe of the univariate local indicators of spatial association (LISA) for this variable, with the clusters of regions with relatively high scores (high share of regional territory under protection both in the region and also in the contiguous territories) represented in red. On the contrary, clusters of regions with low scores for both variables are represented in dark blue. In South Europe, it is still possible to observe the existence of some regions with relatively low scores surrounded by regions with high scores (light blue). Finally, a few regions represented in light-red exhibit relatively high scores, although they are surrounded by regions with small fractions of their territories protected under Natura 2000.

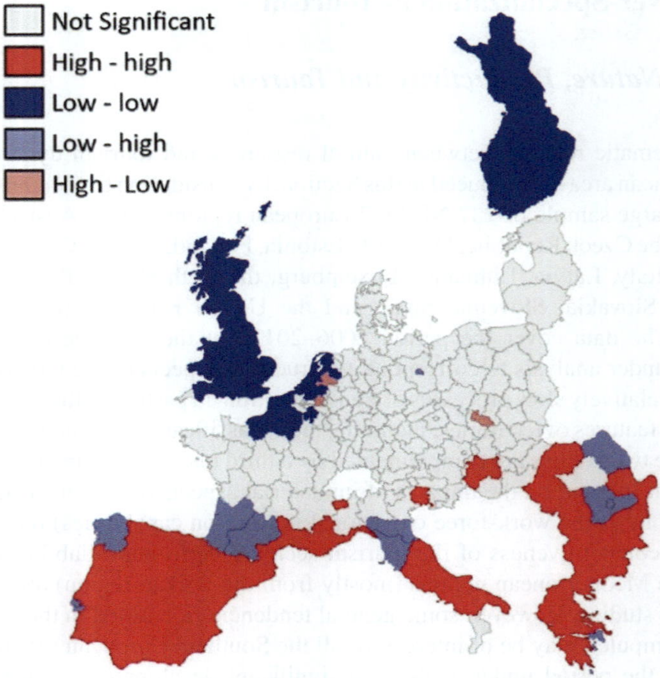

Fig. 15.1 Local indicators of spatial association for natural resources (Share of the Territory Classified as Natura 2000). *Source* European Commission, D.G. Environment

The percentage of the regional territory classified as Natura 2000 would be the indicator used for estimation of the determinants of regional tourism demand (Romão 2015) and regional tourism competitiveness (Romão et al. 2017). In the first case, the regional endowment in natural resources was found to have a positive relation with the nights spent in accommodation establishments, the dependent variable in the model. However, in the second case, natural resources would reveal a negative correlation with the gross value added by the tourism sector, the variable used as a proxy for regional competitiveness. This general tendency was identified by an econometric model, including the overall set of regions under analysis and covering a very large part of Europe. However, Fig. 15.2, representing bivariate local indicators of spatial association (LISA), shows that the problem is mostly observed in Mediterranean regions, where many regions with high scores for the protection of natural resources achieve low levels of value added by tourism activities (represented by light blue colour).

The combined analysis of the results of these two models and related exploratory spatial analyses suggests that many Mediterranean regions assumed forms of mass tourism development in areas with rich and sensitive natural resources, generating relatively low levels of value-added and economic benefits for the local populations,

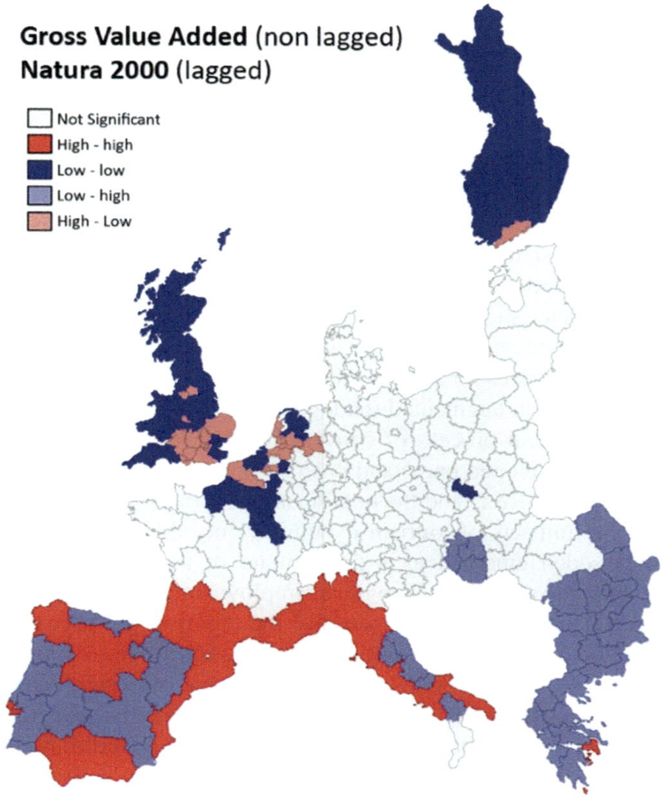

Fig. 15.2 Local indicators of spatial association relating natural resources and economic growth. *Source* Originally published in Romão et al. (2017)

despite the potential negative impacts on ecosystems and landscapes. These results would be complemented by a different type of analysis of the determinants of regional tourism competitiveness, comprising the same regions but focusing on the impacts of both traditional production factors (physical and human capital) and other contextual variables, such as level of specialization in tourism, the intensity of innovation, labour qualification or productivity (Romão and Nijkamp 2019).

In this case, assessing the relations between tourism specialization and competitiveness was particularly relevant: when specialization in tourism was measured by the share of the gross value added by this sector within the regional economy, a positive impact on the competitiveness of tourism activities was observed; however, there was a negative correlation with the gross value added by the tourism sector (the variable used to assess competitiveness) when considering the share of employment in tourism activities. This confirms the results obtained in the previously mentioned studies (Romão 2015; Romão et al. 2017), by clarifying that low levels of productivity are obtained in regions with labour intensive and low value-added tourism services.

15.2.2 Growth and Sustainable Development

By applying similar spatial econometric methods and focusing on the same regions and the same period, Romão and Nijkamp (2018) modelled and analyzed the impacts of territorial characteristics and tourism dynamics on regional economic growth, considering as dependent variable the annual gross of the domestic product per habitant (at constant prices for 2003 and harmonized according to Purchasing Power Standards). By using an augmented (endogenous) production function, the authors could identify a general process of convergence between regions (higher growth levels for the less developed ones) and positive impacts of growth arising from the regional investment in R&D and the qualifications of the work-force.

Regarding the impacts of tourism dynamics, only the share of tourism within regional value-added emerged as a determinant of regional growth, as tourism demand and the share of tourism in regional employment did not reveal a statistically significant correlation. Thus, only partially the "tourism-led-growth hypothesis" (Brida et al. 2016), assuming that tourism dynamics always has a significant impact on economic growth, could be confirmed. However—and most significantly—natural resources were found to be negatively correlated with economic growth. Taking into account the complementary results obtained in the previous studies, this is related to the positive correlation identified by Romão (2015) between natural resources and tourism demand, coexisting with a negative correlation between the endowment in natural resources and the value-added generated by the tourism sector (Romão et al. 2017). The local indicators of the spatial association presented in both studies clearly show that these problems have a particular incidence in the Mediterranean European regions.

In a broader perspective and focusing on a slightly larger sample of 252 European NUTS-2 regions, Romão and Neuts (2017) offered a comprehensive analysis of the previous relationships and impacts assuming the triple bottom line approach to sustainable development. The economic growth was measured according to the regional gross domestic product (GDP) per inhabitant (at current prices), the social dimension of sustainability was assessed by the level of regional unemployment (as a measure of social cohesion), while the environmental dimension was evaluated according to the level of CO2-emissions. Although the sample had some differences, and the method (structural equation modelling) was completely different from the previous approaches, the results were consistent, thus confirming the previous analyses.

In relation to tourism dynamics, the results showed that a higher share of employment in tourism is both related to high tourism demand and also to a low value-added by tourism services. On the other hand, high tourism demand and high value-added by tourism positively affect economic growth. However, the higher importance of employment in tourism appeared as correlated with higher levels of regional unemployment. In the particular case of natural resources, they appeared as positively related to both measures of tourism specialization (share of tourism in the gross value added and regional employment). This shows the importance of these assets as sources of path-dependence (Martin 2014). It was also identified (or confirmed) that higher endowments of natural resources were correlated with lower gross value added by tourism. Moreover, in this case, it was observed a negative correlation between these resources and the levels of employment.

Additionally, the approach proposed by Romão and Neuts (2017) revealed the existence of high levels of tourism demand associated with low value-added by tourism activities, confirming that—for these regions, which are mostly located in Southern Europe—tourism supply relies on mass products and services, related to an unsustainable utilization of resources (as shown by the negative correlations between natural resources, gross value added in tourism and regional economic growth). Moreover, the importance of tourism is correlated with high unemployment. At the same time, the only positive element in terms of the triple bottom line of sustainability is the low level of CO_2-emissions (although this can be related to the low development of other economic activities, like manufactures). For the regions where tourism sector plays a prominent role, it seems clear that its contribution for sustainable development is far from satisfactory, which has particular impacts on the Mediterranean region.

15.2.3 Resilience and Smart Specialization

The previously mentioned studies comprised a particular period of evolution of socio-economic systems, starting with a general tendency of economic growth, followed by an international recession impacting regional economies all over the world (starting in 2007 but with impacts in different moments and with different magnitudes for different regions). As the negative effects of this external shock persisted for several years—and the ability of the economies to avoid and to manage these impacts or to recover a growth path—the concept of socio-economic resilience would be largely adopted from engineering or ecological approaches to social science analyses (Modica and Reggiani 2014). In this context, Romão et al. (2016) analyzed the impacts of this international crisis in a specific Southern European region (Algarve, Portugal, where tourism is primarily based on mass sun-and-sea products with high seasonality, despite the abundance of sensitive ecological resources in large parts of the territory).

This study focused on the inter-sectoral relations within the regional economic structure and noted that the significant reduction in tourism activities observed in 2008 and 2009 would be quickly recovered, with the pre-crisis regional tourism demand being reached in 2010. However, high levels of unemployment would persist for a much more extended period, affecting, in particular, the youngest population. By considering four economic sectors (tradable goods, construction, tourism and non-tradable goods) and their inter-relations over a relatively long period, the analysis identified self-reinforcing mechanisms arising from the interactions between construction, tourism and non-tradable goods. These mechanisms operate both in terms of the inter-sectoral impacts of these activities on regional employment and on the value-added to the regional economy, being noteworthy that the tradable sector (comprising agriculture, fisheries and manufactures) was excluded from this dynamics, clearly losing importance within the regional economic structure over time.

When the international crisis affected tourism dynamics, there was also a significant negative shock on the construction sector, which had assumed large importance in regional employment over the previous two decades. In this context, the persistence of the problems of high unemployment (and lack of resilience) did not appear to be related to the dynamics of tourism, but to the evolution in the construction sector, whose recovery would be much slower. Confirming that regions highly dependent on tourism and construction would reveal lower levels of resilience, as observed by Milio (2014), the study emphasized the importance of the analysis of inter-sectorial relations and specialization patterns for the analysis of the long-term socio-economic impacts of tourism.

Following these concerns and taking into account sectoral priorities assumed within smart specialization strategies (RIS 3—regional innovation strategies 2014–2020) (Foray et al. 2012), Romão (2020) analyzed a broad set of regions over a relatively long period (2006–2017). By focusing on places where tourism was defined as one of the priority sectors for smart specialization, the study includes a large number of Southern European regions (45 territories from Portugal, Spain, Italy, Greece, Cyprus and Malta) among the 55 NUTS-2 regions under analysis. However, some territories from Germany (2 regions), Denmark (4), Austria (2) and Romania (2) were also considered.

As the time-frame for the analysis included a period of growth, an international recession and a process of recovery, it was possible to identify different types of economic impacts of tourism on regional economies: the relation with growth ("tourism-led growth hypothesis") and also different aspects of regional resilience, as defined by Martin et al. (2016)—vulnerability (how to manage the impacts of a recession), shock absorption (how the negative impacts were integrated into the economic structure) and recovery (how to return to a growth path). Additionally, the study explored how different forms of relatedness between tourism and other priority sectors for the regional economies—considering different levels and types of proximity—could have different impacts on their socio-economic performance.

It was interesting to confirm the process of convergence between the most and the least developed regions previously noted by Romão and Nijkamp (2018). However, in this case, it was also possible to observe that the most developed regions are the most resilient ones, revealing lower levels of vulnerability when facing negative impacts, higher ability to absorb a shock and a faster recovery towards a new path of economic growth. On the other hand, different (and eventually contradictory) effects of tourism dynamics were identified: tourism demand and the high importance of tourism within the regional value-added (a measure of specialization) contribute positively to the regional economic performance, both in terms of growth and resilience (when looking at the aspect of vulnerability and recovery). However, the high importance of tourism within the regional employment is negatively correlated with economic growth, while increasing regional vulnerability and slowing down the process of recovery, was associated with higher levels of unemployment.

The results also revealed positive impacts on growth and resilience arising from diverse options of smart specialization priorities, suggesting that different choices based on existing and path-dependent regional capabilities may lead to positive results. For example, the analysis revealed positive impacts from sectors with high proximity (like agriculture and food or mobility and transports) and also less proximity (like manufactures of advanced materials and technologies), suggesting that both related and unrelated variety potentially play a relevant role within regional innovation strategies. Thus, tourism may constitute a central element in a cluster of related activities, within a regional innovation strategy that may also consider other (unrelated) clusters of sectors (Boschma et al. 2016). Moreover, it was observed that a much higher number of unrelated sectors contributes to increasing regional resilience than to promote economic growth, suggesting that unrelated variety offers higher benefits for regional employment than related variety, as proposed by Frenken et al. (2007). However, it was again observed that specialization in the construction sector is correlated with low levels of regional resilience.

15.3 Overtourism Meets No-Tourism: Evolutionary Life-Cycles and Path-Dependent Futures

The characteristics of tourism dynamics described in the different studies presented in the previous Section relate to a situation in which the Southern regions of Europe are mature tourism destinations, well-integrated into international flows, with high demand (at least in the summer season) and the corresponding infrastructures to support tourism activities (for example, accommodation and transport services, entertainment, cultural and recreational facilities). Following the evolutionary conceptualization proposed by Butler (1980) for tourism destinations, Mediterranean regions passed, in general terms, through a long phase of development, in which demand tends to increase relatively fast, substantial investments in infrastructures are made, international companies are attracted, and the specialization in tourism tends to be reinforced in regional economic systems, by mobilizing financial resources for investment, territorial resources to accommodate infrastructures, facilities and services, knowledge and technological applications, along with different types of skills related to tourism activities (Kožić 2019). Eventually reaching a stage of stagnation (due to physical constraints to growth or loss of attractiveness), some regions have passed that development stage. On the other hand, very few territories in the Southern part of Europe can be considered in the initial stages of the cycle (exploration and involvement).

In this sense, overtourism (Dodds and Butler 2019; Sæþórsdóttir and Hall 2020) constitutes an important challenge for tourism in the protected areas of the Mediterranean, thus requiring appropriate solutions. In fact, as the previous studies suggest, current tourism dynamics generally relies on the supply of mass tourism products and services, with potentially high negative impacts on sensitive ecological resources, low value-added, a reduced contribution for economic growth, high vulnerability to external adverse shocks and low levels of socio-economic resilience. Thus, even the slow processes of convergence towards the levels of development of the most advanced European economies can be overturned when a period of recession occurs, with significant social consequences, in particular concerning the levels of unemployment and related inequalities.

The new recession faced by the global economy as a consequence of the COVID-19 pandemic brings the question of resilience back to the centre of the discussions about regional development strategies, in particular those involving a prominent role for the tourism sector. In fact, the systematic, long and generalized lockdowns imposed in many countries had unprecedented consequences for tourism activities, not only in terms of temporary losses of revenues and rising levels of unemployment but also implying the collapse of a large number of companies operating in different types of tourism-related services (Hall et al. 2020). Eventually, sensitive ecosystems could recover from excessive tourism pressure during this period, although these results do not seem clear yet, as pointed out by Corlett et al. (2020). In this context, challenges related to overtourism were suddenly replaced by challenges related to no-tourism—the absence of tourism—in the Mediterranean regions.

Similar structural circumstances are emphasizing the problems related to both overtourism and no-tourism: only if and when the tourism sector achieves a critical socio-economic role in a region, the negative consequences of the excess of tourism (for example, degradation of resources, inflation, distortion of economic structures, deindustrialization, disturbance of lifestyles and loss of quality of life) or related to the absence of tourism (such as economic stagnation, unemployment, risk of poverty, lack of public funds, unused infrastructures, facilities and services) can be observed. Regions where tourism is scarce and not crucial within the regional economic structures surely do not suffer the negative impacts of overtourism or no-tourism.

It must be taken into consideration, however, that regional development (including the aspects related to tourism) is an evolutionary process with different sources of path-dependence, as systematized by Martin (2014). Focusing only on those with direct relation with tourism, several aspects can be generally identified in tourism-dependent regions: sunk costs (large-scale investments, like those related to transportation infrastructures or thematic parks for entertainment), agglomeration economies, different types of knowledge spin-off or technological lock-in (related to high specialization in tourism and concentration of knowledge in this sector) and interregional linkages and dependencies (arising from the territorial dispersion of tourism attractions and resources). With particular importance for the Mediterranean area, natural resources also constitute a strong source of path-dependence, once tourism dynamics is often strongly linked to natural features of the territories.

The constraints imposed by those *sunk costs* as sources of path-dependence are more important when the regions reach the stage of development within the life cycle model proposed by Butler (1980). During that (eventually long) period, tourism demand has fast growth, opening essential business opportunities, which tend to be very attractive to capital investments, the labour market and even to local authorities, as a form of ensuring a relatively fast process of economic prosperity. However, this process may also lead to the concentration of knowledge and innovation capabilities in activities with low value-added and low potential for the integration of advanced technologies, thus constraining the opportunities for development in the future, as exemplified by Kožić (2019) when looking at the qualifications of the work-force in Croatian regions.

In general terms, Southern European regions mostly follow a competitive strategy based on low prices (cost-leadership). These dependencies were historically reinforced, and their importance constrains the opportunities to implement significant changes towards the implementation of differentiation strategies based on the richness of their natural and cultural resources. This different strategic approach could potentially lead to a more sustainable form of tourism development based on the provision of unique experiences supported by local, territorial resources. Such a strategy should also contribute to generating higher value-added for the regional economies, by reinforcing the linkages with other local economic activities, rather than promoting a continuous increase in the number of visitors. In particular, digital technologies can potentially promote the emergence of innovative services enhancing the interrelations between tourism and the sectors with higher proximity, exploring the related variety within the regional economic structures (Neffke et al. 2009).

Diverse creative industries linked to the promotion, understanding and interpretation of local natural and cultural heritage, along with other territorial features, may contribute to tourism diversification while supporting the emergence and consolidation of new activities. On the other hand, activities related to mobility, transports, energy consumption and production or water saving, with an important role within the tourism sector, may also be a source of technological development and innovation with impacts on other industries, while contributing to create and to promote an image of sustainability of the destination.

Taking into account the aspects related to resilience and the different shocks (described in detail by Gössling et al. 2020) severally affecting the tourism industry in the last two decades (9/11 attacks, the international crisis started in 2007 or different pandemic diseases until the recent COVID-19), a strategy of diversification, clearly reducing the regional dependence on tourism activities, seems highly advisable for the Mediterranean regions. Such a strategy should offer positive impacts both in terms of controlling the problems related to overtourism and also the negative impacts related to the absence (or severe reduction) of tourism demand, if and when this occurs. As such, the results obtained by Romão (2020) seem to support a strategic combination of unrelated clusters of related activities (Boschma et al. 2016), where tourism would play a central role within one of them. In fact, strategic development options supported by a diversified regional economic structure seem clearly more adequate to achieve a process of sustainable development, while reinforcing regional resilience.

15.4 Conclusion: How Much Is Too Much?

Supported by a detailed analysis of the relations between natural resources, tourism, specialization patterns, growth, resilience and sustainable development, this work emphasized how an excessive dependence on the tourism sector may lead to a strong regional vulnerability both to overtourism and to no-tourism. In the first case, sensitive territorial resources can be threatened by the excessive presence of tourists, while local communities do not achieve significant long-term economic benefits arising from the provision of mass tourism products and services, supported by labour-intensive production processes and generating low levels of value-added. Although the studies supporting these conclusions do not distinguish between nature-based tourism practices or others, or if they are developed inside or outside sensitive natural areas (which justifies more detailed further research), it is still clear that regions with a higher endowment of natural resources are generally attracting large volumes of tourists, while achieving relatively low levels of value-added for the local economies. On the other hand, they are strongly exposed to negative shocks, potentially affecting tourism demand, thus implying severe socio-economic problems for local communities.

It is also noteworthy that tourism development may be seen as a long-term process, with strong sources of path-dependent, not only including the natural resources that open the opportunities for the supply of very attractive tourism products (like sun and sea), but also all the investments in infrastructures, mobilization of capital resources and investments, local knowledge and skills, technological capabilities or firm creation processes, which tend to concentrate around the tourism sector during the (typically long) stage of development of the destinations. Despite these potential opportunities, overspecialization in tourism may also bring long-term problems related to lack of regional resilience, low levels of technological development, low value-added by the regional economy or degradation of the qualifications of the labour force.

Analyzing the process of local tourism development and identifying if and when tourism starts to be a problem rather than a solution seem crucial policy questions in order to achieve better socio-economic performance and a more sustainable long-term development process. Tourism offers significant benefits in the short-term but severely constraining opportunities in the future. The sector is also highly sensitive to a crisis, as the COVID-19 pandemics emphasized. In this context, supporting the development of clusters of activities not related to tourism (and preferably with higher levels of technological incorporation and value-added), along with the promotion of the potential interactions between tourism and (relatively) related sectors (including food production, information and communication technologies, creative industries, mobility and transports, or energy production and distribution) appear as crucial challenges for smart specialization and regional innovation strategies in the Mediterranean area.

References

Boschma R, Coenen L, Frenken K, Truffer B (2016) Towards a theory of regional diversification. Pap Evol Econ Geogr 16:17

Brida JG, Cortes-Jimenez I, Pulina M (2016) Has the tourism-led growth hypothesis been validated? a literature review. Curr Issues Tour 19(5):394–430

Butler R (1980) The concept of a tourism area life cycle of evolution: implications for management of resources. Can Geogr 24(1):5–12

Corlett RT et al (2020) Editorial—impacts of the coronavirus pandemic on biodiversity conservation. Biol Conserv 246:108571

Dodds R, Butler R (2019) The phenomena of overtourism: a review. Int J Tour Cities 5(4):519–528

Foray D, Goddard J, Beldarrain X, Landabaso M, McCann P, Morgan K, Ortega-Argilés R (2012) Guide to research and innovation strategies for smart specialization. S3P—European Union, Regional Policy, Brussels

Frenken K, Van Oort F, Verburg T (2007) Related variety, unrelated variety and regional economic growth. RegNal Stud 41(5):685–697

Gössling S, Scott D, Hall CM (2020) Pandemics, tourism and global change: a rapid assessment of COVID-19. J Sustain Tour. https://doi.org/10.1080/09669582.2020.1758708s

Hall CM, Scott D, Gössling S (2020) Pandemics, transformations and tourism: be careful what you wish for. Tour Geogr. https://doi.org/10.1080/14616688.2020.1759131

Kožić I (2019) Can tourism development induce deterioration of human capital?. Ann Tour Res 77(C):168–170

Lu J, Nepal SK (2009) Sustainable tourism research: an analysis of papers published in the Journal of Sustainable Tourism. J Sustain Tour 17(1):5–16

Martin R (2014) Path dependence and the spatial economy. In: Fischer M, Nijkamp P (eds) Handbook of regional sciences. Springer, New York, pp 609–629

Martin R, Sunley P, Gardiner B, Tyler P (2016) How regions react to recessions: resilience and the role of economic structure. RegNal Stud 50(4):561–585

Milio S (2014) Impact of the economic crisis on social, economic and territorial cohesion of the European Union, vol. 1. Directorate-general for internal policies. Policy Department B, Luxembourg

Modica M, Reggiani AN (2014) Spatial economic resilience: Overview and perspectives. Netw Spat Econ 15(2):211–233

Neffke F, Henning M, Boschma R (2009) How do regions diversify over time? Industry relatedness and the development of new growth paths in regions. Econ Geogr 87(3):237–265

Romão J (2015) Culture or Nature: a space-time analysis on the determinants of tourism demand in European regions. Discussion Papers Spatial and Organisational Dynamics 15-3

Romão J, Guerreiro J, Rodrigues P (2016) Tourism growth and regional resilience: the 'beach disease' and the consequences of the global crisis of 2007. Tour Econ 22(4):699–714

Romão J, Guerreiro J, Rodrigues P (2017) Territory and sustainable tourism development: a space-time analysis on European regions. Region 4(3):1–17

Romão J, Neuts B (2017) Smart tourism, territorial capital, and sustainable regional development: experiences from Europe. Habitat Int 68:64–74

Romão J, Nijkamp P (2018) Impacts of tourism and territorial capital on regional development: a spatial econometric analysis. Int J Tour Res 20(6):819–829

Romão J, Nijkamp P (2019) Impacts of innovation, productivity and specialization on tourism competitiveness—a spatial econometric analysis on European regions. Curr Issues Tour 22(10):1150–1169

Romão J (2020) Tourism, smart specialization, growth, and resilience. Ann Tour Res 84:102995

Sæþórsdóttir AD, Hall CM (2020) Visitor satisfaction in wilderness in times of overtourism: a longitudinal study. J Sustain Tour. https://doi.org/10.1080/09669582.2020.1817050

Weaver DB, Lawton LJ (2007) Twenty years on: the state of contemporary ecotourism research. Tour Manag 28:1168–1179

Chapter 16
Changing the Growth-Focused Mindset: A Pathway Towards Sustainable Tourism Development

Nicola Camatti, Carolyn Smith, and Jan van der Borg

Abstract Many tourism destinations have long pursued the goal of growth 'at all costs'; the result in many instances has been environmental, socio-cultural and ecological degradation. Amidst mounting evidence of the unsustainability of this approach, we are now witnessing a series of healthy, albeit still timid, attempts to move towards more sustainable tourism development models. Such initiatives experiment with alternative goals and development paths which emphasise the quality rather than the quantity of tourism, alongside the pursuit of wider social and environmental objectives. Yet, despite these positive case studies, change remains slow, and many destinations continue to manage tourism via top-down governance tools—an approach which limits their capacity for sustainable development. Even the most rigid sustainability criteria and the adoption of new governance models have been shown to be insufficient in the absence of a radical change of mindset of tourism stakeholders. A more holistic system is needed, which not only considers but actively engages tourism's wide range of stakeholders to effectively navigate sustainable tourism development.

Keywords Sustainable tourism development · Stakeholder engagement · Transition, mindset · Growth · Knowledge · Awareness-action gap · Sustainability education

---

N. Camatti (✉) · J. van der Borg
Department of Economics (IT), University Ca' Foscari of Venice, Venice, Italy
e-mail: nicola.camatti@unive.it

J. van der Borg
e-mail: jan.vanderborg@kuleuven.be

C. Smith
Department of Geography, University of Cambridge, Cambridge, UK
e-mail: cs932@cam.ac.uk

J. van der Borg
Division of Geography and Tourism (B), KU Leuven University, Louvain, Belgium

A. Mandić and L. Petrić (eds.), *Mediterranean Protected Areas in the Era of Overtourism*, https://doi.org/10.1007/978-3-030-69193-6_16

16.1 Introduction

The aim of this chapter is to outline some of the new ways to frame and deliver sustainable tourism development and highlight key areas for future exploration and elaboration. To this end, the chapter first recalls the well-known consequences of unsustainable tourism growth on the competitiveness and survival of a destination, while highlighting the limitations of the strategies implemented by many destinations to deal with these undesirable effects and their inexorable decline. The contours of an alternative model are then sketched, one which includes the resident population and local entrepreneurs as key tourism stakeholders to determine not only the destination development path but the long-term goals and priorities which will guide its course. The second part of the chapter examines the awareness-action gap, which stymies many sustainability initiatives. The need for educational tourism initiatives, and a consideration of the 'inner dimension' of man and mindfulness, is then examined. The overriding theme of the chapter is the need for tourism destinations to undertake complementary strategies which address sustainable development challenges in a holistic, inclusive and iterative manner. Governance frameworks, business models and socio-cultural values are all key instruments to achieve sustainable development, but all are insufficient alone.

16.2 Unsustainable Tourism: Causes and Consequences

The development of the tourism market knows various, distinct phases. The first phase of this development process is widely known as the period of the Grand Tour: the phenomenon that lies at the very basis of the word 'tourism'. In the seventeenth and eighteenth centuries, young, upper-class men travelled Europe as a rite of passage to read in foreign libraries; study works of art and architecture; admire the landscape and meet their peers. Travel was motivated by a desire to collect both knowledge and experiences; the challenging nature of the journeys was part of the appeal. As Europe's rail network improved, and the preoccupation with neo-classical culture abated, the tradition declined. In its place: 'Cook's Tour' (as associated with Thomas Cook) capitalised on the new railway networks and growing hospitality sector to make travel accessible for the first time—bringing about the birth of contemporary tourism.

After the second world war tourism evolved again. In many industrialised countries, the average income per capita, the amount of paid vacation and car ownership rose quickly, and tourism rapidly became a mass phenomenon. Not only did the number of people travelling rise exponentially, but tourists were inclined to visit the same places at the same time of year: coastal areas during the summer months. Mass tourism was born and with it the mass tourism business model, based on the replication of successful formulas, economies of scale and price competition. Although the tourism market diversified significantly in subsequent decades, this particular business model continues to dominate, both within specific forms of tourism (such as cultural tourism or urban tourism) and for connected industries (such as the cruise industry or the airline industry). This business model focuses on the economic drivers and impacts of tourism; this legacy is evident from the reports of tourism organisations (such as the WTTC or the UNWTO) from the 1990s and the early 2000s.

The exponential growth of tourism demand—and the corresponding growth of the tourism industry—alongside the spatial and temporal concentration of tourism within both high season and vacation hotspots, sparked the first criticisms of this model (see for example Krippendorf 1971), in particular with respect to tourism in natural areas. In the 1990s, a number of authors began to argue that other destination typologies (cities, for example, as Van der Borg (1992) has shown) also suffer from excessive touristic pressure—pressure related to the mass tourism business model they were embracing. Today, cities like Venice (Bertocchi et al. 2020) and Barcelona (Russo and Scarnato 2017) suffer from what is now frequently termed 'overtourism' (UNWTO 2018).

While overtourism is a growing area of research, many directly involved in tourism development processes—stakeholders such as policymakers and those within the tourism industry—still very much foster the traditional business model which favours quantity over quality and the economy above a destination's socio-cultural and environmental and ecological dimensions. The success of a destination is still predominantly measured in the number of overnight stays and the total expenditure of visitors, and this is reflected in the statistical information that is systematically gathered for tourism in countries, regions and places.

This form of development, as Van der Borg (2017) has argued, is partly related to the intrinsic nature of the tourism product. Primary tourism products, or attractions, are often both unique (and therefore extremely scarce), and public goods. This combination of factors often leads to the spontaneous overutilisation of tourism assets, whether they are natural or cultural assets, touristic facilities, infrastructure or public spaces. Spontaneous market forces do not automatically induce optimal allocation of destination assets, but rather maximise their utilisation in order to pursue economic gain. This is what might be called the 'tragedy of tourism commons', to paraphrase Hardin (1968). It has proven to be very difficult to keep popular destinations from following a similar, devastating development path and to embrace a radically different business model.

This form of tourism asset misallocation is evident in the framework presented by Butler in 1980, namely that of the Tourism Area Life Cycle (or TALC, see Fig. 16.1). The life cycle illustrates an early phase of underutilisation (the the 'Exploration' and 'Involvement' stages in Fig. 16.1)—the polar opposite to overtourism, where the non-optimal allocation of tourism assets (or market inefficiency)—where many attractions are already in place, but demand and the associated income is not yet sufficient to cover the costs. This phase is unsustainable as the absence of demand, and hence value discourages entrepreneurs and local policymakers from investing in tourism. In contrast, overtourism is reflected in the 'Stagnation' stage of the life cycle. Butler indicates the existence of a 'critical range of elements of capacity'; he,

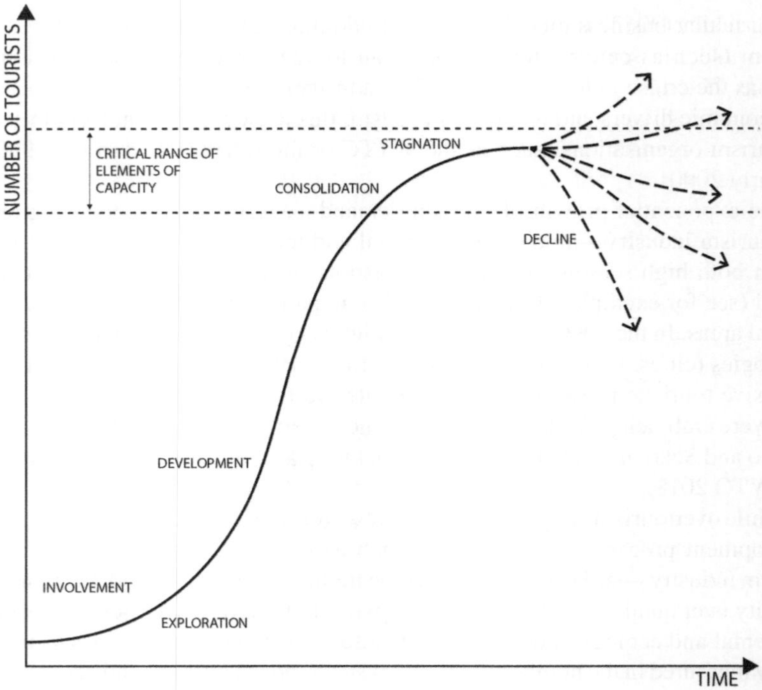

Fig. 16.1 Butler's tourist area life cycle. *Source* Adapted from Butler 1980, p. 7

therefore, suggests that above a certain threshold tourism becomes what now is called overtourism. Surpassing this critical range of elements of capacity, or the Tourist Carrying Capacity (see for example Van der Borg 2017 or Bertocchi et al. 2020), provokes a range of economic, logistical, social and environmental externalities that cause the balance between collective benefits and collective costs to become negative. This not only damages the quality of life of the local population and the business interests of local firms that are not part of the tourism industry, but eventually also degrades the visitor experience and, therefore, the tourism industry itself.

16.3 The Path to Sustainability: What (or Who) Is Missing?

The debates surrounding sustainable tourism development have successfully drawn widespread attention to the need to balance economic interests with socio-environmental concerns; in recent years the concept has become increasingly important within policy frameworks at all levels of governance, adopted within the industry and recognised by consumers. Yet despite this apparent success, the implementation of sustainable tourism development remains relatively limited and highly localised;

both Buckley (2012) and Sharpley (2020) observe little evidence of progress towards sustainable tourism development in recent decades, or even of a more sustainable tourism sector.

Undoubtedly, industrial change has failed to keep pace with academic research; meaningful progress towards sustainable tourism development has become stuck in academic and governance circles (Murphy and Price 2005; Sharpley 2020). This disconnect—between research and policy, and the practical realities of the industry and consumer priorities—is a significant barrier to sustainable tourism development: it situates academics and policymakers as the only far-sighted actors within the tourism system, places the burden of sustainable development on policymakers and largely relies on traditional policy tools to implement and enforce sustainable practices.

16.3.1 The Limitations of Traditional Policy Tools

The challenge of sustainable development, both more generally and with respect to tourism, has highlighted the limitations of traditional governance methods and policy instruments.

The traditional tourism policy domain is primarily concerned with legal and fiscal manipulation via market-based and 'command-control' tools (Hudson and Miller 2012). Market-based instruments are designed to allow governments to incentivise or discourage tourists, companies and other stakeholders in adopting certain patterns of behaviour by adjusting prices. Meanwhile, the introduction of rules, obligations and prohibitions for both tourists and businesses can aim to control access to historic city centres or regulate production standards. Equally widespread are tools that use public spending to ensure the provision of certain services or the protection of public resources which are essential for the sustainable development of a destination (Bramwell 2012). In addition, Zaccai (2012) highlights the recent growth of voluntary tools which have become more commonplace within the tourism policy domain. These voluntary schemes (such as Corporate Social Responsibility, see Taback and Ramanan 2016) have been shown to be more effective for industrial stakeholders than for consumers (McKercher et al. 2010; Zaccai 2012), but their inclusion within the available range of policy instruments and incentives is important for 'beyond compliance' initiatives (Rivera and De Leon 2005; Budeanu et al. 2016).

The limited ability of these tools to address sustainable development has long been recognised (Butler 1991; Dovers 1996; Hunter 2002; Sharpley 2009). The reality is that policy frameworks evolve far more slowly than the current pace of environmental (and related socio-economic and political) change; a result is a reactionary approach to sustainable development which leaves policymakers 'playing catch-up' (Hall 2011, p. 654). However well-designed and implemented, this limits the efficacy of traditional policy-driven approaches to sustainable development; criticism largely centres on the propensity for initiatives to either fail or to result in unforeseen, undesirable consequences (Fodness 2017). Efforts are often undermined by a lack of information: at its most basic, this could concern the appropriate weighting of tariffs or subsidies but, more often, this data gap concerns an understanding of how measures can best be deployed to influence the choices of stakeholders, how and why actions result in unexpected indirect costs, the extent of management costs, and the fundamental efficacy and feasibility of policy aims (Bramwell 2012). The issue is not that these instruments have no value, but alone they are simply unable to address a challenge as complex as sustainable development.

Traditional policy instruments are best suited to specific, spatially defined problems at the local scale (Dovers 2005). This may seem an obvious place to implement sustainable tourism development initiatives, given the industry's dependence on its host destination, but tourism development is an inherently multi-scalar phenomenon; dependent on the external environment and tied to external sectors. It, therefore, cannot be fully understood in isolation. Hall (2011) suggests that as the territorial scale increases (beyond local to regional, national and international levels), tourism sustainability becomes increasingly influenced by factors *beyond* the tourism policy domain. In today's interconnected society, and in the context of global climate change, large-scale dynamics often determine local patterns of development. These large-scale dynamics, therefore, have the potential to undermine the localised approach to policy-making, leading Hall to argue that if tourism policy focuses solely on micro-scale solutions 'it may be inherently doomed to fail' (2011, p. 654).

16.3.2 Innovation, Inclusion and Engagement: A Model for Sustainable Tourism Development

There is growing consciousness of the requirement for innovation to develop new approaches to sustainable tourism development; address the limitations of traditional policy tools and subvert Butler's model of decline, as outlined earlier in this chapter (Smith 2004; Hjalager 2010; Brandão et al. 2019; Hall and Williams 2020). Yet the study of tourism innovation, particularly at the local level, remains in its relative infancy (Costa and Carvalho 2011; Bagiran Ozseker 2019). Both tourism and innovation are highly rooted to place: tourism is entirely dependent on the host destination, while successful innovation systems are embedded within institutional and socio-cultural networks. Authors, therefore, highlight the importance of tourism innovation at the local level and argue that this 'destination point of view' should provide the foundation from which larger-scale innovation is developed (Bagiran Ozseker 2019; Brandão et al. 2019, p. 227).

A destination's innovation capacity hinges on the formation of effective, collaborative stakeholder networks; these are critical to ensure access to the necessary knowledge, infrastructure and financial resources (Jacobsen 2005). Various models have been developed to better articulate the role played by governance, business, research institutions, education and civic society in a networked system of tourism stakeholders (see Costa 1996; Fundeanu 2015; Bagiran Ozseker 2019). Figure 16.2 illus-

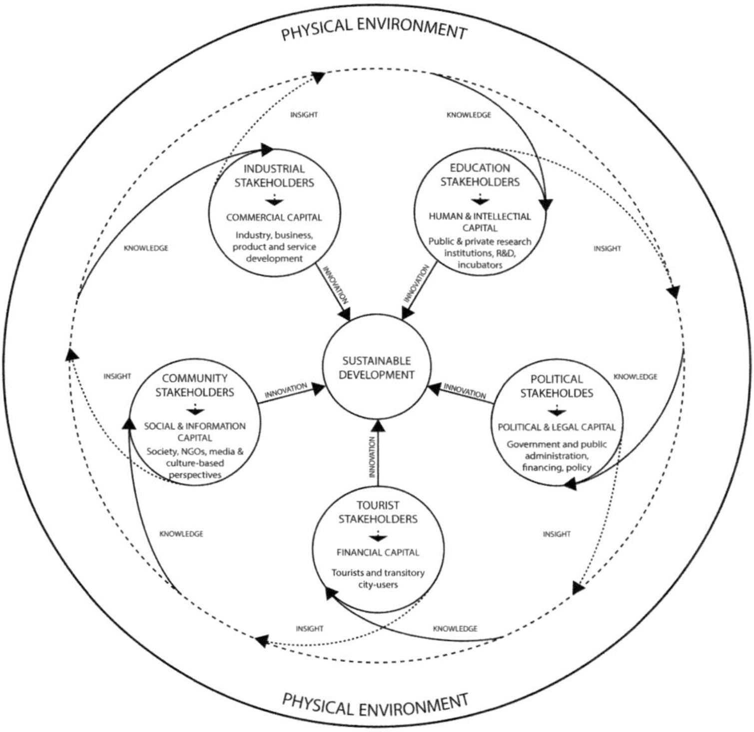

Fig. 16.2 A model for sustainable tourism development based on the collaborative circulation of knowledge. *Source* Drawn after Levett (1998), Laine (2010), and Carayannis et al. (2012)

trates a new conceptual cluster model for sustainable tourism development, which draws inspiration from the general 'Russian doll' model for sustainable development (see Levett 1998), the stakeholder network model proposed by Laine (2010) and the Quintuple Helix model for innovation (see Carayannis et al. 2012).

The model describes a holistic system, composed of five *equally important* stakeholder clusters (or subsystems), linked by a circular process of knowledge sharing and collaboration. Here, the academic and governance spheres are not set apart from the other stakeholders, and the emphasis has been shifted away from economic capital alone by acknowledging that each of the five stakeholder subsystems has a valuable asset (capital) at its disposal which is necessary to effectively navigate sustainable development challenges (Carayannis et al. 2012). The importance of formal knowledge production (within the education subsystem) is recognised but not given precedence over other forms, such as local knowledge from community stakeholders. Innovation occurs as knowledge produced by one set of stakeholders is shared and developed using fresh insight from another group.

- *Political stakeholders*: the governance system remains critically important; it generates 'political and legal capital' in the form of plans, regulations, policies and political leaders.
- *Education stakeholders*: the academic system includes all levels of education and research in universities, institutions and schools; it generates two forms of capital. First, the system develops 'human capital' with the necessary skills and insight for sustainable development; second, it produces 'intellectual capital' through the formal knowledge created within this system.
- *Industrial stakeholders*: the tourism industry is ultimately the subject of sustainable development patterns; it primarily focuses on the generation of 'commercial capital' through its activities.
- *Tourist stakeholders*: as the primary consumers of tourism 'products'; tourists predominantly generate 'financial capital' for the destination or sector.
- *Community stakeholders*: the final subsystem is composed of the local population or wider public at a larger scale; it is concerned with two forms of capital. The first is 'social capital', as associated with culture and social values; the second is 'information capital', derived from local empirical knowledge and communication networks.

Tourism is an inherently spatial phenomenon: it is an industry centred on selling an experience of a specific place; as such, the success and competitiveness of the tourism industry is entirely dependent on the destination itself. In other words, unsustainable tourism development risks undermining its own source of income by degrading the quality of its host destination. Tourism stakeholders are, therefore, mutually dependant on one another to ensure that the industry retains long-term resilience through sustainable tourism development. The model situates all stakeholders within the finite constraints of the 'physical environment' accordingly, and the widest possible participation and inclusion within this framework is necessary for its success.

The model in Fig. 16.2 is generic and can be applied at different scales, as illustrated in Fig. 16.3. The priorities for sustainable development will change according to the territorial scale: the local level is primarily concerned with site capacities (i.e. Tourism Carrying Capacity or waste management), resident participation and industrial engagement, while the international scale is likely to focus more on socio-cultural factors (as explored later in the next section of this chapter) and global climate impacts (e.g. air travel emissions). In each network, for each given destination, a group, actor or agency will serve as a 'broker' to initiate and coordinate the process described by the model. In many contexts, this is likely to be a government agency (as in the model presented by Fundeanu 2015, p. 748) or a working group of key public and private stakeholders, but the precise dynamics will vary considerably according to the specific dynamics of power and trust in a given locality. It is critical, however, that the local networks form the foundations from which the larger networks (regional, national and international) are developed, allowing local dynamics to influence initiatives at larger territorial scales.

Tourism cannot hope to meet the needs and desires of *all* stakeholders *all* of the time; balancing these needs requires communication, empowerment and trust

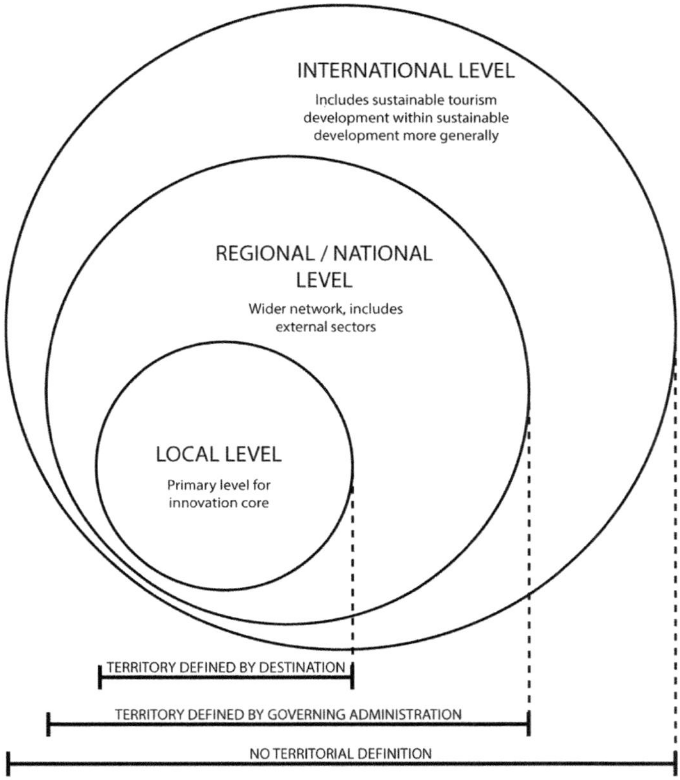

Fig. 16.3 The interdependent territorial scales for sustainable tourism development. *Source* The authors

(Higuchi and Yamanaka 2017; Nunkoo 2017). This must be approached as a *continuous process* to be managed, based on knowledge sharing and active engagement, which will allow specific sustainable development goals and practices to remain resilient, evolving as new challenges, priorities and conflicts emerge.

16.3.3 Locating a Holistic Policy Paradigm

In today's globalised and interconnected society, power structures in most developed democracies have become decentralised, liberalised and market-orientated; this has dramatically decreased the efficacy of 'top-down' policy models (Loorbach 2010). Yet despite dissatisfaction with the 'creeping irrelevance' of tourism policy tools to address sustainable tourism development (Fodness 2017, p. 1673), viable alternatives have been slow to emerge. Hall (2011, p. 656) attributes this to an inability to learn effectively from policy failure, arguing that governance is largely stuck in a 'superficial' pattern of social policy learning, which tends to focus on the incremental modification of existing policy tools. This style of technical learning is comparable to 'single-loop' learning (Argyris 1976; Leeuw et al. 1994; Grin and Loeber 2017), as opposed to 'double-loop' or 'triple-loop' learning, where the failure results in the reassessment of not only the goals and tools but the core framework of ideas and assumptions that were used to define the nature of the problem (Argyris 1976, 2002; Cartwright 2002; Hall 2011).

Transition Management is one response to this challenge: an emerging complexity-based policy paradigm which aims to accelerate the development of a governance framework capable of guiding sustainable development at a systematic scale (Loorbach 2010; Loorbach and Rotmans 2010, p. 239). In the past two decades, Transition Management has been applied at regional and national scales in a range of contexts for various sectors including waste (Paredis 2011), mobility (Kemp and Rotmans 2004; Kemp et al. 2011; Scuttari et al. 2016) and energy (Kern and Howlett 2009); it was first applied to tourism via a Norwegian national initiative in 2010, with 'mixed results' (Gössling et al. 2012, p. 913). The main characteristics of Transition Management are summarised by Rotmans et al. (2001, p. 22) as being concerned with shaping short-term innovation through long-term sustainability considerations (25 years or more); integrating multiple domains, actors and territorial scales; focusing on learning philosophy; aiming for both system innovation and system improvement and maintaining a wide variety of options for future use.

Essentially, Transition Management is a process which empowers a small working group of key stakeholders (usually less than 15) from different backgrounds to actively participate in the formation of new governance tools and approaches (see Loorbach 2010); it is best described as a perspective, rather than a policy instrument (Kemp and Rotmans 2004). The stakeholders are engaged in envisioning exercises to devise long-term sustainable goals, determine potential barriers, define potential pathways and develop new experimental strategies. These strategies are then subject to reflexive monitoring and evaluation, which informs and iterates the aims and approach. The experimental approach to policy development is key; the learning process is as important as the goals and strategies it defines (Loorbach 2010).

At a smaller scale, the Living Lab methodology is gaining popularity across Europe and is of growing interest to policymakers, despite its limited presence in academic literature (Kvisleius et al. 2009; Dell'Era and Landoni 2014). The Living Lab emerged from the intersection between Transition Management, Open Innovation and Collaborative Consumption (Mastelic et al. 2015); the concept has been widely applied to product and technology development, and city design. Essentially, the Living Lab is a participative, learning-focused development of the community consultation, which integrates end-users within design processes. The approach is highly pertinent for the local dimension of tourism development and has been successfully applied to tourism destination management to increase stakeholder engagement, destination innovation capacity and the development and adoption of voluntary sustainability criteria (MacPherson et al. 2008; Guimont and Lapointe 2016).

Both Transition Management and the Living Lab are relatively new methodologies; they are processes rather than prescriptive strategies and therefore offer no guarantees of success. The literature highlights the extent to which both approaches are highly dependent on specific contextual dynamics (place, sector and scale) and active governance (Loorbach and Rotmans 2010; Guimont and Lapointe 2016): Gössling et al. (2012, p. 912) warn that little will emerge from the process itself if stakeholders identify the need for new or modified regulation and this is not followed up by policymakers. The methodologies have been developed in a European context—Transition Management, in particular, has, so far, largely been applied within Nordic political cultures (Loorbach and Rotmans 2010; Gössling et al. 2012)—and questions, therefore, remain as to how adaptable they may be within different political contexts with varying degrees of informality and potentially inconsistent support from distinct levels of governance. Yet both approaches do illustrate the potential role for active (rather than didactic) governance to engage with and empower stakeholders as a pathway for sustainable development.

While the critical importance of tourism stakeholders to the success of sustainable development initiatives is often treated as self-evident in academic, policy and planning literature; little attention has been given to detailing guidance as to how stakeholders might be identified, engaged and included (Moscardo 2011; Budeanu et al. 2016, p. 288). The vast majority of tourism enterprises are small businesses; these firms have a lot to gain from inclusion within a collaborative network such as the system proposed in Fig. 16.2, and processes like Transition Management and the Living Lab (Lynch and Morrison 2007; Brandão et al. 2019), but communicating this immaterial value can be challenging (Hakkarainen and Hyysalo 2013; Mastelic et al. 2015).

In practice, maintaining the engagement of, and managing conflict between, stakeholders remains a significant barrier to sustainable tourism development (Moscardo 2005; Dodds and Butler 2010; Vellecco and Mancino 2010; Hatipoglu et al. 2016). Although many models describe the inclusion of a variety of stakeholders, the groups most often included and empowered in practice tend to be large businesses and government agencies (Moscardo 2005, pp. 33–35; Budeanu et al. 2016, p. 288). Transition Management actively engages only a small group of 'frontrunner' stakeholders (Loorbach 2010); the criteria by which these stakeholders are identified and are judged to be worthy of representation will directly affect the characteristics of the resulting approach to sustainable development. In contrast, the Living Lab tends to operate at the microscale where a wider variety of smaller stakeholders can be more easily engaged, yet questions remain as to how this model might be best scaled up, or included within larger-scale models (Guimont and Lapointe 2016).

While it may make the process of defining shared goals and strategies more challenging, the inclusion of the widest possible range of stakeholder perspectives is necessary to effectively develop a resilient approach to sustainable development. Examining who is empowered by these processes, how key stakeholders are effectively identified, and what impact this selection has on the outcomes are critical elements within the future research agenda, best expanded through case study analysis. What is more, there is a need for long-term assessments, which are able to analyse how these relationships are maintained and evolve over time.

16.4 A Sustainable Mindset: The Socio-Cultural Conditions for Change

While the mitigation of unsustainable development can be framed by policies and regulation, it has become increasingly clear that sustainable development cannot be achieved through the use of new technology and governance alone (Ericson et al. 2014; Wamsler and Brink 2018). Tourism development is a human problem; it, therefore, has a human solution. Yet while the social dimension of sustainability has been accepted for some time, there remains little consensus as to what this means in practice. Our daily choices—and therefore consumer trends and economic spending patterns—are strongly influenced by a mixture of values, emotions, self-concepts, social norms and cultural associations (Sorin 2010); scholars from various disciplines are therefore turning their attention to understanding how these factors relate to sustainable development. Research is focusing increasingly on the socio-cultural factors which underpin long-term, stable changes to both individual and collective behaviour (Burns and Bibbings 2007; Hall et al. 2015).

16.4.1 The Awareness-Action Gap

Many studies indicate that while tourism stakeholders might value sustainability conceptually, it's currently unlikely to influence their choices and behaviour (Budeanu 2007; Timur and Getz 2009; McKercher et al. 2010; Miller et al. 2010). Clearly, awareness alone is not sufficient to catalyse change (Antimova et al. 2012; Mihalic 2016; Lehtonen et al. 2018; Mustapha et al. 2020); overcoming this disconnect between awareness and action requires three fundamental factors: knowledge, a sense of responsibility and a clear strategy.

It is widely accepted that stakeholder participation within sustainable development initiatives is highly dependent on access to knowledge (Moscardo 2005; Rivera and De Leon 2005; Cole 2006; Frisk and Larson 2011). Yet while education has been a core focus of the sustainability agenda for some time, the standard sustainability pedagogy has had limited success in translating knowledge into action (Finger 1994; Stern 2000; Frisk and Larson 2011). Indeed, it has been shown that those with the most knowledge of environmental issues are often the most reluctant to change their behaviour (Hares et al. 2010; McKercher et al. 2010).

Frisk and Larson (2011) argue that the lack of success is due to the focus on information-based knowledge which, while important, is not sufficient to address the awareness-action gap. They draw upon sustainability competency, and behavioural change theory, to outline an educational pedagogy for sustainability action, based on the following four knowledge domains (Frisk and Larson 2011, pp. 4–5):

- *Technical knowledge*: information-based knowledge is usually the central focus of education initiatives. It typically addresses ecological and environmental functions and structures; an understanding of destination-specific dynamics and the socio-economic, environmental impacts of tourism development. This form of knowledge has been shown to have limited effect on direct action, but without it stakeholders cannot make informed decisions.
- *Procedural knowledge*: the fundamental 'how-to information' that allows individuals to take advantage of opportunities, overcome barriers to action and develop strategies.
- *Effectiveness knowledge*: the knowledge that combines factual knowledge with subjective attitudes and self-efficacy beliefs to inform how likely an individual thinks a certain action is to succeed.
- *Social knowledge*: knowledge concerned with the intentions and actions of others through an understanding of social norms. Sustainable development is a particularly normative field as communal values dictate what we decide what is worth sustaining.

Together, these knowledge domains move sustainability education beyond analysis into systems thinking perspective which supports foresighted thinking, stakeholder collaboration and action agency (Frisk and Larson 2011, p. 14); as such, these forms of knowledge are critical components within the circulation of knowledge illustrated in Fig. 16.2.

Yet while the *intention* to behave sustainably is underpinned by knowledge, *actual choices* are often largely based on unconscious or semi-conscious values and beliefs (Antimova et al. 2012); these determine not only our capacity to act, but also our sense of accountability. Frisk and Larson's pedagogical framework highlights the presence of social and value-orientated factors which can be more influential than facts in determining individual choices and actions.

Responsible tourism is a growing area of research, concerned with the moral and ethical dimensions of sustainable tourism development (see Budeanu 2007; Bramwell et al. 2008; Goodwin 2011). Responsibility implies a moral obligation beyond self-interest (Bramwell et al. 2008); moral norms have been shown to have a significant influence on pro-environmental behaviour (Bamberg and Möser 2007). These moral norms are activated once stakeholders understand the impacts of tourism development, recognise how their actions contribute and can connect this with existing behaviours, practices. In lieu of ethical and moral values, a sense of responsibility can be imposed via external incentives and imperatives (i.e. Corporate Social Responsibility and government regulation), but these shallow measures are more limited than intrinsic moral values (Abson et al. 2017). This sense of responsibility then serves as a motivation for stakeholders to make use of available resources to develop strategies for sustainable tourism development (Antimova et al. 2012).

Milhalic (2016) connects sustainable development, responsible tourism and the awareness-action gap with the 'Triple-A Model', which outlines a four-stage, linear progression by which stakeholders move from a state of *Ignorance* to *Action* (Mihalic 2016). The stages can be summarised as follows (Mihalic 2016):

- *Ignorance*: this initial stage is the status of the destination before sustainability issues are first considered; when tourism stakeholders prioritise other values and incentives.
- *Awareness*: later, as stakeholders are exposed to information regarding tourism impacts or destination life-cycle progresses to the point where the negative impacts of tourism development can no longer be denied, the *Awareness* stage is reached.
- *Agenda*: then, as the conceptual understanding is developed into destination-specific knowledge and used to determine goals, sustainable development is on the destination *Agenda*.
- *Action*: this is where the strategies defined within the agenda are manifested.

Mihalic's Triple-A Model does take into account social, economic and environmental considerations (see Mihalic 2016, p. 467), but the pathway between awareness and action is not linear; presenting it as such is potentially unhelpful. While the onset of the Climate Crisis and increasing incidences of overtourism mean that there is now widespread awareness of (un)sustainable development, converting this awareness into meaningful change is a challenging, contested and, crucially, iterative process. Figure 16.4 aims to address this deficiency by acknowledging the role of knowledge, responsibility and strategies in triggering sustainable action. Neither the development of a strategy or the resulting action are endpoints in themselves: a

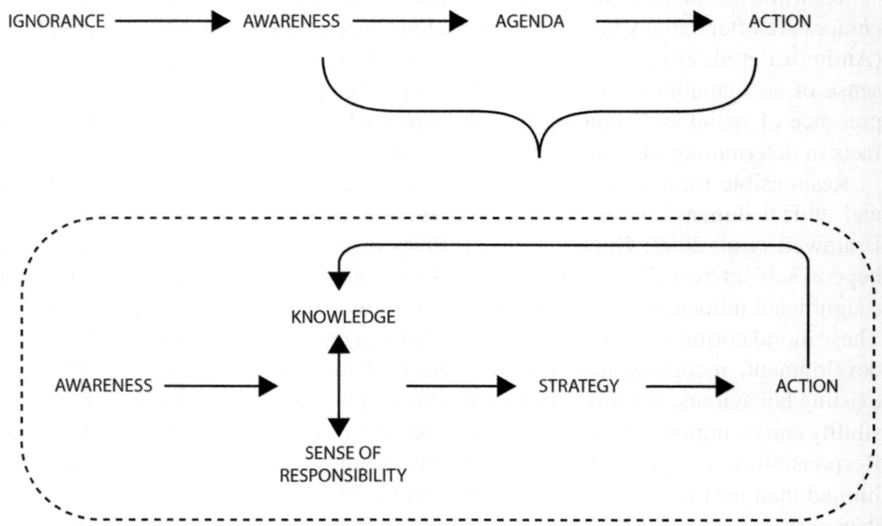

Fig. 16.4 An expansion of the Triple-A Model to better address the awareness-action gap *Source* Drawn after Mihalic (2016, pp. 466–467)

strategy provides the critical sense of self-efficacy necessary to inspire action (Doran et al. 2017), while the action and its impact are subject to evaluation which is used as the basis for the development of future actions.

16.4.2 The Collective Mindset

The challenge of sustainable development concerns not only alterations to prevailing systems and behaviours, but also the wider capacity of both individuals and societies to respond to challenges to their values and beliefs (O'Brien and Hochachka 2010). These unconscious and semi-conscious factors guide our choices, shape our behaviour and define our mindset, which in turn forms the goals and standards of our lives through associations and expectations (Rokeach 1973; Pisters et al. 2019). The mindset is influenced by external factors such as culture, religion, media, social networks and those we trust (Crum and Zuckerman 2017; Blasini et al. 2018), yet changes to the mindset are relatively incremental and unpredictable; attempts to actively influence individuals' mindsets have yielded mixed results (Orosz et al. 2017; Limeri et al. 2020). We tend to avoid uncomfortable, conflicting attitudes and behaviours from challenging our mindset through the mechanism of cognitive dissonance (Festinger 1957), which allows us to hold contrasting beliefs and values simultaneously. This is why O'Brien and Selboe (2015) define sustainability as an 'adaptive challenge' which hinges on our collective ability to reconcile with the uncomfortable reality that sustainable development is not only a pressing concern, but fundamentally our (individual and collective) responsibility.

Cognitive dissonance allows us to value sustainability conceptually without fundamentally challenging our behaviour. This has been shown to be particularly prevalent with regards to air travel (Becken 2007), and is likely to be compounded by the nature of tourism itself: long accepted as a permissive domain where 'anything goes', and prevailing attitudes and behavioural norms are suspended (Shields 1992; Wang 2000). Weaver (2007, 2009) defines the superficial adoption of sustainability as 'veneer environmentalism', emphasising the industrial appeal of the term 'sustainable *development'*—which seems to reinforce the current growth-focused agenda while making the selective use of non-invasive practices (such as recycling, and water- or energy-use reductions) highly marketable. Equally, while global tourism bodies—such as World Tourism Organisation and World Travel and Tourism Council—were quick to adopt the concept of sustainability, this could be seen largely as an attempt to 'green-wash' their growth-focused agendas (Budeanu et al. 2016, p. 289; Mihalic 2016, p. 462; Sharpley 2020, p. 1934). Weaver notes the remarkable similarity between the response of the industry and consumers, 'from strong accordance with the rhetoric of sustainability to an equally strong unwillingness to engage in personal sacrifice to attain the ends espoused by that rhetoric' (Weaver 2009, p. 36; Miller et al. 2010).

The keyword here is a *sacrifice*—and it stems from the negative definition of sustainable development. Sustainability is an ambiguous, and therefore contested, concept (Liu 2003; Hall et al. 2015; Sharpley 2020). It encompasses a variety of subjective value judgements and includes future populations within its range of stakeholders; sustainable development is, therefore, easier to define by what it is *not*, rather than what it *is*. This confusion as to what sustainability is—or should be—is consistently highlighted in tourism literature as a limitation for sustainable development (Smith and Sharicz 2011; Gibson 2012); it often leads to contradictory definitions, policies and strategies (Hall et al. 2015; Fodness 2017). But, crucially, this negative definition represents a more general and collective failure to address sustainable development as a *positive* and *attainable* goal.

Most current sustainable development initiatives rely on shallow, usually government-imposed, incentives or imperatives to prevent or remedy unsustainable practices. These initiatives are based on the fundamental assumption that development is inherently unsustainable—and that this can only be mitigated through (fiscal or legal) manipulation. Even highly effective tools which incentivise sustainable practices imply that damage to the well-being of third parties is an inherent result of tourism development. Equally, while the creation of win-win scenarios for stakeholders is possible—and initiatives such as Corporate Social Responsibility aim to give sustainability currency when it is not—without a broader socio-cultural shift these measures remain fragile and limited (Abson et al. 2017). Given the highly fragmented and dynamic nature of the tourism industry, this leaves collaborative models, such as the system proposed in Fig. 16.2, particularly vulnerable to disintegration.

Attention is therefore shifting from the design of tools to assist stakeholders in pursuing more sustainable growth, towards measures with the capacity to address the deeper socio-cultural values which underpin prevailing behaviour patterns. Scholars from various disciplines are expanding our understanding of the role that our conscience, beliefs, values, culture, social and moral norms, and spirituality take in shaping our mindset—and their ability to uncover new paths for sustainability (Schwartz 1973; Bentley 2000; Stern 2000; Jackson 2005; Thøgersen 2005; McKercher et al. 2010; Edwards 2015). These 'deep leverage points' are those on which it is most difficult to intervene, but also those that can generate the broadest and most enduring impacts (Meadows 1999).

A resilient approach to sustainable tourism development requires a shift away from reactionary strategies towards positive, long-term goals. While we can conceptually value the absence of ecological collapse or the avoidance of a destination's decline, these factors remain abstract and conceptual. The link between sustainability and competitiveness is, by now, well-known (Hassan 2000; Ritchie and Crouch 2003; Cucculelli and Goffi 2016), yet it has been insufficient to incentivise sustainable tourism development, even so, that stakeholders might protect their own long-term source of income by adopting sustainable practices to avoid the decline phase of a destination. Humans struggle to assess the value of future rewards (Bar 2010); this is compounded when these are defined negatively: i.e. 'not failing' rather than a positive gain.

Sustainable tourism development, therefore, requires tangible, measurable goals in place of growth metrics; goals which respect the limitations of planetary, ecological and social boundaries for (natural, physical and cultural) resource consumption. In the field of economics, this shift in values focuses on economic sufficiency as well as economic efficiency and aims to slow the rate and quantity of consumption through a mix of market and regulatory mechanisms. *Time* is, therefore, a fundamental variable and many scholars have begun to associate this approach with the notion of 'slow' consumption, 'de-growth' and, within tourism literature, of Slow Tourism (see, for example, Flipo and Schneider 2008; Hall 2009, 2011; Dickinson and Lumsdon 2010; Martínez-Alier et al. 2010). Hall (2009) defines an approach for 'steady-state tourism': a sustainable system that encourages qualitative tourism development rather than cumulative quantitative growth at the expense of natural capital. This approach is founded on ethical consumption and focuses on living better by consuming less and satisfying non-material needs (Hall 2010, 2015). The success, or health, of a tourism destination, is therefore not measured by only quantitative economic metrics, but these are contextualised and balanced by qualitative factors—such as production processes, wealth distribution, quality of life and social and ecological well-being. In order for this approach to be successful, specific goals must be locally defined through the participative engagement of stakeholders (as outlined in Figs. 16.2 and 16.3, and articulated through the Transition Management envisioning process).

The ability to value economic *sustainability* over economic *expediency* requires education (Frisk and Larson 2011). Education is often seen as imparting value-neutral knowledge, but this is misleading and fails to account for unconscious bias (Sipos et al. 2008, p. 70). It is precisely through acquiring new knowledge that our values change over time; these values inspire local innovation, reshape social norms and define the sustainability mindset (Spence 2012). It is, therefore, necessary to adopt a value-driven stance in education (Kelley and Nahser 2014), which develops the capacity for self-awareness, allowing individuals to situate themselves within a wider understanding of society and nature, as well as understand the complexities of self as an emotional being—and the resulting impacts this human condition has on their choices and behaviour (Rimanoczy 2014; Büchs 2017). Thus, the sustainability mindset is not just a way of thinking, but a way of 'being', which takes into account the emotional aspects of the human experience (Rimanoczy 2014; Kassel et al. 2016; Hermes and Rimanoczy 2018). Adams (2008, p. 63) and Rimanoczy (2014, p. 110) argue that it is this paradigm shift to 'being' rather than 'having' which allows us to value intangibles, qualitative growth and 'the greater good' as opposed to materialistic consumption, quantitative growth and greed. While this may sound lofty, or even idealistic, this mindset follows a much broader movement in contemporary society that places emphasis on reexamining everyday values and our collective position within global ecosystems, such as mindfulness, biocentrism and alternative economics.

To achieve sustainable tourism development, the education of all tourism stakeholders (managers, developers, practitioners, tourists, the general public and local communities) is necessary (Pigram 1990; Spence 2012), as the values, priorities and behaviours of all stakeholders ultimately dictate economic trends, development patterns and participation within collaborative models such as those outlined earlier in this chapter. It was originally hoped that consumer demand would incentivise sustainable tourism practices within the industry, but 'arguably nowhere is the volume of academic research and changed practice so extreme as here' (Budeanu et al. 2016, p. 289). Rather than a failing, this represents a significant opportunity for future improvement. National culture has been shown to influence sustainable decisions (Filimonau et al. 2018); similarly, Spence (2012) underlines the important progress that has been made among Asian populations and developing countries by increasing awareness that sustainability is fundamental for the achievement of long-term goals.

The critical importance of leadership is not a topic which has been addressed in this chapter, but various authors highlight the significance of education at management level for corporate participation in sustainable development initiatives and 'beyond-compliance' performance (Amoah and Baum 1997; Rivera and De Leon 2005), given their weight in the decision-making within the economic system (Shrivastava 1995). This leadership, and the mindset of these key figures, will be crucial in determining the timely success of sustainable development initiatives.

In summary, if we cannot address the values and beliefs which underpin the mindset and dictate our individual and collective choices, the system of incentives and policies designed by economists to address sustainable development is destined to fail in the democratic decision-making process (Pigram 1990; Spence 2012). Furthermore, any intervention which aims to address these socio-cultural 'levers' must aim to align all stakeholder groups towards a new collective mindset for sustainability, so the progress made by some is not undermined by the non-compliance of others. This consideration takes on particular significance, given the intense fragmentation of the tourism sector. This is a critical area for future research, with the potential to garner momentous change.

16.5 Concluding Remarks

This chapter has addressed the flaws in the prevailing, growth-focused approach to tourism development strategies. Many destinations rely on traditional, 'top-down' policy tools to address the unsustainable impacts associated with this business model, but these measures are too limited in scope and scale to result in the necessary paradigm shift for long-term, resilient, sustainable tourism development. The chapter has drawn from various research disciplines to outline some of the key challenges that we face in the pursuit of sustainable tourism.

Sustainability is ambitious, but not an unattainable goal. Overall, the chapter has highlighted the need to treat sustainable tourism development as an iterative, reflexive process, embedded within its locality and directed by its wide range of stakeholders. It requires an innovative, inclusive approach to sustainable development, based on a more holistic understanding of the metabolisms which underpin contemporary tourism. It is critical that the future research agenda addresses the challenges posed by prolonged stakeholder engagement, and that participative governance and management models are used to define positive, shared goals for sustainable development and support a value-driven approach to sustainability education. It will be necessary to investigate these themes on a case study basis due to the variety of local inflexions, dynamic interdependencies and global interrelationships which characterise contemporary tourism. It is perhaps even more important that these observations and conclusions are disseminated beyond the academic sphere. The challenge of sustainable tourism development involves all stakeholders; each group has a role to play and significant potential for intervention.

The final section of the chapter addressed the socio-cultural dimensions of sustainable development; a growing area of interdisciplinary investigation. Breaking down the awareness-action gap hinges on a combination of various forms of knowledge, the sense of responsibility devised from moral/ethical imperatives and clear strategies. The section also highlighted the need to reframe the collective mindset both in reference to our perceptions of what sustainability really means and our worldview and value systems more generally. These 'deep leverage points' are notoriously difficult to influence and progress in this area has lagged behind political and economic developments— but this means that there is significant scope for radical change. The role of education and knowledge building is critical to these processes, and the advancement of sustainable development pedagogies is a significant area for future research.

This chapter has addressed the collective approach to sustainable tourism development as if all involved are inherently good-natured, assuming that if sufficiently enlightened with knowledge and instilled with an understanding of their responsibility stakeholders will be compelled to make sustainable choices. At best, this could be described as optimistic, at worst, naïve. Often the economic returns for a few stakeholders today are seen to be worth more than the decline of an entire destination which will fall on the following generation. In time, as sustainable choices become the social norm; policymakers will be required to maintain the threshold of what is acceptable and reflect the values of the majority. In other words, social change can pre-empt policy; in democratic nations, governance is a reflection of the collective mindset.

References

Abson DJ, Fischer J, Leventon J, Newig J, Schomerus T, Vilsmaier U, Von Wehrden H, Abernethy P, Ives CD, Jager NW, Lang DL (2017) Leverage points for sustainability transformation. Ambio 46:30–39. https://doi.org/10.1007/s13280-016-0800-y

Adams JD (2008) Six dimensions of mental models. In: Wirtenberg J, Russell W, Lipsky D (eds) The sustainable enterprise fieldbook: when it all comes together. Greenleaf Publishing, Sheffield, UK, pp 60–70

Amoah VA, Baum T (1997) Tourism education: policy versus practice. Int J Contemp Hosp Manag 9(1):5–12. https://doi.org/10.1108/09596119710157531

Antimova R, Nawijn J, Peeters P (2012) The awareness/attitude-gap in sustainable tourism: a theoretical perspective. Tour Rev 67(3):7–16. https://doi.org/10.1108/16605371211259795

Argyris C (1976) Single-loop and double-loop models in research on decision making. Adm Sci Q 21(3):363–375. https://doi.org/10.2307/2391848

Argyris C (2002) Double-loop learning, teaching, and research. Acad Manag Learn & Educ 1(2):206–218. https://doi.org/10.5465/amle.2002.8509400
Bagiran Ozseker D (2019) Towards a model of destination innovation process: an integrative review. Serv Ind J 39(3–4):206–228. https://doi.org/10.1080/02642069.2018.1491970
Bamberg S, Möser G (2007) Twenty years after Hines, Hungerford, and Tomera: a new meta-analysis of psycho-social determinants of pro-environmental behaviour. J Environ Psychol 27(1):14–25. https://doi.org/10.1016/j.jenvp.2006.12.002
Bar M (2010) Wait for the second marshmallow? future-oriented thinking and delayed reward discounting in the brain. Neuron 66(1):4–5. https://doi.org/10.1016/j.neuron.2010.04.001
Becken S (2007) Tourists' perception of international air travel's impact on the global climate and potential climate change policies. J Sustain Tour 15(4):351–368. https://doi.org/10.2167/jost710.0
Bentley MD (2000) Consumer trends and expectations: an international survey focusing on environmental impacts. Ind Environ 23:57–59
Bertocchi D, Camatti N, Giove S, Van der Borg J (2020) Venice and overtourism: simulating sustainable development scenarios through a tourism carrying capacity model. Sustainability 12(2):512. https://doi.org/10.3390/su12020512
Blasini M, Peiris N, Wright T, Colloca C (2018) The role of patient–practitioner relationships in placebo and nocebo phenomena. In: Colloca L (ed) International review of neurobiology. Academic Press, Cambridge, MA, pp 211–231. https://doi.org/10.1016/bs.irn.2018.07.033
Bramwell B, Lane B, McCabe S, Mosedale J, Scarles C (2008) Research perspectives on responsible tourism. J Sustain Tour 16(3):253–257. https://doi.org/10.1080/09669580802208201
Bramwell B (2012) 21 interventions and policy instruments for sustainable tourism. In: Theobald WF (ed) Global tourism. Elsevier, Burlington, pp 406–425
Brandão F, Breda Z, Costa C (2019) Innovation and internationalisation as development strategies for coastal tourism destinations: the role of organisational networks. J Hosp Tour Manag 41:219–230. https://doi.org/10.1016/j.jhtm.2019.10.004
Büchs M (2017) The role of values for voluntary reductions of holiday air travel. J Sustain Tour 25(2):234–250. https://doi.org/10.1080/09669582.2016.1195838
Buckley R (2012) Sustainable tourism: research and reality. Ann Tour Res 39(2):528–546. https://doi.org/10.1016/j.annals.2012.02.003
Budeanu A (2007) Sustainable tourist behaviour—a discussion of opportunities for change. Int J Consum Stud 31(5):499–508. https://doi.org/10.1111/j.1470-6431.2007.00606.x
Budeanu A, Miller G, Moscardo G, Ooi C (2016) Sustainable tourism, progress, challenges and opportunities: an introduction. J Clean Prod 111(Part B):285–294. https://doi.org/10.1016/j.jclepro.2015.10.027
Burns P, Bibbings LJ (2007) Mitigation: potential for emission reduction through behavioral change. In: E-CLAT technical seminar: policy dialogue on tourism transport and climate change: stakeholders meet researchers. UNESCO, Paris, France. Available at: https://research.brighton.ac.uk/en/publications/mitigation-potential-for-emission-reduction-through-behavioral-ch
Butler RW (1980) The concept of a tourist area cycle of evolution: implications for management of resources. Can Geogr/Le Géographe canadien 1:5–12
Butler RW (1991) Tourism, environment, and sustainable development. Environ Conserv 18(3):201–209. https://doi.org/10.1017/S0376892900022104
Carayannis EG, Barth TD, Campbell DF (2012) The Quintuple Helix innovation model: global warming as a challenge and driver for innovation. J Innov Entrep. https://doi.org/10.1186/2192-5372-1-2
Cartwright S (2002) Double-loop learning: a concept and process for leadership educators. J LeadShip Educ 1(1):68–71. https://doi.org/10.12806/V1/I1/TF1
Cole S (2006) Information and empowerment: the keys to achieving sustainable tourism. J Sustain Tour 14(6):629–644. https://doi.org/10.2167/jost607.0
Costa CMM da (1996) Towards the improvement of the efficiency and effectiveness of tourism planning and development at the regional level: planning, organisations and networks; the case

of Portugal, Thesis (Doctoral). University of Surrey, UK. Available at: http://epubs.surrey.ac.uk/855568/1/27598746.pdf
Costa T, Carvalho L (2011) Tourism innovation: a literature review complemented by case study research In: International conference on tourism and management studies, vol 1. Algarve, pp 23–33. Available at: http://www.tmstudies.net/index.php/ectms/article/viewFile/173/211
Crum A, Zuckerman B (2017) Changing mindsets to enhance treatment effectiveness. JAMA—J Am Med Assoc 317(20):2063–2064. https://doi.org/10.1001/jama.2017.4545
Cucculelli M, Goffi G (2016) Does sustainability enhance tourism destination competitiveness? Evidence from Italian destinations of excellence. J Clean Prod 111(Part B):370–382. https://doi.org/10.1016/j.jclepro.2014.12.069
Dell'Era C, Landoni P (2014) Living lab: a methodology between user-centred design and participatory design. Creat Innov Manag 23(2):137–154. https://doi.org/10.1111/caim.12061
Dickinson JE, Lumsdon L (2010) Slow travel and tourism. Earthscan, London
Dodds R, Butler R (2010) Barriers to implementing sustainable tourism policy in mass tourism destinations. Tourismos 5(1):35–53
Doran R, Hanss D, Larsen S (2017) Intentions to make sustainable tourism choices: do value orientations, time perspective, and efficacy beliefs explain individual differences? Scand J Hosp Tour 17(3):223–238. https://doi.org/10.1080/15022250.2016.1179129
Dovers SR (1996) Sustainability: demands on policy. J Public Policy 16(3):303–318
Dovers SR (2005) Environment and sustainability policy: creation, implementation, evaluation. Federation Press, Alexandria
Edwards AR (2015) The heart of sustainability: restoring ecological balance from the inside out. New Society Publishers, Gabriola Island
Ericson T, Kjønstad BG, Barstad A (2014) Mindfulness and sustainability. Ecol Econ 104:73–79. https://doi.org/10.1016/j.ecolecon.2014.04.007
Festinger L (1957) A theory of cognitive dissonance. Stanford University Press, Stanford, CA
Filimonau V, Matute J, Mike M, Faracik R (2018) National culture as a driver of pro-environmental attitudes and behavioural intentions in tourism. J Sustain Tour 26(10):1804–1825. https://doi.org/10.1080/09669582.2018.1511722
Finger M (1994) From knowledge to action? exploring the relationships between environmental experiences, learning, and behavior. J Soc Issues 50(3):141–160. https://doi.org/10.1111/j.1540-4560.1994.tb02424.x
Flipo F, Schneider F (2008) Economic De-growth for ecological sustainability and social equity. In: Proceedings of the 1st international conference on economic de-growth for ecological sustainability and social equity, Paris, France, 18th–19th April
Fodness D (2017) The problematic nature of sustainable tourism: some implications for planners and managers. Curr Issues Tour 20(16):1671–1683. https://doi.org/10.1080/13683500.2016.1209162
Frisk, E. and Larson, K. (2011) 'Educating for Sustainability: Competencies & Practices for Transformative Action', *Journal of Sustainability Education*, 2
Fundeanu DD (2015) Innovative regional cluster, model of tourism development. Procedia Econ Financ 23:744–749. https://doi.org/10.1016/s2212-5671(15)00501-8
Gibson K (2012) Stakeholders and sustainability: an evolving theory. J Bus Ethics 109:15–25. https://doi.org/10.1007/s10551-012-1376-5
Goodwin H (2011) Taking responsibility for tourism. Goodfellow Publishers, Oxford
Gössling S, Hall CM, Ekström F, Brudvik Engeset A, Aall C (2012) Transition management: a tool for implementing sustainable tourism scenarios? J Sustain Tour 20(6):899–916. https://doi.org/10.1080/09669582.2012.699062
Grin J, Loeber A (2017) Theories of policy learning: agency, structure, and change. In Fischer F, Miller GJ (eds) Handbook of public policy analysis: theory, politics, and methods. Routledge, New York. https://doi.org/10.4324/9781315093192-25

Guimont D, Lapointe D (2016) Empowering local tourism providers to innovate through a living lab process: does scale matter? Technol Innov Manag Rev 6(11):18–25. https://doi.org/10.22215/timreview1031

Hakkarainen L, Hyysalo S (2013) How do we keep the living laboratory alive? learning and conflicts in living lab collaboration. Technol Innov Manag Rev 3(12):16–22. https://doi.org/10.22215/timreview749

Hall CM (2009) Degrowing tourism: décroissance, sustainable consumption and steady-state tourism. Anatolia 20(1):46–61. https://doi.org/10.1080/13032917.2009.10518894

Hall CM (2010) Changing paradigms and global change: from sustainable to steady-state tourism. Tour Recreat Res 35(2):131–143. https://doi.org/10.1080/02508281.2010.11081629

Hall CM (2011) Policy learning and policy failure in sustainable tourism governance: from first- and second-order to third-order change? J Sustain Tour 19(4–5):649–671. https://doi.org/10.1080/09669582.2011.555555

Hall CM (2015) Economic greenwash: on the absurdity of tourism and green growth. In: Reddy MV, Wilkes K (eds) Tourism in the green economy. Routledge, London, UK, pp 361–380

Hall CM, Gössling S, Daniel S (2015) The Routledge handbook of tourism and sustainability. Routledge, Abingdon, UK. https://doi.org/10.4324/9780203072332

Hall CM, Williams AM (2020) Tourism and innovation. Routledge, Abingdon. https://doi.org/10.4324/9780203938430

Hardin G (1968) The tragedy of the commons, Science, 162. J Nat Resour Policy Res 162(13):3

Hares A, Dickinson J, Wilkes K (2010) Climate change and the air travel decisions of UK tourists. J Transp Geogr 18(3):466–473. https://doi.org/10.1016/j.jtrangeo.2009.06.018

Hassan SS (2000) Determinants of market competitiveness in an environmentally sustainable tourism industry. J Travel Res 38(3):239–245. https://doi.org/10.1177/004728750003800305

Hatipoglu B, Alvarez MD, Ertuna B (2016) Barriers to stakeholder involvement in the planning of sustainable tourism: the case of the Thrace region in Turkey. J Clean Prod 111(Part B):306–317. https://doi.org/10.1016/j.jclepro.2014.11.059

Hermes J, Rimanoczy I (2018) Deep learning for a sustainability mindset. Int J Manag Educ 16(3):460–467. https://doi.org/10.1016/j.ijme.2018.08.001

Higuchi Y, Yamanaka Y (2017) Knowledge sharing between academic researchers and tourism practitioners: a Japanese study of the practical value of embeddedness, trust and co-creation. J Sustain Tour 25(10):1456–1473. https://doi.org/10.1080/09669582.2017.1288733

Hjalager AM (2010) A review of innovation research in tourism. Tour Manag 31(1):1–12. https://doi.org/10.1016/j.tourman.2009.08.012

Hudson S, Miller G (2012) 13 Ethical considerations in sustainable tourism. In: Theobald WF (ed) Global tourism. Elsevier, Burlington, pp 248–266

Hunter C (2002) Sustainable tourism and the touristic ecological footprint. Environ Dev Sustain 4:7–20. https://doi.org/10.1023/A:1016336125627

Jackson T (2005) *Motivating* sustainable consumption: a review of evidence on consumer behaviour and behavioural change, a report to the sustainable development research network. London

Jacobsen D (2005) Processes influencing innovation in the tourism system in Woodburn, new South Wales. In: Carson D, MacBeth J (eds) Regional tourism cases: innovation in regional tourism. Common Ground Publishing, Altona, pp 131–136

Kassel K, Rimanoczy I, Mitchell SF (2016) The sustainable mindset: connecting being, thinking, and doing in management education. In: Academy of management proceedings. Academy of Management, New York, p 16659. https://doi.org/10.5465/ambpp.2016.16659abstract

Kelley S, Nahser R (2014) Developing sustainable strategies: foundations, method, and pedagogy. J Bus Ethics 123(4):631–644. https://doi.org/10.1007/s10551-013-2014-6

Kemp R, Avelino F, Bressers N (2011) Transition management as a model for sustainable mobility. European Transport—Trasporti Europei 47:25–46

Kemp R, Rotmans J (2004) Managing the transition to sustainable mobility. In: Elzen B, Geels FW, Green K (eds) System innovation and the transition to sustainability. Edward Elgar, Cheltenham, p 237. https://doi.org/10.4337/9781845423421.00019

Kern F, Howlett M (2009) Implementing transition management as policy reforms: a case study of the Dutch energy sector. Policy Sci 42:391–408. https://doi.org/10.1007/s11077-009-9099-x
Krippendorf J (1971) Marketing et tourisme. Herbert Lang, Berne
Kvisleius N et al (2009) Living labs as tools for open innovation: open innovation. Commun & Strat (Montpellier) 74(2):75–94
Laine M (2010) The nature of nature as a stakeholder. J Bus Ethics 96:73–78. https://doi.org/10.1007/s10551-011-0936-4
Leeuw FL, Rist RC, Sonnischen RC (1994) Can governments learn? comparative perspectives on evaluation and organizational learning. Transaction Publishers, New Brunswick
Lehtonen A, Salonen A, Cantell H, Riuttanen L (2018) A pedagogy of interconnectedness for encountering climate change as a wicked sustainability problem. J Clean Prod 199:860–867. https://doi.org/10.1016/j.jclepro.2018.07.186
Levett R (1998) Sustainability indicators—integrating quality of life and environmental protection. J R Stat Society Ser A: Stat Soc 161(3):291ˆ302. https://doi.org/10.1111/1467-985x.00109
Limeri LB. et al. (2020) Growing a growth mindset: characterising how and why undergraduate students' mindsets change. Int J STEM Educ 7(35). https://doi.org/10.1186/s40594-020-00227-2
Liu Z (2003) Sustainable tourism development: a critique. J Sustain Tour 11(6):459–475. https://doi.org/10.1080/09669580308667216
Loorbach D (2010) Transition management for sustainable development: a prescriptive, complexity-based governance framework. Governance 23(1):161–183. https://doi.org/10.1111/j.1468-0491.2009.01471.x
Loorbach D, Rotmans J (2010) The practice of transition management: examples and lessons from four distinct cases. Futures 42(3):237–246. https://doi.org/10.1016/j.futures.2009.11.009
Lynch P, Morrison A (2007) The role of networks. In: Michael EJ (ed) Micro-clusters and networks: the growth of tourism. Elsevier Science, Oxford, pp 43–60
MacPherson R, Wilson R, Foote L (2008) Voluntary standards as a tool for increasing the sustainability of the marine recreation industry and improving MPA effectiveness in Hawaii and Mesoamerica. In: Proceedings of the 11th international coral reef symposium. Ft. Lauderdale, Florida, pp 1170–1172
Martínez-Alier J, Pascual U, Vivien FD, Zaccai E (2010) Sustainable de-growth: mapping the context, criticisms and future prospects of an emergent paradigm. Ecol Econ 69(9):1741–1747. https://doi.org/10.1016/j.ecolecon.2010.04.017
Mastelic J, Sahakian M, Bonazzi R (2015) How to keep a living lab alive? Info 17(4):12–25. https://doi.org/10.1108/info-01-2015-0012
McKercher B, Prideaux B, Cheung C, Law R (2010) Achieving voluntary reductions in the carbon footprint of tourism and climate change. J Sustain Tour 18(3):297–317. https://doi.org/10.1080/09669580903395022
Meadows D (1999) Leverage points: places to intervene in a system. The Sustainability Institute, Hartland. Available at: http://drbalcom.pbworks.com/w/file/fetch/35173014/Leverage_Points.pdf. Accessed 24 September 2020
Mihalic T (2016) Sustainable-responsible tourism discourse—towards "responsustable" tourism. J Clean Prod 111:461e470. https://doi.org/10.1016/j.jclepro.2014.12.062
Miller G, Rathouse K, Scarles C, Holmes K, Tribe J (2010) Public understanding of sustainable tourism. Ann Tour Res 37(3):627–645. https://doi.org/10.1016/j.annals.2009.12.002
Moscardo G (2005) Peripheral tourism development: challenges, issues and success factors. Tour Recreat Res 30(1):27–43. https://doi.org/10.1080/02508281.2005.11081231
Moscardo G (2011) Exploring social representations of tourism planning: issues for governance. J Sustain Tour 19(4–5):423–436. https://doi.org/10.1080/09669582.2011.558625
Murphy PE, Price GG (2005) Tourism and sustainable development. In: Theobald WF (ed) Global tourism, 3rd edn. Elsevier, Burlington, MA, USA, pp 167–193
Mustapha M, Zulkifli FZ, Awang KW (2020) Enhancing sustainability through implementation of balanced scorecard: a case study of beach resorts. J Sustain Sci Manag 15(1):136–147

Nunkoo R (2017) Governance and sustainable tourism: what is the role of trust, power and social capital? J Destin Mark Manag 6(4):277–285. https://doi.org/10.1016/j.jdmm.2017.10.003
O'Brien K, Hochachka G (2010) Integral adaptation to climate change. J Integr Theory Pract 5(1):89–102
O'Brien K, Selboe E (2015) The adaptive challenge of climate change. Cambridge University Press, Cambridge. https://doi.org/10.1017/CBO9781139149389
Orosz G, Péter-Szarka S, Bőthe B, Tóth-Király I, Berger R (2017) How not to do a mindset intervention: learning from a mindset intervention among students with good grades. Front Psychol 8(311). https://doi.org/10.3389/fpsyg.2017.00311
Paredis E (2011) Transition management as a form of policy innovation. A case study of Plan C, a process in sustainable materials management in Flanders. 26. Ghent. Available at: https://www.researchgate.net/publication/228453295_Transition_management_as_a_form_of_policy_innovation_A_case_study_of_Plan_C_a_process_in_sustainable_materials_management_in_Flanders. Accessed 24 September 2020
Pigram JJ (1990) Sustainable tourism-policy considerations. J Tour Stud 1(2):2–9
Pisters SR, Vihinen H, Figueiredo E (2019) Place based transformative learning a framework to explore consciousness in sustainability initiatives. Emot, Space Soc 32:100578. https://doi.org/10.1016/j.emospa.2019.04.007
Rimanoczy I (2014) A matter of being: developing sustainability-minded leaders. J Manag Glob Sustain 2(1):95–122. https://doi.org/10.13185/jm2014.02105
Ritchie JRB, Crouch GI (2003) The competitive destination: a sustainable tourism perspective. CABI Publishing, Wallingford. https://doi.org/10.1079/9780851996646.0000
Rivera J, De Leon P (2005) Chief executive officers and voluntary environmental performance: Costa Rica's certification for sustainable tourism. Policy Sci 38:107–127. https://doi.org/10.1007/s11077-005-6590-x
Rokeach M (1973) The nature of human values. The Free Press, New York
Rotmans J, Kemp R, Van Asselt M (2001) More evolution than revolution: transition management in public policy. Foresight 3(1):15–31. https://doi.org/10.1108/14636680110803003
Russo AP, Scarnato A (2017) "Barcelona in common": a new urban regime for the 21st-century tourist city? J Urban Aff 40(4):455–474. https://doi.org/10.1080/07352166.2017.1373023
Schwartz SH (1973) Normative explanations of helping behavior: a critique, proposal, and empirical test. J Exp Soc Psychol 9(4):349–364. https://doi.org/10.1016/0022-1031(73)90071-1
Scuttari A, Volgger M, Pechlaner H (2016) Transition management towards sustainable mobility in Alpine destinations: realities and realpolitik in Italy's South Tyrol region. J Sustain Tour 24(3):463–483. https://doi.org/10.1080/09669582.2015.1136634
Sharpley R (2009) Tourism development and the environment: beyond sustainability?. Earthscan, London
Sharpley R (2020) Tourism, sustainable development and the theoretical divide: 20 years on. J Sustain Tour 28(11):1932–1946. https://doi.org/10.1080/09669582.2020.1779732
Shields R (1992) Places on the margin: alternative geographies of modernity. Psychology Press, Hove
Shrivastava P (1995) The role of corporations in achieving ecological sustainability. Acad Manag Rev 20(4):936–960. https://doi.org/10.2307/258961
Sipos Y, Battisti B, Grimm K (2008) Achieving transformative sustainability learning: engaging head, hands and heart. Int J Sustain High Educ 9(1):68–86. https://doi.org/10.1108/14676370810842193
Smith MK (2004) Seeing a new side to seasides: culturally regenerating the English seaside town. Int J Tour Res 6(1):17–28. https://doi.org/10.1002/jtr.467
Smith PAC, Sharicz C (2011) The shift needed for sustainability. the learning organization, n. 1. Cit CIIcs J Bus Ethics: Celeb First Thirty Years Publ 18(1):73–86. https://doi.org/10.1007/978-94-007-4126-3_32
Sorin D (2010) Sustainability, self-identity and the sociology of consumption. Sustain Dev 18(3):172–181. https://doi.org/10.1002/sd.457

Spence M (2012) The sustainability mindset, project syndicate 17. Available at: http://www.lsf-lst.ca/media/The_Sustainability_Mindset_-_Michael_Spence_-_Project_Syndicate.pdf. Accessed 24 September 2020

Stern PC (2000) Toward a coherent theory of environmentally significant behavior. J Soc Issues 56(3):407–424. https://doi.org/10.1111/0022-4537.00175

Taback H, Ramanan R (2016) Ethics and corporate social responsibility for spirituality and sustainability. In: Dhiman S, Marques J (eds) Spirituality and sustainability. Springer, Cham, Switzerland, pp 73–88. https://doi.org/10.1007/978-3-319-34235-1_5

Thøgersen J (2005) How may consumer policy empower consumers for sustainable lifestyles? J Consum Policy 28:143–177. https://doi.org/10.1007/s10603-005-2982-8

Timur S, Getz D (2009) Sustainable tourism development: how do destination stakeholders perceive sustainable urban tourism? Sustain Dev 17:220–232. https://doi.org/10.1002/sd.384

UNWTO (2018) 'Overtourism'? understanding and managing urban tourism growth beyond perceptions. UNWTO, Madrid. https://doi.org/10.18111/9789284420070

Van der Borg J (1992) Tourism and urban development: the case of Venice, Italy. Tour Recreat Res 17(2):46–56. https://doi.org/10.1080/02508281.1992.11014649

Van der Borg J (2017) Sustainable tourism in venice: what lessons for other fragile cities. Cambridge Scholars Publishing, Cambridge

Vellecco I, Mancino A (2010) Sustainability and tourism development in three Italian destinations: stakeholders' opinions and behaviours. Serv Ind J 30(13):2201–2223. https://doi.org/10.1080/02642060903287500

Wamsler C, Brink E (2018) Mindsets for sustainability: exploring the link between mindfulness and sustainable climate adaptation. Ecol Econ 151:55–61. https://doi.org/10.1016/j.ecolecon.2018.04.029

Wang N (2000) Tourism and modernity: a sociological analysis. Pergamon, Oxford

Weaver DB (2007) Towards sustainable mass tourism: paradigm shift or paradigm nudge? Tour Recreat Res 32(3):65–69. https://doi.org/10.1080/02508281.2007.11081541

Weaver DB (2009) Reflections on sustainable tourism and paradigm change. In: Gössling S, Hall CM, Weaver DB (eds) Sustainable tourism futures: perspectives on systems, restructuring and innovations. Routledge, New York, NY, USA, pp 33–40

Zaccai E (2012) Over two decades in pursuit of sustainable development: influence, transformations, limits. Environ Dev 1(1):79–90. https://doi.org/10.1016/j.envdev.2011.11.002

Chapter 17
Technological Solutions to Overtourism: Potential and Limits

Ulrike Gretzel

Abstract Overtourism is a pressing issue for tourism destinations, including those whose attractiveness and experiential offerings are based on natural resources. The relationship between overtourism and technology is complicated, but technological solutions are generally seen as critical for overcoming or at least mitigating overtourism and its negative consequences for destinations. This chapter discusses a variety of smart technologies and illustrates how they can support smart tourism initiatives aimed at avoiding or solving overtourism issues and increasing overall sustainability at the destination. In addition to highlighting various solution potentials, this chapter also discusses the possible drawbacks of smart technology use in light of the specific characteristics of nature-based destinations. It concludes that comprehensive and holistic strategies based on a combination of technological and governance-related solutions are needed to combat the potentially detrimental effects of overtourism in nature-based destinations.

Keywords Technological solutions · Overtourism · Smart tourism · Sustainable destinations · Nature-based destinations

17.1 Introduction

Solving or mitigating overtourism problems has become a growing concern for destinations (Dodds and Butler 2019), especially for those that need to protect invaluable cultural and natural resources. Technologies like social media and smartphones have been partly blamed for their role in creating overtourism (Alonso-Almeida et al. 2019), but have also been identified as means to combat overtourism in addition to or replacing traditional forms of sustainable management and demarketing (Gretzel 2019).

U. Gretzel (✉)
University of Southern California, Los Angeles, CA, USA
e-mail: gretzel@usc.edu

A. Mandić and L. Petrić (eds.), *Mediterranean Protected Areas in the Era of Overtourism*, https://doi.org/10.1007/978-3-030-69193-6_17

Greater adoption of smart technologies in the provision and consumption of tourism experiences (Kabadayi et al. 2019), advances in tourist tracking technologies and methodologies (Shoval and Isaacson 2009), greater analytical capabilities of tourism service providers and destinations (Buhalis et al. 2019), and a more evident focus on achieving social and environmental sustainability through the implementation of smart tourism principles (Gretzel et al. 2015a) open up new opportunities for creating and managing tourist flows and influencing specific tourist behaviors. At the same time, such "smart" or advanced technological approaches also create new challenges (Gretzel et al. 2015b) and do not provide holistic solutions unless accompanied by sustainability-driven governance at the destination level. This chapter therefore critically analyzes the advantages and limitations of technological solutions to overtourism in light of their long-term viability and their ability to create change instead of shifting the problem to other areas or postponing impacts.

17.2 Overtourism

Overtourism is a perceptual concept that encompasses negative feelings by destination residents or tourists in response to diminishing quality of life or a deteriorating touristic experience caused by an excessive presence of tourists (Goodwin 2017). Veríssimo et al. (2020) highlight the political dimension of overtourism by drawing attention to the state of conflict it represents and the lack of coordination among stakeholders that leads to uncontrolled tourism development. Mihalic (2020) identifies acceleration in the growth of tourism demand and supply as the main root of overtourism. Capocchi et al. (2019) situate overtourism at the intersection of tourism growth, spatial and temporal concentration, and insufficient governance. Similarly, Milano et al. (2019a) describe overtourism as a reduction in general well-being caused by excessive/ill-managed tourism growth. Dodds and Butler (2019) also highlight the role of governance and policy-making, notably the lack thereof, in fostering overtourism. This is echoed by Koens et al. (2018) and Eckert et al. (2019) who see overtourism as fundamentally a destination management issue. Others (e.g., Phi 2020; Pasquinelli and Trunfio 2020) describe overtourism as a media phenomenon through which tourism-related issues are framed, and public opinion about tourism is influenced.

Based on a review of overtourism-focused literature, Nilsson (2020) identifies the following common negative consequences of overtourism: difficulties to move around, price increases, difficulty finding available services/facilities, sense of insecurity, tourist-related disturbances (noise, waste, etc.), a growing supply of unregulated accommodations, and rapid gentrification/touristification of local areas. Koens et al. (2018) highlight that such negative perceptions are typically judged against perceived benefits and that overtourism sentiment only emerges when negative perceptions outweigh what might be gained from large numbers of tourists (e.g., jobs or short-term rental income for residents or discounts, safety-in-numbers, and superior touristic infrastructure for tourists). Phi (2020) demonstrates through a content analysis of news media articles that most often it is the tourists who are blamed for overtourism and that therefore, reporting on overtourism phenomena typically focuses on tourist numbers.

Perceptions of overtourism often lead to antitourism (Martín et al. 2018), meaning public expressions of discontent with tourism impacts and demands for degrowth, or even tourismphobia (Milano et al. 2019b), which can involve a variety of actions from outright discrimination and attacks against tourists to vandalism against touristic infrastructure (Zerva et al. 2019). This contentious relationship between tourists and residents and between growth-oriented industry providers and politically motivated local governments forms a core theme in contemporary overtourism literature (Capocchi et al. 2020).

Overtourism has been mostly studied in urban contexts (Nilsson 2020; Phi 2020; Koens et al. 2018), although, as a phenomenon, it is clearly not limited to cities. Indeed, Clark and Nyaupane (2020) find lots of nature-based destination examples in their analysis of overtourism reports in the media. There are a handful of overtourism studies that acknowledge its relevance for nature-based destinations. Pecot and Ricaurte-Quijano (2019), for example, report on overtourism in the Galapagos Islands, one of the most significant nature-based destinations globally. Sæþórsdóttir and Hall (2021) report on visitor perceptions in a wilderness area in Iceland, a nature-based destination that is often mentioned in conjunction with overtourism.

In relation to nature-based destinations, however, the more established concept of carrying capacity (Navarro Jurado et al. 2012) seems to be seen as more applicable as it encompasses not only the social carrying capacity and perceptual sustainability (i.e., perceptions of residents and tourists, respectively) but also the physical, environmental, economic, and infrastructural capacities of destinations (Eckert et al. 2019). In contrast, overtourism as a concept mostly captures the often tremendous fluctuations in visitor demand, the spatial conflicts that result from it, as well as the media attention that regularly follows (Phi 2020). Because of its urban focus and emphasis on perceptions, effects on animals or plants are largely absent from the overtourism discourse. This chapter argues that a consideration of both concepts is necessary in the context of nature-based destinations. When overtourism is added to carrying capacity considerations, then the issues are no longer just management issues but require more comprehensive development considerations and new models of governance.

The 2020 COVID-19 pandemic has put a damper on discussions of overtourism in the industry. Only the early, and later debunked as fake, stories of animals returning to otherwise overcrowded destinations like Venice directly linked to overtourism (Daly 2020). In contrast, measures to overcome "undertourism" started to emerge at the forefront of industry and media reports. This is very short-sighted as reports from China clearly illustrate how fast demand can bounce back and lead to the same old problems. It also ignores the fact that during the pandemic, a lot of the domestic tourism demand in various countries shifted to nature-based destinations and created bottlenecks and negative sentiments there. In the United States, for example, large numbers of tourists flocked to the national parks during the Memorial Day, Independence Day, and Labor Day holiday weekends, leaving parks completely overwhelmed, with some reportedly nearing their breaking points (Coren and Kopf 2020). Therefore, addressing overtourism issues, and especially addressing them in the context of nature-based destinations is even more pressing of an issue as in past years. Technological solutions are often hailed as possible ways to at least mitigate if not eradicate overtourism problems (Veríssimo et al. 2020; Camatti et al. 2020). Pasquinelli and Trunfio (2020) specifically refer to smart technologies as potentially transformative because of their potential to not only manage symptoms but address the causes of overtourism.

Capocchi et al. (2019) show that overtourism literature addresses technologies to some extent. As discussed in the Introduction section of this chapter, the relationship between technology and overtourism is a complicated one. In many ways, information and communication technologies fuel the erratic demand spikes that put a strain on all types of carrying capacity (Gretzel 2019; Alonso-Almeida et al. 2019). Online platforms make unregulated accommodation not only more accessible but also more profitable and therefore, more attractive as a real estate management option. Smartphones and their apps make it easy and convenient for tourists to conquer spaces beyond the tourist precincts. Camera innovations and social media create new needs and new performativities that change tourist behavior (Gretzel 2020; Dinhopl and Gretzel 2016a, b). Social media influencers encourage large audiences to imitate their travel behavior (Pasquinelli and Trunfio 2020). Online travel reviews and bucket lists widely distributed online concentrate demand in critical ways. Recent reports from China show that so-called "daka tourists" (tourists who "punch the card," meaning they travel to be able to post evidence of having been to popular destinations on social media) continue to overwhelm smaller locales (Hutton 2020). Social media also allow for negative sentiment to spread quickly and develop into antitourism activism (Gretzel 2019).

At the same time, the same technologies also provide unprecedented opportunities for a different kind of tourism development that explores new governance approaches aimed at encouraging and coordinating technology development in light of clearly defined sustainable development goals. These efforts are usually referred to as smart tourism (Gretzel et al. 2015a). Smart tourism takes advantage of so-called smart technologies to address sustainability issues, including overtourism. Moreno-Izquierdo et al. (2018) see smart tourism as an instrumental step in the process of overcoming overtourism and achieving responsible tourism.

17.3 Smart Technologies

Smart technology is a summary term for software, hardware, and infrastructure technology that enables new ways of supporting decision-making processes, sometimes making human input redundant. Harrison et al. (2010) define smart as exploiting operational, near-real-time real-world data, integrating and sharing data, and using complex analytics, modeling, optimization, and visualization to make better operational decisions. Gretzel (2011:759) describes two main functions of smart technology: "(1) the ability to sense the environment; and, (2) the ability to learn from actions to maximize success in achieving particular objectives." Höjer and Wangel (2015) argue that it is not so much individual technologies but the interconnection, synchronization, and concerted use of different technologies that constitutes smartness. Buhalis (2019) stresses the pervasive applicability and the adaptive capacity of smart technology solutions. Gretzel et al. (2015a) highlight the opportunities different smart technologies provide in (co-)creating new value propositions within phygital (digital + physical) ecosystems. Indeed, bridging the physical and digital realms is one of the main aspects that characterize smart technologies. As such, smart technologies extend the capabilities of e-tourism beyond digital representation, communication, and transaction.

At the infrastructure level, smart technologies provide new opportunities for connectivity. Creating free, public wireless networks, for instance, is the main feature of smart development efforts (Gretzel et al. 2018). Near-field communication (NFC) is another example of smart technology that supports communication and data exchange (Egger 2013). Controversial next-generation infrastructure technologies like 5G, stratospheric balloons (see for example loon.com), and Elon Musk's satellite program (Starlink.com) to further facilitate high-speed connectivity and support the transfer of immense volumes of data are also typically mentioned in conjunction with smart technology development. Magasic and Gretzel (2020) discuss the essential role connectivity plays in facilitating tourism. The potential to extend high-speed, broadband connectivity to rural and even remote areas in cost-effective ways has huge ramifications for nature-based destinations.

At the phygital level, smart technology in the form of sensors, beacons, or radio frequency identification (RFID) tags enable the Internet of Things (IoT) that can be exploited for both tourism experience creation and tourism management (Cavada et al. 2018). The IoT collects and communicates data at rates that allow for new touristic value propositions, new efficiencies, and new levels of automated control. Energy grids, transportation networks, and waste management systems can automatically adjust to demand levels or trigger alerts. Entry barriers can become sensitive to tourist flows. Natural and built environments can communicate their states as well as their experiential potential to managers or tourists.

Mobile applications and social media platforms serve as the communication backend of smart ecosystems. They are typically accessed through smart devices, which are mostly portable (with smart speakers being the exception) and increasingly wearable (e.g., smart watches and smart glasses). They enable a feedback loop between smart objects and human users. While smartphones and tablets continue to play a dominant role in this context and remain the focus of research in this context (Dorcic et al. 2019), new interfaces like robots and smart contact lenses continuously emerge. Technology that facilitates augmented or virtual reality experiences (Egger and Neuburger 2020) further complements the array of ways in which smart ecosystems communicate with human users. Thus, a multiplicity of interfaces and an eventual disappearance of separate devices needed to access them further characterize smart technology development.

Smart technology also encompasses systems, applications, and approaches needed to store, transfer, transform, and display vast amounts of data. Data flows constitute the lifeblood of smart ecosystems and happen among machines, among humans, and between humans and machines. Here, data processing and analytics supported by machine learning and artificial intelligence algorithms and applications need to be highlighted as they lead to new opportunities for interacting with, understanding and managing/controlling a smart environment. Smart technologies not only collect and process large amounts of data at unprecedented speeds but also handle data that was not traditionally used in tourism, for example, sensor data and biometric data. Smart technologies further enable new forms of accessing such data, for instance, through real-time dashboards, and, thus, facilitate new insights and create action potentials.

Together, this assortment of ever more sophisticated smart technologies available at increasingly manageable cost provides opportunities to develop, manage, and experience destinations in new ways. When coupled with specific development goals, such efforts are referred to as smart tourism.

17.4 Smart Tourism

Smart tourism has gained momentum globally over recent years and is continuously fueled by new technological developments. Smart tourism efforts encompass the use of smart technology in the form of individual intelligent systems (Gretzel 2011) or complex phygital ecosystems to achieve a variety of tourism development goals (Gretzel 2021). These goals are aligned with the United Nations Sustainable Development Goals (https://sdgs.un.org/goals), from which the original smart development ideas developed (Joss et al. 2019). In the tourism context, value creation opportunities for the tourism industry, enhanced experiences for tourists, and quality of life for destination residents have been imposed on these original social and environmental sustainability goals (Gretzel et al. 2015a; Buonincontri and Micera 2016). The conceptual pillars on which most smart tourism initiatives are built can be summarized as effective use of advanced technology, mobility/accessibility, sustainability, and knowledge development/innovation/creativity (Gretzel 2018).

Boes et al. (2016) point out that achieving smart tourism goals requires the development of soft smartness (knowledge and governance aspects) in addition to technology and infrastructure components, so-called hard smartness factors. Gretzel (2021) also argues that smart technologies alone do not define smart tourism and that smart tourism governance is instrumental to the success of smart tourism development. In accordance, Ivars-Baidal et al. (2019a) propose a tight relationship between destination management and smart tourism implementation.

Perles Ribes and Ivars-Baidal (2018) emphasize the important link between smart tourism and sustainability. However, Rafael (2020) demonstrates that data, technology, and innovation are at the center of smart tourism conceptualizations in the academic literature; sustainability is mentioned but remains insufficiently addressed. As a result, the sustainability pillar of smart tourism remains under-conceptualized and under-researched. This is especially problematic for nature-based destinations for which sustainability concerns stand at the forefront.

Further, just like overtourism, smart tourism has been mostly conceptualized and researched in urban contexts. Gretzel (2018) illustrates that an urban bias permeates all aspects of smart tourism, which hinders its successful implementation in nature-based destination contexts. This is not just a matter of scale. Many assumptions related to smart tourism infrastructure, digital business ecosystems, and tourist behaviors simply do not apply to nature-based destinations. In terms of smart technology implementation, issues range from lack of connectivity that is often prominent in nature-based destinations (Magasic and Gretzel 2020) to negative implications of technology use, such as distraction or the encouragement of risky behaviors as in the case of taking selfies with wildlife.

Concerning overtourism in relation to smart tourism, only Ivars-Baidal et al. (2019b) and Gretzel (2021) have explicitly addressed the two concepts together. Examining the smart tourism development plans of Spanish and Portuguese destinations, Ivars-Baidal et al. (2019b) find that while combating overtourism is not a priority for the investigated destinations, their smart tourism plans still include technological solutions related to the deconcentration of tourist flows, the deseasonalization of tourism and the avoidance of general stress factors such as mobility bottlenecks. Thus, despite the seemingly obvious connection between overtourism mitigation and smart tourism development efforts because of the need to rely on smart technologies for both, concrete discussions on the overlap of the concepts and the central role but also limits of technology in moving sustainable tourism agendas forward are currently missing from the literature. The next section therefore presents a rudimentary framework that seeks to sketch out how smart tourism principles and technologies can help address overtourism, with a specific focus on nature-based destinations.

17.5 Conceptual Framework

Given that overtourism is a complex phenomenon fundamentally based on communication, management, planning, and regulation issues, technological solutions need to address a variety of aspects to help overcome it. Based on strategies put forward by Koens and Postma (2017), Pasquinelli and Trunfio (2020), and Camatti et al. (2020), the most common avenues for dealing with overtourism involve:

1. Dispersion of tourist flows
2. Stricter and more widespread regulation
3. Optimizing tourism
4. Increasing resident benefits from tourism
5. Increasing capacity

6. Reducing conflict

These strategies are listed, ranging from more short-term solutions to long-term initiatives. For instance, the dispersion of tourist flows involves marketing alternative tourism products, limiting access through ticketing, dynamic pricing, or incentivizing off-peak visitation times for tourists who are at the destination. Regulating tourism means implementing limits to noise, requiring permits for tourist access or for specific tourism business operations, imposing restrictions on unregulated accommodation offerings, creating pedestrian or no alcohol zones, restricting access for tour busses or cruise ships, etc. Optimizing tourism encompasses more long-term marketing strategies that seek to increase the yield from tourism. These can include rebranding efforts, de-marketing campaigns, targeting different visitor segments, or de-seasonalizing tourism through the creation of new touristic offerings. Increasing the benefits residents derive from tourism can be achieved through incentivizing local employment, compensating in the form of taxes, or creating tourism experiences that are attractive for locals. Increasing capacity is typically a long-term strategy because it usually involves changes to the overall and the touristic infrastructures at the destination. However, in the medium term, improving infrastructure could also be achieved through higher efficiency in use. Last but not least, given the perceptual dimension of overtourism, reducing conflict is a necessity. This can range from short-term detection of negative sentiment to more long-term education programs for residents to help them understand the value of tourism and opening up pathways for resident and tourist participation in the governance of a destination. The following section will illustrate how smart tourism supports the achievement of these varied strategies applied in combating overtourism.

17.5.1 Potential Technological Solutions

The various smart technologies discussed earlier in the chapter together enable five critical smart tourism functions that support the specific overtourism eradication/mitigation strategies explained above. These functions are:

1. Interpretation/Education
2. Monitoring/Management
3. Targeted Persuasion/Nudging
4. Innovation/Value Co-Creation
5. Governance/Planning

The Interpretation/Education function is enabled through smart technologies that bring the destination alive for tourists and facilitate deep engagement and understanding, for example, through augmented reality mobile applications. Artificial intelligence-supported translation can not only enhance tourist experiences but also reduce resident–tourist conflicts. Interpretation/Education-related technologies can further help residents experience the destination from a touristic point of view.

Monitoring/Management technologies are central to smart tourism development efforts and support the fight against overtourism in multiple, critical ways. Regulation, governance, tourist dispersion, and capacity management all need data. Mihalic (2020) specifically discusses the importance of monitoring and diagnosing overtourism risk. The data collected through smart technologies embedded in destination infrastructure, such as sensors and camera networks, and transformed into insights through big data analytics provide real-time management support and long-term planning capacity. Roaming data from mobile phones and other tourist tracking approaches (Shoval and Isaacson 2009) further feed into smart tourism data observatories.

Targeted Persuasion/Nudging support dispersion and optimization strategies by providing behavioral data from social media, web analytics, or mobile phone data and translating these into dynamic offerings or real-time recommendations. Gamification plays an important role here, too, because of the persuasive power of these approaches (Xu et al. 2017). Beacons that push information to smart devices via mobile applications further open up opportunities for behavioral change.

Innovation/Value Co-Creation refers to efforts implemented at smart destinations to encourage the dynamic interplay between actors. Innovation labs that bring different stakeholders together to creatively envision overtourism solutions are an example of how this could be realized. Additionally, platforms that foster microentrepreneurship and interactions with locals (Ditta-Apichai et al. 2020) can create opportunities for residents to become involved in tourism and can foster the kinds of exchanges that reduce antitourism sentiment.

Governance/Planning as a central tenet of smart tourism uses smart technologies to facilitate participatory governance approaches, e.g., through mobile applications that allow residents or tourists to upload user-generated contents they would like to bring to the attention of governing bodies. Smart technologies can also create the simulations necessary to make informed strategic decisions.

Figure 17.1 summarizes the dynamic interaction between smart tourism functions enabled by the phygital smart technology ecosystem and strategic overtourism management needs. It illustrates how both aspects can build on the smart tourism pillars and can ultimately lead to the central smart tourism outcomes, namely greater well-being at the destination exemplified by an increased quality of life for residents and enhanced experiences for tourists.

Camatti et al. (2020) use the case of Dubrovnik to illustrate how concrete technological solutions can be implemented within a smart tourism framework. They list, among other things, a smart visitor counting system, a destination smartcard, a sensor-based noise monitoring system, and an interactive web-based platform that enables direct communication between residents and administrative bodies as examples of smart tourism initiatives that help the destination combat overtourism. Their case study shows the many possibilities that emerge from smart tourism innovations and the general guidance smart tourism provides for implementing them.

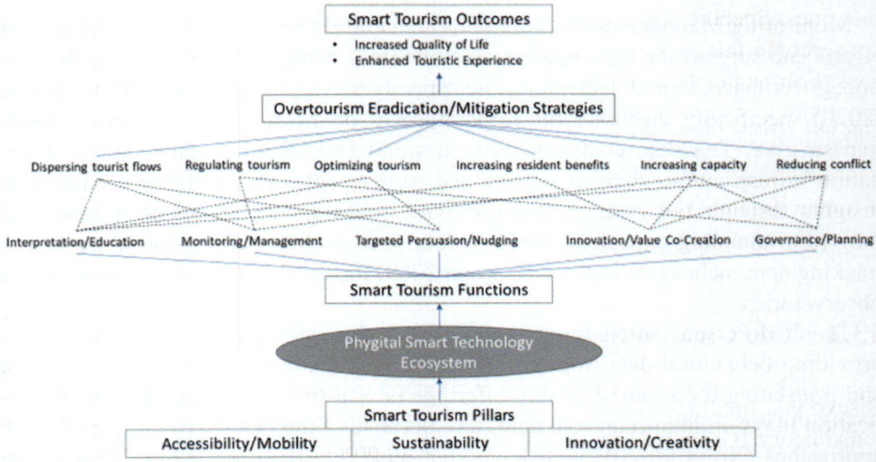

Fig. 17.1 Relationship between smart tourism functions and overtourism strategies *Source* Author

17.5.2 Limits to Technological Solutions

It is important to note that smart technologies per se are not a panacea for overtourism problems. First, the discussion on smart tourism presented in this paper clearly outlined the necessity for smart tourism technology solutions to be accompanied by strong and innovative governance. Second, Gretzel et al. (2015b) highlight a number of ethical problems, such as privacy and diversity concerns that directly emerge from the use of smart technologies. Third, an overreliance on smart technologies creates dependencies, security issues, and other vulnerabilities that need to be carefully considered. Technologies are far from perfect and human judgment should not be replaced. Fourth, stakeholder buy-in and coordination are essential for smart tourism success, and in many destinations, the tourism organizations in charge do not have the necessary mandates, funding, or capacities to gain this buy-in or implement solutions that require major infrastructure or regulatory changes. Consequently, there are potentially many perceptual and political barriers that need to be overcome when envisioning and actioning technological solutions to overtourism.

There are also limits to applying smart tourism frameworks in nature-based destinations that struggle with overtourism. As mentioned already, connectivity is critical for smart technologies, and nature-based destinations typically do not have the necessary infrastructure. Widespread connectivity might also interfere with nature due to the negative effects of electromagnetic frequencies or might take away from the experiences tourists seek from such destinations. Even when connectivity is possible, other environmental issues such as energy consumption or e-waste need to be considered. Further, while overall sustainability concerns are deeply integrated into smart tourism, it still prioritizes human needs, as exemplified by the desired smart tourism outcomes. Specific smart goals for nature-based destinations should be formulated to overcome this and other urban biases inherent in smart tourism conceptualizations. From a practical perspective, compact and accessible spaces in urban settings make monitoring as well as redirecting of tourist flows much easier. Nature-based destinations have unique physical and spatial characteristics and might not be able to offer a large diversity of experiences. It could be that providing virtual alternatives, at least temporarily, represents a more sensible solution (Guttentag 2020).

17.6 Conclusion

This chapter explored the relationship between overtourism and smart tourism by discussing how smart tourism pillars and smart technologies enable functions that can serve destinations well when trying to deal with problems created by overtourism. It provided an overview of the diverse technological solutions possible but also stressed that technological solutions alone are not enough to tackle a problem as complex and wide-ranging as overtourism. It further highlighted that both overtourism and smart tourism have been conceptualized and researched in urban contexts. More theorizing and empirical research are needed that to take the specific characteristics and needs of nature-based destinations into account.

References

Alonso-Almeida MDM, Borrajo-Millán F, Yi L (2019) Are social media data pushing overtourism? The case of Barcelona and Chinese tourists. Sustainability 11(12):3356

Boes K, Buhalis D, Inversini A (2016) Smart tourism destinations: ecosystems for tourism destination competitiveness. Int J Tour Cities 2(2):108–124

Buhalis D (2019) Technology in tourism-from information communication technologies to etourism and smart tourism towards ambient intelligence tourism: a perspective article. Tour Rev 75(1):267–272

Buhalis D, Harwood T, Bogicevic V, Viglia G, Beldona S, Hofacker C (2019) Technological disruptions in services: lessons from tourism and hospitality. J Serv Manage 30(4):484–506

Buonincontri P, Micera R (2016) The experience co-creation in smart tourism destinations: a multiple case analysis of European destinations. Inf Technol Tour 16(3):285–315

Camatti N, Bertocchi D, Carić H, van der Borg J (2020) A digital response system to mitigate overtourism. The case of Dubrovnik. J Travel Tour Mark: 1–15. https://doi.org/10.1080/10548408.2020.1828230

Capocchi A, Vallone C, Amaduzzi A, Pierotti M (2020) Is 'overtourism' a new issue in tourism development or just a new term for an already known phenomenon? Curr Issues Tour 23(18):2235–2239

Capocchi A, Vallone C, Pierotti M, Amaduzzi A (2019) Overtourism: a literature review to assess implications and future perspectives. Sustainability 11(12):3303

Cavada D, Elahi M, Massimo D, Maule S, Not E, Ricci F, Venturini A (2018) Tangible tourism with the internet of things. In: Information and communication technologies in tourism 2018. Springer, Cham, pp. 349–361

Clark C, Nyaupane GP (2020) Overtourism: an analysis of its coverage in the media by using framing theory. Tour Rev Int 24(2–3):75–90

Coren MJ, Kopf D (2020) Once again, a pandemic has stoked Americans' love for national parks. Accessed online 2 Dec 2020 at: https://qz.com/1908674/covid-19-has-americans-visiting-national-parks-in-record-numbers/

Daly N (2020) Fake animal news abounds on social media as coronavirus upends life. Accessed online 15 Sep 2020 at: https://www.nationalgeographic.com/animals/2020/03/coronavirus-pandemic-fake-animal-viral-social-media-posts/#close

Dinhopl A, Gretzel U (2016a) Selfie-taking as touristic looking. Ann Tour Res 57:126–139

Dinhopl A, Gretzel U (2016b) Conceptualizing tourist videography. Inf Technol Tour 15(4):395–410

Ditta-Apichai M, Kattiyapornpong U, Gretzel U (2020) Platform-mediated tourism micro-entrepreneurship: implications for community-based tourism in Thailand. J Hosp Tour Technol 11(2):223–240

Dodds R, Butler R (2019) The phenomena of overtourism: a review. Int J Tour Cities 5(4):519–528

Dorcic J, Komsic J, Markovic S (2019) Mobile technologies and applications towards smart tourism–state of the art. Tour Rev 74(1):82–103

Eckert C, Zacher D, Pechlaner H, Namberger P, Schmude J (2019) Strategies and measures directed towards overtourism: a perspective of European DMOs. Int J Tour Cities 5(4):639–655

Egger R (2013) The impact of near field communication on tourism. J Hosp Tour Technol 4(2):119–133

Egger R, Neuburger L (2020) Augmented, Virtual, and Mixed Reality in Tour. In: Xiang Z, Fuchs M, Gretzel U, Höpken W (eds) Handbook of e-tourism. Springer, Cham, pp 1–25

Goodwin H (2017) The challenge of overtourism. Responsible tourism partnership working paper 4. Accessed online 15 Sep 2020 at: https://haroldgoodwin.info/publications/

Gretzel U (2011) Intelligent systems in tourism: a social science perspective. Ann Tour Res 38(3):757–779

Gretzel U (2018) From smart destinations to smart tourism regions. J Reg Res 42:171–184

Gretzel U (2019) The role of social media in creating and addressing overtourism. In: Dodds R, Butler R (eds) Overtourism: issues, realities and solutions. De Gruyter, Berlin, pp 62–75

Gretzel U (2020) The growing role of social media in city tourism. In: Morrison AM, Coca-Stefaniak JA (eds) Routledge handbook of tourism cities. Routledge, New York, pp 389–399

Gretzel U (2021) Smart tourism development. In: Dieke P, King B, Sharpley R (eds) Tourism in development: Reflective essays. CABI, Oxford, UK, pp 159–168

Gretzel U, Ham J, Koo C (2018) Creating the city destination of the future – The case of smart Seoul. In: Wang Y, Shakeela A, Kwek A, Khoo-Lattimore C (eds) Managing Asian destinations. Springer, Cham, Switzerland, pp 199–214

Gretzel U, Reino S, Kopera S, Koo C (2015a) Smart tourism challenges. J Tour 16(1):41–47

Gretzel U, Sigala M, Xiang Z, Koo C (2015b) Smart tourism: foundations and developments. Electron Markets 25(3):179–188

Guttentag D (2020) Virtual reality and the end of tourism? A substitution acceptance model. In: Xiang Z, Fuchs M, Gretzel U, Höpken W (eds) Handbook of e-tourism. Springer, Cham, pp 1–19

Harrison C, Eckman B, Hamilton R, Hartswick P, Kalagnanam J, Paraszczak J, Williams P (2010) Foundations for smarter cities. IBM J Res Dev 54(4):1–16

Höjer M, Wangel J (2015) Smart sustainable cities: Definition and challenges. In: Hilty LM, Aebischer B (eds) ICT innovations for sustainability, advances in intelligent systems and computing. Springer, New York, pp 333–349

Hutton M (2020) China's 'daka' tourists show that post-pandemic travel will be no more meaningful. Accessed online 2 Dec 2020 at: https://www.scmp.com/magazines/post-magazine/travel/article/3093122/chinas-daka-tourists-show-post-pandemic-travel-will

Ivars-Baidal JA, Celdrán-Bernabeu MA, Mazón JN, Perles-Ivars ÁF (2019a) Smart destinations and the evolution of ICTs: a new scenario for destination management? Curr Issues Tour 22(13):1581–1600

Ivars-Baidal JA, Hernández M.G., de Miguel SM (2019b) Integrating overtourism in the smart tourism cities agenda. E-Rev Tour Res 17(2)
Joss S, Sengers F, Schraven D, Caprotti F, Dayot Y (2019) The smart city as global discourse: Storylines and critical junctures across 27 cities. J Urban Technol 26(1):3–34
Kabadayi S, Ali F, Choi H, Joosten H, Lu C (2019) Smart service experience in hospitality and tourism services. J Serv Manage 30(3):326–348
Koens K, Postma A (2017) Understanding and managing visitor pressure in urban tourism: A study to into the nature of and methods used to manage visitor pressure in six major European cities. Research Report. Centre of Expertise Leisure, Tourism & Hospitality, Breda, The Netherlands
Koens K, Postma A, Papp B (2018) Is overtourism overused? Understanding the Impact of Tourism in a City Context. Sustainability 10(12):4384
Magasic M, Gretzel U (2020) Travel connectivity. Tour Stud 20(1):3–26
Martín JMM, Martínez JMG, Fernández JAS (2018) An analysis of the factors behind the citizen's attitude of rejection towards tourism in a context of overtourism and economic dependence on this activity. Sustainability 10(8):2851
Mihalic T (2020) Concpetualising overtourism: A sustainability approach. Ann Tour Res 84:103025
Milano C, Cheer J, Novelli M (eds) (2019a) Overtourism: Excesses, discontents and measures in travel & tourism. CABI, Wallingford
Milano C, Novelli M, Cheer J (2019b) Overtourism and tourismphobia: A journey through four decades of tourism development, planning and local concerns. Tour Plan Dev 16(4):353–357
Moreno-Izquierdo L, Ramón-Rodríguez A, Such-Devesa MJ (2018) The challenge of long-term tourism competitiveness in the age of innovation: Spain as a case study. J Reg Res 42:13–24
Navarro Jurado E, Tejada Tejada M, Almeida García F, Cabello González J, Cortés Macías R, Delgado Peña J, Solís Becerra F (2012) Carrying capacity assessment for tourist destinations. Methodology for the creation of synthetic indicators applied in a coastal area. Tour Manag 33(6):1337–1346
Nilsson JH (2020) Conceptualizing and contextualizing overtourism: the dynamics of accelerating urban tourism. Int J Tour Cities 6(4):657–671
Pasquinelli C, Trunfio M (2020) Overtouristified cities: an online news media narrative analysis. J Sustain Tour 28(11):1805–1824
Pecot M, Ricaurte-Quijano C (2019) '¿Todos a Galapagos?': Overtourism in wilderness areas of the Global South. In: Milano C, Cheer J, Novelli M (eds) Overtourism: Excesses, discontents and measures in travel and tourism. CABI, Wallington, UK, pp 70–79
Perles Ribes JF, Ivars-Baidal J (2018) Smart sustainability: a new perspective in the sustainable tourism debate. J Reg Res 42:151–170
Phi GT (2020) Framing overtourism: A critical news media analysis. Current Issues Tour 23(17):2093–2097
Rafael C (2020) Analysis of scientific production-smart tourism destination, technology and sustainability. In: Rocha R, Abreu A, Vidal de Cavalho J, Liberato D, Gonzalez EA, Liberato P (eds) Advances in tourism, technology and smart systems. Singapore, Springer, pp. 599–613
Sæþórsdóttir AD, Hall CM (2021) Visitor satisfaction in wilderness in times of overtourism: a longitudinal study. J Sustain Tour 29(1):123–141
Shoval N, Isaacson M (2009) Tourist mobility and advanced tracking technologies. Routledge, New York
Veríssimo M, Moraes M, Breda Z, Guizi A, Costa C (2020) Overtourism and tourismphobia: A systematic literature review. Tour Int Interdisc J 68(2):156–169
Xu F, Buhalis D, Weber J (2017) Serious games and the gamification of tourism. Tour Manag 60:244–256
Zerva K, Palou S, Blasco D, Donaire JAB (2019) Tourism-philia versus tourism-phobia: Residents and destination management organization's publicly expressed tourism perceptions in Barcelona. Tour Geogr 21(2):306–329

Chapter 18
Resilience Conceptualisation and Protected Areas in the Jadranska Hrvatska Region

Blanka Šimundić, Zvonimir Kuliš, and Vinko Muštra

Abstract Protected areas (PA) are complex socio-ecological systems where socio-cultural, economic and ecological perspectives intervene. They have a challenging mission which involves the conservation of nature, provision of ecosystem services and creating an opportunity for the development of the local community. Resilience thinking appears as a novel approach which might foster the understanding of how these missions and goals interact. It refers to a specific model of how socio-ecological systems respond to disturbances, or what attributes shape their response to stress (resistance, adaptability, vulnerability). Within this chapter, we discuss the concept of resilience thinking in national parks (NP) in the Jadranska Hrvatska region, Croatia. The analysis involves the investigation of the aspects of disturbances in NPs, and the adaptation of resilience thinking in the management plans of the selected NPs. The research revealed that the selected NPs are facing the increasing visitors' use which is emphasising the role of good governance and efficiency of management in applying innovative solutions aiming to advance the resilience of these socio-ecological systems.

Keywords Resilience · Sustainable tourism · National parks · Croatia · Jadranska Hrvatska region

18.1 Introduction

Resilience research extends since the late 1970s and is characterised by conceptual transformations and a different approach to the meanings of the concept. Its definition has evolved and proliferated as resilience thinking has witnessed the transition from

B. Šimundić (✉) · Z. Kuliš · V. Muštra
Faculty of Economics, Business and Tourism, University of Split, Split, Croatia
e-mail: blans@efst.hr

Z. Kuliš
e-mail: zvonimir.kulis@efst.hr

V. Muštra
e-mail: vinko.mustra@efst.hr

A. Mandić and L. Petrić (eds.), *Mediterranean Protected Areas in the Era of Overtourism*, https://doi.org/10.1007/978-3-030-69193-6_18

ecological to the social sciences (Anderies et al. 2006; Folke 2006). As a conceptual approach, resilience deals with change (Berkes and Ross 2013; Brouder and Saarinen 2018) and in this context, it involves the recovery and resistance abilities of environmental and socio-economic structures affected by a change. Recovery means the "bounce back" to the pre-shock state of a socio-economic system. Resistance, on the other hand, is a broader concept and emphasises the ability of a socio-economic system to process a transition from one socio-economic structure to another (Hill et al. 2008; Simmie and Martin 2010). Resilience is understood as the ability of a system to maintain its identity and to adapt its vital structure and function in the face of disturbance (Orchiston et al. 2016). Currently, there is no joint agreement upon the definition of resilience. Consequently, multiple typologies of resilience definitions exist (Brand and Jax 2007; Olsson et al. 2014; Berbés-Blázquez and Scott 2017), mostly based on either different approaches to shock definitions or different socio-economic system resilience dimensions (for example, persistence, adaptive capacity, return time to recover, vulnerability, economic response capacity, organisation and disaster management) (Pimm 1991; Holling 1996; Adger 2000; Folke 2006; Walker et al. 2012; Mancini et al. 2012; Berkes and Ross 2013). Also, the appearance of an evolutionary turn within the economic geography approach over the last two decades has brought a significant step forward in regional development studies and tourism research (Brouder 2017; Brouder and Saarinen 2018). Evolutionary economic geography is a framework, which enables a deeper understanding of the relationship between the tourism economy and overall economic development of different places (Brouder and Saarinen 2018). In that context, tourism is taken out of its tourist-centric focus into comprehensive regional and relational contexts, which are vital for understanding the resilience of different places (Brouder and Saarinen 2018).

The resilience concept offers a valuable tool for understanding interactions among different stakeholders in complex systems, such as the PAs (Cochrane 2010, 2017). PAs are a mainstay of biodiversity conservation, contributing to local communities livelihoods. As a consequence of increasing nature-based tourism development, tourists are becoming significant stakeholders for PAs (Strickland-Munro 2017). The interactions among various stakeholders and interest groups make a vital aspect of a PA ecosystem, as they occur over multiple scales (from local, to regional and national, or even global) and through multi-level governance structures (Janssen et al. 2007; Cumming et al. 2015; Strickland-Munro et al. 2010; Strickland-Munro 2017).

Ensuring the resilience in withstanding the shocks imposed by the growing visitor use is a major challenge for the management of PAs. There is a long-standing pressure on PAs to increase revenues from market-based sources, including tourism development. Unfortunately, in many cases, such efforts are followed by the deterioration of protected features. Enhancing PA resilience requires recognising and managing spatial connections with its surroundings aiming to improve biodiversity conservation in situations of conflict and poor governance and to make people aware of the importance of PAs for their well-being (Cumming et al. 2015). This approach to management would be conscious and anticipative of complexity, uncertainty and dynamism. In its essence resilience thinking involves developing adaptation strategies so that systems can respond to the disturbances and reorganise without losing their core features (Walker et al. 2004) in the continually changing complex social and ecological systems (Walker and Salt 2006). This is in line with the adaptive management approach, which is often discussed as a feasible solution for sustainable and resilient PAs.

This chapter discusses the concept of resilience thinking in PAs and analyses its adaptation in the Croatian NPs. To do so, we reflect on the nexus between resilience and tourism development within PAs. Through iterative, reflective phases of research, the chapter applies modified Cochrane's (2017) framework, which aims to facilitate the understanding of dynamics and governance in the Croatian NPs facing extensive tourism development. The empirical analysis aiming to ascertain the adaptation of resilience thinking in PA management involved an in-depth analysis of eight NPs in Croatia.

18.2 Resilience Theory in Tourism and Protected Area Tourism Systems

18.2.1 Resilience in Tourism

The concept of resilience emphasises the connections between social and ecological systems and focuses on the identification of factors that cause vulnerability in systems along with the factors that enhance system capacity to absorb or withstand disturbances (Strickland-Munro 2017). Situations such as climate change, depopulation, natural disasters and market fluctuations are examples of a variety of disturbances (factors) that impose long- and short-run pressures on society. Since resilience is the capacity to withstand the disturbance, learn and develop (Folke et al. 2002), resilient systems are those which can better absorb social, political and economic changes. The resilience concept expresses the complexity of the linked socio-ecological system in that they do not evolve in linear progression but in cycles, composed of four stages: (i) reorganisation—rapid change after a destabilising event; (ii) exploitation—creation of new systems through the exploitation of social and other forms of capital; (iii) conservation—the gradual construction of a new stable state and (iv) release—a disturbance event (or events) which destabilises the existing systems (Holling 2001; Cochrane 2017).

The everyday experience shows that this world is chaotic and requires an understanding of how environments and societies operate as complex adaptive systems (Calgaro et al. 2014; Farrell and Twining-Ward 2005; Schianetz and Kavanagh 2008). Resilience is about adaptation, including building human resource capacities to change efficiently, creating learning institutions that can address changing circumstances while maintaining core values, understanding feedbacks in dynamic social and environmental systems and generally encouraging flexibility, creativity and innovation in the culture of a community (Lew et al. 2016). Evolutionary resilience, in particular, suggests that all systems are in a constant state of adaptation within an ever-flowing field of change (Davoudi 2012; Simmie and Martin 2010; Lew et al. 2016). The level of resilience depends on political, social, organisational and institutional characteristics of the system under the analysis. It is a contextual phenomenon, which according to Ruiz-Ballesteros (2011, p. 665) should not be used only to understand the course of the past events but to predict them. The local knowledge is required to make it fully operational. Building the efficient resilience empowers the system to adjust to confrontational environmental conditions through flexibility and adaptability in the face of external pressures, based on predicting the impacts of global change and the complex and dynamic nature of individuals, organisations and society in response to these impacts (Gallopín 2006; Marshall 2010).

The resilience came relatively late to tourism as a term with significant growth after the mid-2000s. Early papers on resilience were focused on its use in an economic context (Hall 2018). Holder (1980) uses the term in the context of strengthening the resilience of the Caribbean economy through tourism, adding to the diversity of economic structure and income generation. The first paper to induce tourism into the ecological dimensions of resilience regarding the value of biodiversity was Lovejoy (1994). Tyler and Dangerfield (1999) were the first to bring ecological resilience as a means of tourism-oriented resource management in ecotourism context. As stated by Lew et al. (2016), Tyler and Dangerfield's paper is to "blame" for contributing to the substantial and ongoing confusion that exists in tourism studies concerning the relationships between resilience and sustainable development. According to Derissen et al. (2011), sustainability mitigates change by maintaining resources above normative safe levels, whereas resilience adapts to change by building capacities to return to the desired state following a disruption. Faced with the modern challenges of climate change and natural disasters, economic and cultural globalisation and numerous other predictable and unpredictable drives of change, communities are faced with two questions, one dealing with the sustainability and the second with the resilience. Hence, concerning sustainability they need to answer what exactly they want to protect and conserve, and to keep from changing, whereas concerning resilience they have to be sure what exactly they want to adapt and to change into something new, preferably better than before (Lew et al. 2016). The answers will, of course, vary, depending on a community's goals and local circumstances.

Researchers have discussed the concept of resilience in a context of rural tourism (e.g. Perpar and Udovc 2007), ecotourism (e.g. Sakuma 2020) and wine tourism (e.g. Alebaki and Ioannides 2018). Furthermore, the concept has been used to identify the actions to reduce vulnerability and increase resilience in coastal resorts (e.g. Calgaro and Cochrane 2010), to analyse the relationships between stakeholders in a tourism destination (e.g. McDonald 2009) and to analyse interactions between tourism and environmental and social processes (e.g. Lew 2014; Bec et al. 2016; Butler 2018; Cheng and Zhang 2020; Haisch 2020; Heslinga et al. 2020). The concept is useful for understanding how tourism industry and related enterprises could respond effectively and adapt to increasing global changes and disturbances (Farrell and Twining-Ward 2004; Tyrrell and Johnston 2008; Biggs et al. 2012).

The previously mentioned studies support the conclusion that the resilience approach could be useful to analyse the interactions in complex socio-ecological systems (Cochrane 2017). Additionally, Ruiz-Ballesteros (2011) indicated that such an approach might be of value to communities, particularly governance agencies, interested in developing community-based tourism, as it enables the analysis of attitudes, behaviour and activities essential to understand how to achieve socio-ecological sustainability. Holladay and Powell (2013) and Hamzah and Hampton (2013) reasoned that local leadership, local control and robust social networks enhanced the resilience of community-based tourism. Shepard (2017) argues that change is good, particularly if the response to it assists in building the resilience and sustainability of communities, whereas this capacity takes place mostly at the governance level, especially in the tourism-focused community. Cochrane (2010) suggested that a healthy and resilient tourism system able to provide livelihood benefits and manage resources sustainably has three crucial elements: (i) the ability to connect market forces; (ii) cooperation between different stakeholders; and (iii) strong leadership (from the private or public sector). In the case of PAs, the list should be advanced by considering the contribution of tourism to the conservation of natural assets as the fourth element (Cochrane 2017). Conclusively, the importance of strong leadership, the ability to self-organise through robust networks and the need for tourism to support conservation are perceived as the crucial principles in resilience thinking (Cochrane 2017).

18.2.2 Resilience and Protected Area Tourism

International Union for Conservation of Nature (IUCN) states PAs as a clearly defined geographical space, recognised, dedicated and managed, through legal or other effective means, to achieve the long-term conservation of nature with associated ecosystem services and cultural values (Dudley and Stolton 2008). Defining the formal boundaries of PAs is impossible without support from external institutions. This means that the creation and maintenance of PAs is heavily dependent on their compatibility with institutions in the broader social and economic system. Each PA has social and ecosystem characteristics, often including stated management goals, that influence (and are influenced by) governance, affecting economic outputs and social outcomes in the socio-ecological system (Ostrom 2009). An expansion of prioritising community involvement and benefit alongside biodiversity conservation now drives the management of many PAs (Strickland-Munro 2017). According to Worboys et al. (2015), this focus on human communities often occurs in the essence of equity compensation and social justice, as decision-makers seek to move away from historic "fortress conservation" practices.

PAs and tourism have a close relationship as tourism is an increasingly important source of revenue for nature conservation. Consequently, tourists and representatives of the tourism industry become essential stakeholders when it comes to PA management planning. Cumming et al. (2015) argue that global environmental change and growing cross-scale anthropogenic influences mean that PAs can no longer be thought of as ecological islands that function independently of the broader social-ecological system in which they are located. These conclusions are supported with evidence of increasing interest in nature-based tourism and associated positive and adverse impacts on PAs and surrounding communities (Cochrane 2010).

The key stakeholders and their interactions are integral components of the broader PA tourism system. Their interactions are highly dynamic and complicated mainly because they reflect (i) trade-offs between sustainability of resource uses and provision of recreational opportunities; (ii) resource or economically dependent surrounding communities of PAs; and (iii) diverse and often unpredictable PAs institutional arrangements linked to multi-scale governance and high levels of state influence (Strickland-Munro 2017). All of this, along with inherent complexities, suggest that resilience thinking as a novel approach could be useful for understanding interactions within PA tourism systems. Cumming et al. (2015) noted that the dominant processes that shape and alter PA resilience are primarily social and economic, at a broader scale. The same authors argue that the concept of the socio-ecological system is useful for PA management because it explicitly implies that the manager, other stakeholders and related institutions are part of a cohesive unite, which in turn, may identify opportunities to enhance the resilience of systems that would otherwise be overlooked.

Based on the previous discussion, in the following sections, we consider and adapt Cochrane's (2017) approach. Cochrane explored the management issues in the context of ensuring the resilience of the natural systems to withstand the shocks of tourism visitation and related infrastructure. She evaluated the engagement with resilience in management plans of the UK's NPs and revealed that there was a weak leadership or lack of clarity over the roles of different stakeholders, which lead to confusion over responsibility for tourism development in the parks. It is essential to accentuate that this approach raises the awareness of the system approach and the significance of organisation and planning in PA tourism systems. The author argues that developing and using practical instruments to manage tourism in natural areas and to maximise its contribution to conservation is essential for engagement with resilience.

Following the above-presented framework, in the next section, the PA tourism system is introduced, and its elements are explored to reason the dynamics and challenges it is facing (for example, visitor use and seasonality). Subsequently, based on the methodology proposed by Cochrane (2017), the engagement with resilience in eight NPs in the Jadranska Hrvatska region, in Croatia is assessed. This analysis involved several stages. First, the latest version of management plans available for each of the NPs was accessed. The key terms associated with resilience as revealed by the literature review were used as a framework for the plans' content analysis. This was accomplished to reveal to what extent the resilience concept was embedded within the specific NP's management. The framework applied in this research broadens the understanding of the capacity of PA governance to respond to and manage change positively. We aimed to emphasise the importance of building such resilience that leads to greater long-term economic, socio-cultural and environmental sustainability.

18.3 Tourism Development and National Parks in Jadranska Hrvatska Region

According to the EU NUTS2016 (valid until 31 December 2020) classification scheme, Jadranska Hrvatska and Kontinentalna Hrvatska are corresponding to two NUTS 2 regions in Croatia. Jadranska Hrvatska is a coastal region of Croatia and covers a territory of 24,705 km^2 with 1,374,071 inhabitants, thus comprising 44% of the country's territory and 33% of the population. According to the Eurostat (2020), the list of the EU regions with the highest numbers of tourist nights in 2018 is dominated by coastal regions around the Mediterranean Sea, counting for eight Mediterranean regions in top 10, and one of them is Jadranska Hrvatska. The highest number of nights spent in tourist accommodation in 2018 was recorded in Spain's Atlantic island destination of Canarias (99.9 million), the capital region of Île-de-France (which had the second-highest number of nights spent in tourist accommodation at 86.0 million) and the Adriatic region of Jadranska Hrvatska (Croatia; 84.8 million). Also, Jadranska Hrvatska accounts for 86.5% of all tourist arrivals and 94.7% of all overnights realised in Croatia in 2018 (Croatian Bureau of Statistics, CBS 2019).

Between 2010 and 2018, the number of nights spent in EU-27 tourist accommodation increased by 28.7% overall (Eurostat 2020), while in Jadranska Hrvatska this increase was by 150% (CBS 2019). Annual increases in EU-27 ranged between 1.5 and 5.0% per year over this period, with the latest annual growth rate for 2018 equals to 2.6%. In Jadranska Hrvatska the annual rates of change for the same period have ranged between 2.43 and 61.52% (for 2012), and in 2018 the increase was by 3.65%. This area is highly dependent on tourism, characterised by high seasonality (67.1% of total inbound tourist arrivals in 2018 was realised in July, August and September) and dependence on foreign markets (92%, mostly from Germany, Austria, Slovenia, Italy). The presented figures indicate the state of tourism performance in Jadranska Hrvatska with progressive annual growth on average by 7% in the analysed period. At the same time, it is essential to notice that the problems of imbalanced tourism development between Jadranska and Kontinentalna Hrvatska are persistent and massive, causing the presence of overtourism in Jadranska Hrvatska, especially on its coastal strip (see Peeters et al. 2018).

While tourism is an important and vibrant economic activity for the Croatian economy with many positive effects (for example, income earnings in local communities and export benefits at the national level), it is important to accentuate its cost and benefits in terms of socio-ecological and environmental impacts, especially concerning the PAs. The role of PAs in tourism development has ever-growing importance (Petrić 2008; Cumming et al. 2015). According to the recent poll research of the Zagreb Institute of Tourism (2020), 55% of the Croatian tourist are motivated to come to Croatia because of its nature. Protected natural and cultural sites are important resources for tourism development in Jadranska Hrvatska, as there are eight national parks, seven nature parks and seven UNESCO's world heritage sites. Consequently, the importance of understanding and managing the nexus of tourism and PAs is increasing. Although the NP tourism system is comprised of both, supply and demand-side features, as well as of management issues, in this chapter we deal only with demand-side features and managing issues of the NP systems in Jadranska Hrvatska.

NPs are one of the six-categories of IUCN category system of protected areas identified by their primary management objective. They present category II PAs in the IUCN system. They are large natural or near-natural areas set aside to protect large-scale ecological processes, along with the complement of species and ecosystems characteristic of the area, which also provide a foundation for environmentally and culturally compatible, spiritual, scientific, educational, recreational and visitor opportunities (Cumming et al. 2015).

All PAs in Croatia are registered in the *Register of Protected Areas,* maintained by the Ministry of Environment Protection and Energy. According to the 2020 data, there are 412 permanently PAs in nine national categories of protection, representing 8.68% of the entire territory of the Republic of Croatia, or 7,476.28 ha. With the accession of the Republic of Croatia to the European Union, areas of the Natura 2000 ecological network has been defined, covering 36.73% of the terrestrial area and 15.42% of the sea (internal marine waters and territorial waters), i.e. 29.08% of the total surface of the Republic of Croatia. The Natura 2000 ecological network consists of 781 areas in total, out of which 38 Special Protection Areas and 743 Special Areas of Conservation.

All of the eight Croatian NPs are situated in Jadranska Hrvatska region and comprise approximately 1.1% of the total surface of Croatia (Table 18.1).

They diverge in terms of surface, a number of associated local administrative units (LAUs) and population number. Half of them have a surface area larger than 10.000 ha. The smallest NP in terms of the surface is Brijuni NP, whereas Mljet NP is the smallest in terms of the population, both being maritime PAs. The oldest and the largest NP is Plitvice Lakes, established in 1949. The process of tufa formation, which resulted in the building of the tufa, or travertine barriers and resulted in the creation of the lakes, is the outstanding universal value, for which the Plitvice Lakes were internationally recognised on 26 October 1979 with their inscription onto the UNESCO World Heritage List. This NP is the only one situated in the continental part of Jadranska Hrvatska, while the seven of them are located on the Adriatic coast.

According to the official data, every fourth Croatian tourist visits one of the eight NPs and 11 nature parks in Croatia. The analysis of the arrivals to the NPs in the period 2006–2018 reveals the average annual rise by 5.8%, while the overall percentage of growth in the same period was 92.8% (2018 vs. 2006). The highest overall growth of visitors in 2018 compared to 2006 was recorded in Northern Velebit (216%), followed by the Kornati and Plitvice Lakes, with 207% and 107% growth rate, respectively. The first half of the analysed period is characterised with the decline in a number

Table 18.1 Jadranska Hrvatska National Parks (NP)

NP	Area of NP (ha)	Date of designation	Local administrative units (LAU)	Population in LAU
Brijuni	3.400,46	9 November 1983	Pula	56.388
Kornati	21.571,14	13 August 1980	Murter-Kornati, Sali	4.025
Krka	11.063,68	20 February 1985	Drniš, Ervenik, Kistanje, Knin, Promina, Skradin, Šibenik	69.783
Mljet	5.287,53	12 November 1960	Mljet	1.155
Paklenica	9.507,56	19 October 1949	Gospić, Lovinac, Starigrad	14.573
Plitvička jezera	29.630,77	8 April 1949	Plitvička Jezera, Rakovica, Saborsko, Vrhovine	7.082
Risnjak	6.340,29	15 September 1953	Bakar, Čabar, Čavle, Delnice, Jelenje, Lokve	30.314
Sjeverni Velebit	11.157,29	17 June 1999	Senj	6.283

Source Data retrieved from the official sites of Bioportal—Nature Protection Information System, Croatian Agency for the Environment and Nature and Croatian Bureau of Statistics CBS (2020)

of visitors in only one NP, Risnjak. Hence, in 2018 it recorded an overall decline of visitors by 47%, as compared to the number of visitors in 2006 (Table 18.2).

The number of visitors must be observed taking into the account uneven annual distribution of tourist, with extraordinary pressures on the surface and infrastructures of the NP in the period from April to October, while over than 50% of all visits to NPs occurs in July and August (Ružić and Šutić 2014). Thus, the NPs in Jadranska Hrvatska, especially Plitvice lakes, experience the pressures discussed under the discourse of overtourism (Peeters et al. 2018). The impacts of overtourism are associated with the type of destination. Social impacts prevail in urban destinations, environmental impacts in rural, while all three impact categories are relevant in coastal and islands and heritage and attractions (Peeters et al. 2018). Peeters et al. (2018) additionally revealed environmental issues related to pollution and waste, social issues related to overcrowding of transport infrastructure and tourism sites. The authors concluded that there were no common economic impacts for the regions under their study and that every region has its specific economic issues.

Given the conclusions delivered by Peeters et al. (2018), within this analysis attention was paid to PA management in terms of enhancing the process of planning

Table 18.2 Number of visitors in eight national parks in Jadranska Hrvatska

	Brijuni	Kornati	Krka	Mljet	Paklenica	Plitvice Lakes	Risnjak	Northern Velebit	Total
2006	168.431	77.096	660.128	83.832	127.111	866.218	29.310	9.704	2.021.830
2007	178.073	94.605	700.823	87.816	110.338	927.661	24.831	11.949	2.136.096
2008	180.276	71.780	696.699	91.788	115.943	948.891	21.738	13.644	2.140.759
2009	162.664	89.300	632.378	88.455	110.350	939.747	21.416	16.043	2.060.353
2010	145.152	86.130	668.027	96.391	112.898	962.322	17.005	15.168	2.103.093
2011	156.549	91.780	683.739	95.498	118.288	1.083.451	18.212	19.372	2.266.889
2012	150.943	n/a	732.999	97.148	114.321	1.157.019	16.359	16.620	2.285.409
2013	151.007	94.257	786.635	120.464	114.381	1.188.798	13.725	15.777	2.485.044
2014	153.086	105.000	804.411	100.787	122.189	1.184.449	11.338	14.360	2.495.620
2015	160.010	157.574	951.106	112.156	119.686	1.357.304	12.715	16.471	2.887.022
2016	181.560	165.200	1.071.561	126.699	127.848	1.429.228	14.346	16.913	3.133.355
2017	169.299	229.061	1.284.720	140.329	140.561	1.720.331	16.575	22.919	3.723.795
2018	171.794	237.435	1.354.802	145.751	144.624	1.796.670	16.816	30.638	3.898.530

Source Author's calculation based on the official data from Parkovi Hrvatske (2020), and Croatian Bureau of Statistics, CBS (2006–2011)

and activities to mitigate conflicts over nature protection and tourism development. The most common measures by destination management organisations and local governments to relax the adverse effects of overtourism are related to spatial and time distribution of visitors (i.e. directing to a greater number of attractions over a prolonged season); pursuing inappropriate visitor behaviour; or improving the capacity of infrastructure, accommodation and facilities (Peeters et al. 2018). Since the ability to deal with change depends on good governance (Cochrane 2017), the resilience thinking needs to be applied in each stage of programming the development of socio-ecological systems. By doing so, the socio-ecological systems can anticipate the change, avoid the adverse effects of the anticipated change and transform into a new desirable state.

18.4 Resilience Thinking and National Parks in Croatia

The legal framework for the management of nature protection in Croatia (RH) includes three main documents, the Nature Protection Act (from now on Act, Official Gazette (OG) no. 80/13, 15/18), Regulation on the Ecological Network and the Jurisdiction of Public Institutions for Management of Sites of Ecological Network (OG no. 80/19) and the Nature Protection Strategy and Action Plan of the Republic of Croatia for the period 2017–2025 (from now on Strategy, OG no. 72/2017). The Act defines the principles for nature protection, while Strategy refers to capacities needed (NPs management authority) to ensure maximum contribution to nature conservation. The Strategy includes five strategic goals, which encompass specific objectives and activities stemming from them. First two goals target the increase in the effectiveness of crucial nature protection mechanisms and the reduction of the direct pressures on nature and the promotion of sustainable use of natural resources. The Strategy also aims to strengthen the capacities of the nature protection system, increase the knowledge and availability of data on nature and to raise the level of knowledge, understanding and support for nature protection among the critical stakeholders of the PA.

For the management of NPs in the Jadranska Hrvatska region, eight park-level public institutions, i.e. NPs' management authorities (NPAs), were established. They are primarily focused on the conservation of nature, cultural heritage and traditional values; visitors management (where visitation is allowed) and cooperation with the local community to ensure the long-term conservation of natural values within the area. The administrative council of the NPA is obliged to adopt a ten-year management plan (MP) and the annual programme for the protection, maintenance, conservation, promotion and use of the PA. The first generation of PA MPs was adopted in 2007 for NP Risnjak, NP Northern Velebit, NP Plitvice Lakes and NP Paklenica. In 2020, a new planning cycle for most public institutions in Croatia began, and by the end of 2022, all PAs should adopt new MPs. Currently, there are five NPs with valid MPs, including NP Brijuni (until 2026), NP Krka (until the end of 2020), NP

Kornati (until 2023), NP Mljet (until 2026) and NP Plitvice Lakes (until 2028). The above-listed MPs make the subject of research in this chapter.

The valid MPs were firstly analysed according to the overall ten-year objectives they define. The main goals in all five MPs refer to conservation and enhancement of the wildlife, natural resources and cultural heritage; improvement of the visitor management system to reduce high season tourism pressures; and promotion of partnership with local communities and knowledge-based management. Furthermore, the contribution of the NPs to the development of the local community is acknowledged. We noticed the transition of the traditional NP management, which now involved the interconnected resource management and participation of relevant stakeholders. These findings suggest that NPs were implementing the system approach when defining their objectives and activities. Considering the content analysis of the objectives within selected MPs, we concluded that visitor management is one of the critical challenges addressed. Within MPs, tourism is recognised as one of the key threats for the protected ecosystem. MPs provide the assessment of both positive and negative impacts of tourism development. The tourism-related benefits mostly refer to the number of employed persons and revenues from the entry fees. Cooperation between NPAs and relevant stakeholders, particularly the local community, is stressed as a precondition for the achievement of sustainable development goals.

In the next step, an examination is performed by the keywords most usually cited in the academic publications discussing the concept of the resilience in the protected natural areas. It revealed the following words: “resilience” (and its variations, such as resiliency), “adaptation/adaptability”, “vulnerability”, “resistance” and “ecosystem”. The term “climate change” was also included to ascertain the level of concern about this. Following was the analysis of their appearance in the MPs. Although it is ambiguous to search for English words in documents written in Croatian, the highest challenge was to translate the word resilience to Croatian. In order to keep its necessary broad scope of meaning besides the term resilience, we added to the search term “resistance”. Usually, resilience translation to Croatian refers to resistance, and only a few translations keep the basic meaning of the term resilience (Croatian—*rezilijentnost*). Therefore, we have searched for the frequency of both terms in the MPs, not to omit the broader scope of the resilience meaning when compared to the resistance meaning.

Table 18.3 gives the results of the frequency of use of the searched term within selected MPs. We considered the resulting numbers to be a proxy indicator of the recognition of the resilience concepts and climate change. We noticed that the most frequently used term in MPs was an *adaptation*, followed by *climate change* and *vulnerability*. The term “resilience” was not present, which we anticipated due to the already explained problems concerning a translation from English to the Croatian language. Still, the term “resistance” appeared, but only once in the MP for NP Plitvice Lakes and in the context of the *ability to resist the shocks in tourism markets*. As mentioned, the most frequently used term was “adaptability”. In the context of resilience, “adaptability” once referred to the *adaptation of infrastructure due to increasing visitors demand* (in MP for NP Mljet) and mostly on the *adaptation of*

Table 18.3 Resilience concepts appearing in Jadranska Hrvatska National Park (NP) management plans

NP	Pages in the management plan	Year of publication	Period	Frequency term appeared in documents					
				Resilience	Resistance	Adaptation	Climate change	Vulnerability	Ecosystem
Brijuni	161	2016	2016–2025	0	0	7	0	6	12
Mljet	120	2017	2017–2026	0	0	6	8	9	13
Krka	124	2011	2011–2020	0	0	10	0	2	11
Kornati	54	2014	2014–2023	0	0	3	0	0	4
Plitvice Lakes	360	2019	2019–2028	0	2	126	24	32	91
Total				0	2	152	32	49	131

Source Authors, the data are retrieved from management plans available at the websites of Public Institution National Park Brijuni, Public Institution National Park Mljet, Public Institution National Park Krka, Public Institution National Park Kornati, Public Institution National Park Plitvice Lakes *management activities when specific PA and species are concerned*, but without clarity how to adapt management activities. This could be discussed in the Annual Action Plan; however, they were not included in the analysis. The "vulnerability" term showed to be the most associated to the resilience thinking since in MPs it referred not only to *the protection of vulnerable species* but also to *the protection of the vulnerable areas in NPs due to the increasing urbanisation*. This comprehensive analysis suggests that most of the NPAs do not have a strong awareness of resilience thinking, while the presence of the system approach to NPs tourism system was confirmed.

To analyse the adaptation of a system approach to address the tourism-related challenges, we have decided to include additional keywords, namely "tourism shock", "shock", "tourism" and "visitors". The terms "tourism shock" and "shock" appeared only once, and both of them in the MP for NP Plitvice Lakes. "Tourism shock" referred to the already explained reference with the term "resistance" while the term "shock" was used in the context of *negative demographic changes affecting the Plitvice Lakes administrative unit*. The search for the additional term "tourism" and "visitor", on the other side, showed high frequency and in each MP it was referred to at least once per page of the document. A closer analysis of the context within which the term is used revealed that NPs develop the objectives related to visitors experience, tourist infrastructure and marketing to respond to changing nature of tourism demand. By doing so, NPs work towards building resilience without actually recognising resilience as their objective in MPs.

In addition, we searched for term "ecosystem", for two reasons, i.e. to reveal biodiversity management issues and to detect if there was the notion of the need to develop ecosystem services to gain social and economic benefits within NPs. We found out that the term was frequently used in the context of *management activities dealing with the preservation of nature and species*, while the term did not refer to the holistic perception of interconnected systems. Nonetheless, we confirm the adaptation of a system approach since all MPs emphasised interdisciplinary planning processes for problem-solving and integration of recreation into land management planning and conservation.

Although climate change is an important topic on a global scale, the term appeared only twice in the analysed MPs (NP Mljet; NP Plitvice Lakes). In both MPs the use of the term mostly referred to the *preservation of the habitat conditions in the face of climate change*, but with loose aspirations and questionable implementation of the actions mentioned. In addition, the role of the NPs' territories in contributing to the reduction of negative effects on climate was not referred to in the analysed NPs.

18.5 Conclusion

According to IUCN World Park Congress (2014), PAs should be considered as "an investment in the world's future as a matter of world security and as affordable and scalable solutions to leapfrog to more resilient ecosystems, societies and economies".

The socio-ecological system approach defines the objectives of their management which should determine multiple, simultaneous benefits while ensuring that realisation of one benefit does not degrade other benefits, or net harm the other beneficiaries (Cochrane 2017). Resilience assessment guides the development of a conceptual model of an integrated social-ecological system where key actors, ecological structures and their interactions are identified in relation to the larger context in which they are embedded (Quinlan et al. 2016).

The increasing importance of understanding and managing the nexus of tourism and socio-ecological system such as a PA is supported by the growth of international tourism, increased demand for natural and cultural heritage visitation, and international conservation efforts (McCool and Spenceley 2014). Additionally, increasing and diversifying demands from society on PAs, accelerate the need for a greater institutional capability to manage visitors and tourism development in PAs (Strickland-Munro 2017). Cumming et al. (2015) argue that the concept of the socio-ecological system is useful for the PA management because it explicitly implies that managers, stakeholders and related institutions are a part of a cohesive unite, which in turn, may identify opportunities to enhance the resilience of systems that would otherwise be overlooked. The question we set out in this research was whether tourism systems in NPs in the Jadranska Hrvatska region, Croatia, are currently a part of a cohesive unit, which may enhance their resilience.

The analysis involved the investigation of the aspects of disturbances in NPs, and the adaptation of the resilience thinking in the management plans of the selected NPs. The research evidenced that NPs are facing increasing tourism demand. To cope with these challenges, they have implemented a variety of interventions fostering resilience and have integrated them in recently adopted ten-year MPs. We noted the significant efforts to improve the "experience and enjoyment" of the visitors within the parks, throughout the involvement of the local communities and the improvement of infrastructure (facilities). Additionally, the analysis of visitor demand was performed in each MP. Diversification of tourism-related activities, and tourism products were proposed, accentuating the need for visitors' dispersion in time and space. Given the above, it may be concluded that MPs help developing capacities to respond to adverse effects, thus leading to enhancement of resilience in several NPs.

The research aimed to provide the evidence on the resilience thinking in the MPs for five NPs in Jadranska Hrvatska region. Overall, we conclude that the resilience approach in analysed MPs encompasses the reactive actions to disturbances, among others related to the excessive visitors and associated adverse impacts on biodiversity. Hence, the MPs could be categorised as reactive and adaptable to the system changes and disturbances, which is fundamental to resilience. In addition, we revealed that in some cases, MPs' goals and monitoring of their achievement foster resilience thinking more than it was shown by the very presence of the "resilience-associated terms" referred in the MP document. This is particularly the case with the goals related to tourism demand and visitor use adaptation to change. With this regard, adaptive management as a feasible solution for reaching resilience was confirmed in most of the NPs under the investigation. Furthermore, we noticed the use of the system approach in several MPs to analyse the interactions within the NPs. This leads to the conclusion that NPs managing agencies increase the adaptive capacities to react to the changes. However, an anticipative approach, which is vital to prepare proactive strategies to respond to future challenges, still lacks.

The discussion above yields several proposals for the improvement of the NPs' management plans by involving resilience thinking in the process. Hence, more intense engagement is needed in sharing up-to-date knowledge and best practices during the process of MP preparation. By doing so, the future management policies by the NP authorities would be more anticipative and innovative to advance the resilience of the NP's socio-ecological system. This research, focused on the Jadranska Hrvatska region, revealed that the nature-based destinations under analysis are currently a part of a cohesive unit, which fosters their resilience. Considering the growing demand for nature-based tourism, similar research in Croatia and the Mediterranean area is needed to develop a comprehensive framework for future enhancement of the PAs' resilience.

References

Adger WN (2000) Social and ecological resilience: are they related? Prog Hum Geogr 24(3):347–364

Alebaki M, Ioannides D (2018) Threats and obstacles to resilience: insights from Greece's wine tourism. In Cheer JM, Lew AA (eds) Tourism, resilience and sustainability: adapting to social, political and economic change. Routledge, London

Anderies JM, Walker BH, Kinzig AP (2006) Fifteen weddings and a funeral: case studies and resilience-based management. Ecol Soc 11(1)

Bec A, McLennan CL, Moyle BD (2016) Community resilience to long-term tourism decline and rejuvenation: a literature review and conceptual model. Curr Issues Tourism 9(5):431–457

Berbés-Blázquez M, Scott D (2017) The development of resilience thinking. In: Butler R (ed) Tourism and resilience. CAB International, London, UK, pp 9–22

Berkes F, Ross H (2013) Community resilience: toward an integrated approach. Soc Nat Resour 26(1):5–20

Biggs D, Hall CM, Stoecki N (2012) The resilience of formal and informal tourism enterprises to disasters: reef tourism in Phuket, Thailand. J Sustain Tourism 20(5):645–665

Brand FS, Jax K (2007) Focusing the meaning (s) of resilience: resilience as a descriptive concept and a boundary object. Ecol Soc 12(1)

Brouder P (2017) Evolutionary economic geography: reflections from sustainable tourism perspective. Tourism Geographies 19(3):438–447

Brouder P, Saarinen J (2018) Co-evolution and resilient regions: moving towards sustainable tourism development. In: J Sarrinen AM Gill (eds) Resilient destinations and tourism: governance strategies in the transition towards sustainability in tourism. Routledge, New York, pp 67–76

Butler R (2018) Resilience in the face of changing circumstances. In: Cheer JM, and Lew AA (eds) Tourism, resilience and sustainability: adapting to social, political and economic change. Routledge, London

Calgaro E, Cochrane J (2010) Comparative destination vulnerability assessment for Thailand and Sri Lanka. Stockholm Environment Institute, Stockholm

Calgaro E, Lloyd K, Dominey-Howes D (2014) From vulnerability to transformation: a framework for assessing the vulnerability and resilience of tourism destinations. J Sustain Tourism 22:341–360

Cheng L, Zhang J (2020) Is tourism development a catalyst of economic recovery following natural disaster? An analysis of economic resilience and spatial variability. Curr Issues in Tou*rism* 23:1–22
Cochrane J (2010) The sphere of tourism resilience. Tourism Recreation Res 35(2):173–185
Cochrane J (2017) Tourism resilience in UK National Parks. In: Butler R (ed) Tourism and resilience. CAB International, London, UK, pp 121–134
Croatian Bureau of Statistics, CBS (2006–2011) Visitors to main tourist sights and attractions. First release, different numbers
Croatian Bureau of Statistics, CBS (2019) Tourist arrivals and nights in 2018. First release, CBS, (online). Available at: https://www.dzs.hr/Hrv_Eng/publication/2018/04-03-02_01_2018.htm
Croatian Bureau of Statistics, CBS (2020) Population estimate by sex, by towns/municipalities, 31 December (online). Available at: https://www.dzs.hr/PXWeb/Menu.aspx?px_db=Stanovni%c5%a1tvo&px_language=en
Cumming GS, Allen CR, Ban NC, Biggs D, Biggs HC, Cumming DH, De Vos A, Epstein G, Etienne M, Maciejewski K (2015) Understanding protected area resilience: a multi-scale, social-ecological approach. Ecol Appl 25(2):299–319
Davoudi S (2012) Resilience: a bridging concept or a dead end? Plann Theor Pract 13(2):299–333
Derissen S, Quaas MF, Baumgärtner S (2011) The relationship between resilience and sustainability of ecological-economic systems. Ecol Econ 70(6):1121–1128
Dudley N, Stolton S (eds) (2008) Defining protected areas: an international conference in Almeria, Spain. IUCN, Gland, Switzerland, 220 p. (online). Available at: https://portals.iucn.org/library/sites/library/files/documents/2008-106.pdf
Eurostat (2020) Tourism statistics at regional level (online). Available at: https://ec.europa.eu/eurostat/web/tourism/publications
Farrell BH, Twining-Ward L (2004) Reconceptualising tourism. Ann Tourism Res 31(2):274–295
Farrell BH, Twining-Ward L (2005) Seven steps towards sustainability: tourism in the context of new knowledge. J Sustain Tourism 13(2):109–122
Folke C (2006) Resilience: the emergence of a perspective for social–ecological systems analyses. Glob Environ Change 16(3):253–267
Folke C, Carpenter S, Elmqvist T, Gunderson L, Holling CS, Walker B (2002) Resilience and sustainable development: building adaptive capacity in a world of transformations. AMBIO J Hum Environ 31(5):437–440
Gallopín GC (2006) Linkages between vulnerability, resilience, and adaptive capacity. Glob Environ Change 16(3):293–303
Haisch T (2020) Working towards resilience: collective agency in a tourism destination in the Swiss Alps. In: Sarrinen J, Gill AM (eds) Resilient destinations and tourism: governance strategies in the transition towards sustainability in tourism. Routledge, New York, pp 121–135
Hall MC (2018) Resilience in tourism: development, theory, and application. In: Cheer JM, Lew AA (eds) Tourism, resilience and sustainability: adapting to social, political and economic change. Routledge, New York
Hamzah A, Hampton MP (2013) Resilience and non-linear change in island tourism. Tourism Geographies 15(1):43–67
Heslinga J, Groote P, Vanclay F (2020) Examining social-ecological resilience: using content analysis to assess changes in public perceptions of the synergetic interactions between tourism development and landscape protection. In: Sarrinen J, Gill AM (eds) Resilient destinations and tourism: governance strategies in the transition towards sustainability in tourism. Routledge, New York
Hill E, Wial H, Wolman H (2008) Exploring regional economic resilience. Institute of Urban and Regional Development Working Paper 2008/4. University of California, Berkeley, CA. http://doi.org/10.13140/RG.2.1.5099.4000
Holder JS (1980) Buying time with tourism in the Caribbean. Int J Tourism Manage 1(2):76–83
Holladay PJ, Powell RB (2013) Resident perceptions of social–ecological resilience and the sustainability of community-based tourism development in the Commonwealth of Dominica. J Sustain Tourism 21(8):1188–1211

Holling CS (1996) Engineering resilience versus ecological resilience. In: Schulze PC (ed) Engineering within ecological constraints. National Academy Press, Washington, DC
Holling CS (2001) Understanding the complexity of economic, ecological, and social systems. Ecosystems 4(5):390–405
International Union for Conservation of Nature (IUCN) World Parks Congress (2014) Available at: www.worldparkscongress.org. Accessed 5 Nov 2020
Janssen MA, Anderies JM, Ostrom E (2007) Robustness of social-ecological systems to spatial and temporal variability. Soc Nat Resour 20(4):307–322
Lew AA (2014) Scale, change and resilience in community tourism planning. Tourism Geographies 16(1):14–22
Lew AA, Ng PT, Ni CC, Wu TC (2016) Community sustainability and resilience: similarities, differences and indicators. Tourism Geographies 18(1):18–27
Lovejoy TE (1994) The quantification of biodiversity: an esoteric quest or a vital component of sustainable development? Philos Trans Roy Soc Lond Series B Biol Sci 345(1311):81–87
Mancini A, Salvati L, Sateriano A, Mancino G, Ferrara A (2012) Conceptualising and measuring the economic issues in the evaluation of socio-ecological resilience: a commentary. Int J Latest Trends Fin Econ Sci 2(3):190
Marshall NA (2010) Understanding social resilience to climate variability in primary enterprises and industries. Glob Environ Change 20(1):36–43
McCool S, Spenceley A (2014) Tourism and protected areas: a growing nexus of challenge and opportunity. Koedoe - African Protected Area Conservation and Science. 56
McDonald JR (2009) Complexity science: an alternative world view for understanding sustainable tourism development. J Sustain Tourism 17(4):455–471
Nature Protection Act, Official Gazette no. 80/13, 15/18
Nature Protection Strategy and Action Plan of the Republic of Croatia for the period 2017–2025 Official Gazette no. 72/2017
Olsson P, Galaz V, Boonstra WJ (2014) Sustainability transformations: a resilience perspective. Ecol Soc 19(4)
Ostrom E (2009) A general framework for analysing sustainability of socio-ecological systems. Science 325(5939):419–422.
Orchiston C, Prayag G, Brown C (2016) Organisational resilience in the tourism sector. Ann Tourism Res 56:145–148
Parkovi Hrvatske (2020) Broj posjetitelja u nacionalnim parkovima Hrvatske, 2012–2020
Peeters P, Gössling S, Klijs J, Milano C, Novelli M, Dijkmans C, Eijgelaar E, Hartman S, Heslinga J, Isaac R, Mitas O, Moretti S, Nawijn J, Papp B, Postma A (2018) Research for TRAN committee—overtourism: impact and possible policy responses. European Parliament, Policy Department for Structural and Cohesion Policies, Brussels
Perpar A, Udovč A (2007) Rural tourism and protected areas-factors to increase resilience of rural areas. Acta agriculturae Slovenica 89(1):115–128
Petrić L (2008) Kako turizam razvijati na održiv način u zaštićenim obalnim prostorima? Primjer: “Park prirode Biokovo”. Acta turistica nova 2(1):5–24
Pimm SL (1991) The balance of nature? Ecological issues in the conservation of species and communities. University of Chicago Press, Chicago
Public Institution National Park Brijuni (2016) Plan upravljanja (2016–2025), (online). Available at: https://www.np-brijuni.hr/download/6f67e492-4a50-4136-93fd-89f06287a549/28-02-2019/plan_upravljanja_np_brijuni_2016_2025g.pdf
Public Institution National Park Kornati (2014) Plan upravljanja 2014–2023, (online). Available at: https://www.np-kornati.hr/images/plan_upravljanja/Plan_upravljanja_NP_Kornati_tekst_%20i_zone.pdf
Public Institution National Park Krka (2011) Plan upravljanja, (online). Available at: http://npkrka.hr.win9.mojsite.com/buk/plan-upravljanja/files/assets/downloads/publication.pdf
Public Institution National Park Mljet (2017) Plan upravljanja 2017–2026 (online). Available at: https://np-mljet.hr/wp-content/uploads/2020/dokumenti/Plan-upravljanja_2.pdf

Public Institution National Park Plitvička jezera (2019) Plan upravljanja Nacionalnim parkom Plitvička jezera 2019–2028 (online). Available at: https://np-plitvicka-jezera.hr/wp-content/uploads/2019/10/Plan-upravljanja-NP-Plitvi%C4%8Dka-jezera.pdf
Quinlan AE, Berbés-Blázquez M, Haider LJ, Peterson GD (2016) Measuring and assessing resilience: broadening understanding through multiple disciplinary perspectives. J Appl Ecol 53(3):677–687
Regulation on the Ecological Network and the Jurisdiction of Public Institutions for Management of Sites of Ecological Network Official Gazette no. 80/19
Ruiz-Ballesteros E (2011) Social-ecological resilience and community-based tourism: an approach from Agua Blanca, Ecuador. Tourism Manage 32(3):655–666
Ružić V, Šutić B (2014) Ecological risks of expansive tourist development in protected areas-case study: Plitvice Lakes National Park. Collegium Antropologicum 38(1):241–248
Sakuma S (2020) Revitalising the nation: building resilience through ecotourism in Okinawa, Japan. In: Sarrinen, J, Gill AM (eds) Resilient destinations and tourism: governance strategies in the transition towards sustainability in tourism. Routledge, New York, pp 171–186
Schianetz K, Kavanagh L (2008) Sustainability indicators for tourism destinations: a complex adaptive systems approach using systemic indicator systems. J Sustain Tourism 16(6):601–628
Shepard VA (2017) Resilience and destination governance. In: Butler R (ed) Tourism and resilience. CAB International, London, UK, pp 53–68
Simmie J, Martin R (2010) The economic resilience of regions: towards an evolutionary approach. Camb J Reg Econ Soc 3(1):27–43
Strickland-Munro JK (2017) Resilience and protected area tourism in Purnululu National Park. In: Butler R (ed) Tourism and resilience. CAB International, London, UK, pp 137–149
Strickland-Munro JK, Allison HE, Moore SA (2010) Using resilience concepts to investigate the impacts of protected area tourism on communities. Ann Tourism Res 37(2):499–519
Tyler D, Dangerfield JM (1999) Ecosystem tourism: a resource-based philosophy for ecotourism. J Sustain Tourism 7(2):146–158
Tyrrell TJ, Johnston RJ (2008) Tourism sustainability, resiliency and dynamics: towards a more comprehensive perspective. Tourism Hospitality Res 8(1):14–24
Walker BH, Salt D (2006) Resilience thinking: sustaining ecosystems and people in a changing world. Island Press, Washington, DC
Walker BH, Holling CS, Carpenter SR, Kinzig A (2004) Resilience, adaptability and transformability in social–ecological systems. Ecol Soc 9(2)
Walker BH, Carpenter SR, Rockstrom J, Crépin AS, Peterson GD (2012) Drivers, "slow" variables, "fast" variables, shocks, and resilience. Ecol Soc 17(3)
Worboys GL, Lockwood M, Kothari A, Feary S, Pulsford I (eds) (2015) Protected area governance and management. ANU Press, Canberra
Zagreb Institute for Tourism (2020) Stavovi i potrošnja turista u Hrvatskoj – Tomas Hrvatska 2019., (online). Available at: http://www.iztzg.hr/files/file/RADOVI/KNJIGE/TOMAS-Hrvatska-2019.pdf

Chapter 19
Concentrate Mass Tourism to Promote Sustainable Development

Tomaz Ponce Dentinho

Abstract Mass tourism has a bad reputation. Nevertheless, once established, it mobilises the specialisation of the place in the touristic activities and promotes economic growth. Parallel to this, other places remain protected due to the backwash effects of the touristic activities. That stated this chapter aims to evaluate the sustainability and resilience of places regarding the development of touristic activities. To do that spatial density of tourism in Portugal is first analysed, and its effect on demographic growth is elaborated. Then, a demographic growth model for the Portuguese municipalities was created to include the effect of tourism and to test its impact on a destination's sustainability concerning its life cycle stage. Results indicate that places with more intense tourism might protect the sustainability and resilience of neighbouring places, including those rich with natural assets, and, at the same time, provide the means for fostering its sustainability.

Keywords Tourism · Sustainability · Resilience · Life cycle · Demographic growth model

19.1 Introduction

Tourism is booming around the world, illuminating the minds of regional developers. Nevertheless, due to the global crises of COVID-19 and local, distresses and catastrophes, the awareness of the need for developing tourism sustainably becomes a primary concern (Richards and Derek 2000). Although bigger cities are often more resilient than the smaller ones, they can also suffer from overtourism that colonises their centres and removes traditional activities. On the other, hand smaller, sparsely populated rural places are focused on increasing productivity of former agricultural land by attracting tourists and eventually evolve into integrated communities consisting of both, tourists and locals. Nevertheless, after being distressed by tourists or defeated by other 'colonial' activities such as sport, mineral extraction, etc., such

---

T. P. Dentinho (✉)
Centro de Estudos de Economia Aplicada do Atlântico (CEEAplA), University of the Azores, Ponta Delgada, Portugal
e-mail: tomas.lc.dentinho@uac.pt

A. Mandić and L. Petrić (eds.), *Mediterranean Protected Areas in the Era of Overtourism*, https://doi.org/10.1007/978-3-030-69193-6_19

areas may be abandoned again. The opportunities and threats of this rapidly emerging and booming reality linked to the expansion of economic development, low fares of liberalised air transportation, and broad public support, need to be carefully examined to avoid the destruction and 'prostitution' of cultural and creative ambiences and, instead, to use the opportunity to further develop and integrate peoples and places. The growing numbers of abandoned hotels in the middle of nowhere, degraded and shrinking villages and displacement of populations from old city centres simply can't be overlooked.

This research departs from two assumptions. On the one hand, space is limited and tends to be intensely specialised depending on the site-specific economic basis. On the other hand, places tend to keep their rank in the urban network, unless being impacted by a sudden interest of external demand provoked by some natural and/or cultural assets.

Based on a temporal perspective delineating tourism development in a place in various phases, i.e. discovery, growth, success, maturity and decline (Butler 1980), two scenarios may be outlined. In the so-called 'serious' scenario, a place can avoid the substitution of its former activities by touristic activities thus keeping the ratio between tourists and residents relatively low, while keeping at the same time its rank in the city/region hierarchy unchanged. In the so-called 'prostituted/redeemed' scenario, tourism activities support higher land use rents, push non-touristic activities to other places, and, leaning on the community's competitive advantages promote tourism-led growth. Specific scenarios can be explained by the 'spread' and 'backwash' effects of spatial interaction. Hence, in the 'spread effect scenario', a tourism-oriented place may abandon some touristic activities, which consequently move (spread) towards nearby locations. In the second scenario, some places facing a general growth of the touristic activities in a country or a region they are located, experience backwash effects associated with the competitive advantage of nearby locations.

The spatial analysis of tourism development creates new perspectives on the major worries of sustainable tourism. Destination's sustainability, associated with its natural and cultural environments (Taylor 1995; Butler 1999), refers not only to its territory but also to the surrounding areas influenced by spread and backwash effects. Furthermore, overall development strategies should consider the strong spatial specialisation of tourism (Kappert 2000; Edgell 2016) and the constant need for redemption of invaded spaces, to protect important attractions. Regarding the spatial specialisation assumption, there are two issues to address, the first one referring to the potential of tourism to be the focus of an area's specialisation and the second one to its impact on urban hierarchy. The aforementioned is to be investigated firstly by positioning each particular Portuguese city-region in the national city-regions' hierarchy and then by searching the extent tourism specialisation impacts on this ranking. It is also important to investigate if touristic activities in a specific place involve spread and backwash spatial effects affecting not only the sustainability of the place itself but also of the surrounding places.

Various approaches to measuring tourism sustainability based on different types of indicators have been employed so far. Hence, by investigating residents' perceptions, Woo et al. (2016) research how tourism affects their life quality in terms of both material and non-material life and life satisfaction. The overall measurement model consists of five major constructs and 20 observed indicators. Lee and Jan (2018) examined, based on the tourism area life cycle stage, residents' perceptions of the sustainability of community-based tourism in six Taiwanese communities, and found that results significantly differ between the pre-development and post-development stage. Asmelash and Kumar (2019) analyse tourist satisfaction with sustainable heritage tourism in Tigrai, the inception of ancient Ethiopian civilisation. Perceived overall sustainable heritage tourism was operationalised as the average of four dimensions and sixteen sub-dimensions with the proper set of indicators. Blancas et al. (2018) present a vectorial composite indicator, called Differential Dynamic Index (DDI) that is defined via two components: one dynamic, to graduate the evolution of the destination regarding its sustainability, and the other static, to relativise its position concerning other territories, evaluating the status achieved in the social, economic and environmental parameters, which affect the degree of sustainability. Based on a system of indicators calculated in municipalities that are representative of the different tourist areas and environments of Catalonia (Spain), Torres-Delgado and Palomeque (2018), identified the key variables of sustainable local tourism and aggregated them into a single global score called the ISOST index that enables the definition of sustainable tourism thresholds. Given the lack of available data to produce indicators for cities, Navarro et al. (2019) propose an index to measure sustainable tourism at the European NUTS 2 level, while Modica et al. (2018) produced a sustainable tourism performance measurement system (STPMS) on the case of South Sardinia (Italy), to provide guidelines and inputs for other European countries and tourist destinations that are currently in the process of implementing the European Tourism Indicator System (ETIS) toolkit or similar methodologies. Pan et al. (2018) propose a joint analysis of human perceptions confronted with environmental indicators of sustainability while Edgell (2016) associates sustainability with morality.

Despite their valuable contributions, none of the above papers investigates the spatial interaction effect of sustainability that is crucial for understanding the resilience and sustainability of peoples and places. With this regard, the following chapter aims firstly to analyse touristic data for Portuguese municipalities to illustrate the spatial specialisation of tourism and its relevance in the urban hierarchy. Following is the proposal of a demographic growth model that includes spatial interaction sustainability, to understand the relations between tourism and economic growth. Finally, the results are discussed, and the conclusions are outlined with the regional policy implications highlighted.

19.2 Tourism and City/Region Hierarchy in Portugal

Portugal has 308 municipalities in the Main Land, Madeira and Azores. In the last decades, there is a substantial increase in the suburbs of the administrative, political centres of Lisbon, Porto, Coimbra, Funchal, and Ponta Delgada and a considerable increase in the population of the Algarve. The rest of the country experienced a substantial decrease in the population with some resilience in the capitals of former districts and the municipalities along the coast.

On the other hand, tourism becomes a very concentrated activity with only six municipalities, out of 308, with more than 4.000 tourists per 100 residents. Hence, it is possible to identify a small group of competitive tourism destinations in Portugal (Fig. 19.1), developed around different attractions, confirming the idea that the market itself groups destinations into various clusters. With this regard, the coastal municipalities of Algarve and Alentejo, in Nazaré and on the island of Porto Santo are focused on the 'sea and sun' tourism. Cultural tourism is developed in Lisbon, Porto and Óbidos. Nature tourism is well developed in the mountains of São Mamede, Gerês and on the islands of São Miguel and Madeira. Finally, Fátima is known as a destination for religious tourism.

All the other municipalities, although opting to attract tourists and reach a competitive advantage on the tourism market, do not succeed to achieve these goals.

Additionally, there is a relative and absolute population growth only in those municipalities that are in the mature stage of touristic development, with more than 4.000 tourists per 100 residents (Fig. 19.2). Other tourist destinations, not showing such a tourist intensity level, are maintained at an infancy stage of their tourism

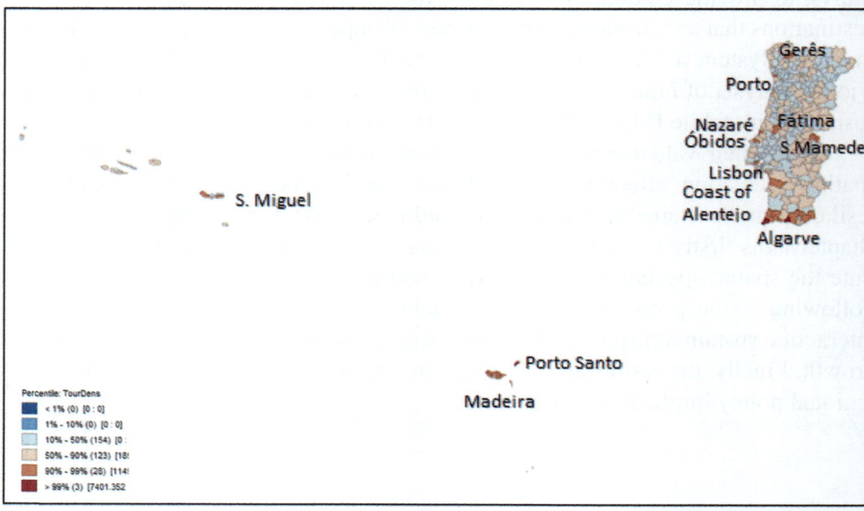

Fig. 19.1 Tourism per 100 residents in Portuguese Municipalities (2011). *Source* Retrieved from https://www.pordata.pt/Portugal

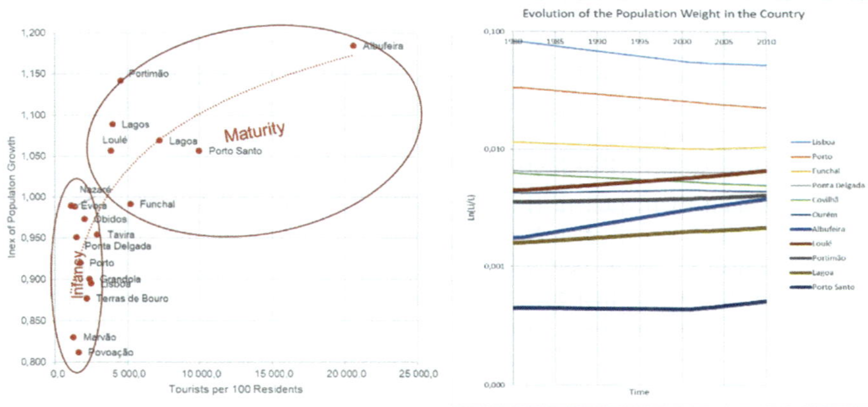

Fig. 19.2 Tourism stage and demographic relative growth. *Source* Authors

development life cycle, experiencing at the same time a decrease in the population, either in absolute terms or regarding its weight within the country.

This leads to a conclusion that only if the tourist intensity indicator exceeds 4.000 tourists per 100 residents, there is a positive relation between touristic specialisation and demographic growth. In other words, only those destinations that are in the maturity stage of their life cycle have an impact on urban hierarchy. With this regard, maturity should be used to redeem the 'prostituted' place by keeping it competitive in the global market and, considering backwash effects from the surrounding areas, wisely protecting them from being devastated.

19.3 Tourism in a Demographic Growth Model

This section elaborates a demographic growth model of a spatial network of city-regions, highlighting the role of tourism. The aim is to understand how tourism affects urban hierarchy and sustainability.

19.3.1 The Model Elaboration

Each city-region is the outcome of a productive space that depends on labour, capital, land, spatial interaction and respective elasticities. A Cobb Douglas production function is assumed for each city-region (i) in a specific period (t) with constant returns on the scale for capital and labour.

$$Y_{it} = Q_{it}^{\sigma} K_{it}^{\alpha} L_{it}^{\beta} \tag{1}$$

where α = elasticity of capital; β = elasticity of labour in period (t); μ_i = elasticity of land in city-region (i); Y_{it} = the product of city-region (i) in time (t); K_{it} = the capital of city-region (i) in time (t); and L_{it} = labour of city-region (i) in time (t). Q_{it} = the spatial interaction factor that redistributes product among city-regions and includes the potential of the city-region ($\boldsymbol{T}_i$) as a limiting factor, representing the territorial capacity of the place, usually taken as land.

The spatial interaction factor $Q_{it} = \sum_{jt}^{nt} (\boldsymbol{T}_j/T_{jt})\exp^{\left(-\nu d_{ij}\right)}$ depends on the relation between the potential of each city-region ($\boldsymbol{T}_j$) relative to its use (T_{jt}) weighted by a function of distances between city-regions (d_{ij}). By internalising the factor $(\boldsymbol{T}_j/T_{jt})$ in the exponential form by defining $\left[p_{it} = -\boldsymbol{ln}\,(\boldsymbol{T}_j/T_{jt})\right]$, the spatial interaction takes the following form:

$$Q_{it} = \sum_{jt}^{nt} \exp^{(-p_{it-1} - \nu d_{ij})} \tag{2}$$

where p_{it-1} = the bid-rent of city-region (i) in period (t −1).

For any time (t), optimality requires that the Marginal Productivity of Capital and Labour equals respectively to r = return on capital—considered fixed for all the periods and all the city-regions, and w_{it} = wages for each period t in city-regions (i).

$$w_{it} = \beta Q_{it}^{\sigma} K_{it}^{\alpha} L_{it}^{\beta-1} \tag{3}$$

$$r = \alpha Q_{it}^{\sigma} K_{it}^{\alpha-1} L_{it}^{\beta} \tag{4}$$

$$\frac{w_{it}}{r} = (\alpha/\beta)\frac{K_{it}}{L_{it}} \tag{5}$$

The model assumes that economic growth comes from investment fed by savings in income $(s_t Y_{1t} - \delta(K_{1t}))$, where s_t = marginal propensity to save in time (t) and that labour follows the capital needs based on expression (4). With some calculations, the change in labour between (t) and ($t+1$)$\left(\frac{L_{1t+1}}{L_{1t}} - (1-\delta)\right)$ results from labour in the previous period (L_{1t}) weighted by the production factor costs and elasticities $((\omega_{it}; r; \sigma; \alpha; \beta)$, and by the scale factor (Q_{it}^{σ}) influenced by spatial interaction and spatial income distribution.

$$\frac{L_{1t+1}}{L_{1t}} - (1-\delta) = s_t(\beta r/\alpha\omega_{it})^{1-\alpha} Q_{it}^{\sigma} (L_{1t})^{(\alpha+\beta-1)} \tag{6}$$

The city increases with savings (s), decreases with wages $(\beta r/\alpha\omega_{it})^{1-\alpha}$, rises with spatial interaction Q_{it} of expression (2), and grows in population as related to the previous period. The logarithm of expression (6) leads to the econometric model of expression (7).

$$Ln\left(\frac{L_{1t+1}}{L_{1t}} - (1-\delta)\right) = Ln(s_t) + (1-\alpha)Ln(\beta r/\alpha\omega_{it}) + \sigma Ln Q_{it} + (\alpha+\beta-1)Ln(L_t) + \varepsilon_{it} \quad (7)$$

The identification of the model parameters results from the following operations:

- First, by defining the relation $(\boldsymbol{T}_j/T_{jt})$ of formula (2) as the relation between the long-term weight of the city-region (i) in the country and the weight of the city-region in the country in time (t). Therefore the bid-rent of each city-region (i) in time (t) is $[p_{it} = -\boldsymbol{ln}(T_j/T_t)]$ as presented above;
- Second, the parameters $(\alpha_t + \beta_t)$, associated with the elasticity of capital and labour, relate to the coefficients $(\alpha_t + \beta_t - 1)$ of $(Ln(L_{it}))$ in expression (7).
- Third, the parameter of the distance (ν) in expression (2) is calibrated to secure the scale consistency of (2) and the robustness of parameter σ of $(Ln(Q_{it}))$.
- Finally, δ, or the depreciation of capital, is defined exogenously.

How to include tourism? Tourism can be included with an extra population ($L_{it}^E = L_{it} + X_{it}$), where $X_{it} =$ number of residents equivalent of tourists in the region (i) in time (t). The effect of Maturity Tourism (from Fig. 19.2) can also be analysed with a dummy variable for municipalities with mature tourism (D_{mt}) that have more than 4.000 tourists per 100 residents.

$$Ln\left(\frac{L_{1t+1}}{L_{1t}} - (1-\delta)\right) = Ln(s_t) + (1-\alpha)Ln(\beta r/\alpha\omega_{it}) + \sigma Ln Q_{it} + (\alpha+\beta-1)Ln\left(L_{it}^E\right) + +\gamma D_{mt} + \varepsilon_{it} \quad (8)$$

19.3.2 Data for the Model

The data set for the model involved 616 observations correspondent to 308 municipalities and two periods of change: 1981–2001 and 2001–2019. Population comes from PORDATA Data Base; the distances between municipalities—adjusted when involving crossing the sea—come from ARCGIS Database. T_j/T_{jt} is the median population weight of the municipality in the whole country, adjusted for a total sum of 1.00, divided by the population weight of the municipality in time (j). Finally, the indices of the product per capita per municipality are the ones estimated by Pedro Ramos (1998) for 2001 and extrapolated for 2019 based on the estimates of the EUROSTAT for NUTS II regions. For the Azores, the spatial profile of the product per capita per island (SREA) served to estimate the product per capita for all the municipalities of the same island for 2001 and 2019. The municipalities of Madeira got the same indices for the product per capita, being changed only from 2001 to 2019.

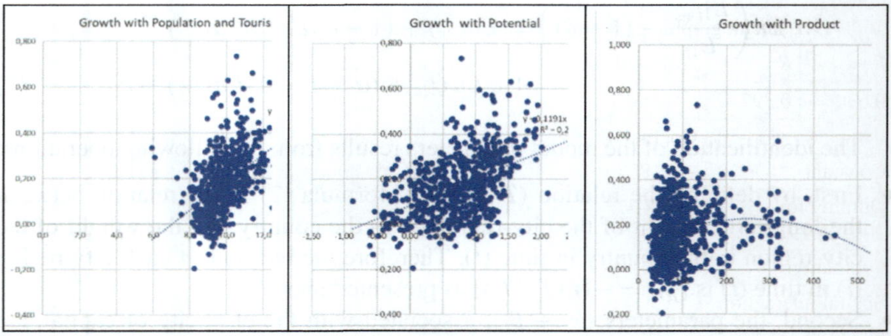

Fig. 19.3 The relation between population change and, respectively, population with tourism, potential and product per capita. *Source* Authors

Figure 19.3 indicates, as expected from the theoretical model (Expression 8), that Population and Tourism, on one hand, and Spatial Interaction Potential, on the other hand, influence directly population growth.

Nevertheless, the relation between Product per Capita, as a proxy of regional wages (noted as the one derived theoretically in expression [6], with a fixed return on capital [r]), influence population growth negatively. This may be explained by the situation that at a lower wage level, regional growth attracts more people as compared to the situation when at a higher wage level investments move elsewhere, ultimately decreasing economic growth.

19.3.3 Model Estimates

Table 19.1 presents the results of the demographic growth model designed to estimate the impact of tourism on the economic and demographic growth of places within a networked system of city/region. Results show quite robust coefficients for the variables Population and Tourism, Spatial Interaction Potential and the Dummy for Mature Tourism. As expected, Product Per Capita transformations show collinearity but still indicate interesting results. The model estimated with a depreciation factor of $(1-\delta) = 0$ shows better results than the model with $(1-\delta) = 0,25$, which shows to be logical considering that periods of change last for 20 years.

Results in Tables 19.1 and 19.2 reveal interesting interpretations:

- The saving rate implicit in the constant and in the dummy for the period 2001/2019 indicates that there was a decrease in the model savings rate from 0.84 in the period 1981–2001 to 0.77 in the period from 2001 to 2019.
- The variable 'Population with tourist/resident equivalent' presented in Table 19.2 shows slightly better results indicating that the econometric model in expression (8) is better than the one presented in expression (7).

Table 19.1 Demographic growth model using population

R	R^2	Adjusted R^2	Standard error	Durbin-Watson	F	Sig.
0.630	0.397	0.391	0.10716	1.505	66.814	0.000
		Coefficient	T	Sig.	Tolerance	VIF
Constant		−0.18	−3.904	0.000		
Dummy for 2001/2019		−0.077	−7.982	0.000	0.808	1.238
Population		0.024	4.762	0.000	0.668	1.498
Spatial interaction potential		0.104	10.575	0.000	0.659	1.518
Dummy for mature tourism		0.171	5.813	0.000	0.968	1.033
Product per capita Index		0.001	4.763	0.000	0.102	9.808
Product per capita Index 2		−1.41E-06	−2.988	0.003	0.109	9.185

Source Author

Table 19.2 Demographic growth model using population with tourism

R	R^2	Adjusted R^2	Standard error	Durbin-Watson	F	Sig.
0.632	0.399	0.393	0.10695	1.510	66.814	0.000
		Coefficient	T	Sig.	Tolerance	VIF
Constant		−0.196	−4.171	0.000		
Dummy for 2001/2019		−0.077	−8.037	0.000	0.812	1.232
Population with tourism		0.026	5.022	0.000	0.664	1.506
Spatial interaction potential		0.103	10.472	0,000	0.660	1.516
Dummy for mature tourism		0.164	5.583	0.000	0.963	1.038
Product per capita Index		0.001	4.717	0.000	0.102	9.816
Product per capita Index 2		−1.403E-6	−2.984	0.003	0.109	9.178

Source Author

- Spatial interaction potential that changes only with the bid rends (because distance and spatial attrition are the same for both periods), plays an important role in the demographic growth of municipalities, indicating that land markets expressed in the bid rends play a crucial role in the dynamic spatial equilibrium of a city's/region's system.

- Mature Tourism present in the Algarve—Portimão, Lagos, Lagoa, Loulé, and Albufeira—and in Madeira—Funchal and Porto Santo (Fig. 19.2) has an important role in demographic growth when compared with municipalities that are in the tourism infancy stage or those that are not touristically developed. Mature Tourism exists in so-called 'prostituted' places redeemable through growth. Infant tourism is on edge between becoming prostituted and eventually redeemed, or it may stay 'serious' avoiding touristic growth.
- Finally, the inverted U-curve relating wages with the growth of city/regions indicates the balancing factor between wages and rural–urban migration. For some time people are attracted by the cities growing in size and wages, but when wages become too high, the flow of investments to other places reduces the growth of the city/region with higher income.

19.4 Discussion

The questions posed in the introductory part of the chapter may now be answered more competently:

- Thus, the research confirmed the existence of a tourism spatial specialisation. On the one hand, touristic activities are highly concentrated (as presented in Fig. 19.1). On the other hand, tourism affects demographic growth only if, as a dominant activity in a place, brings it to the maturity stage.
- Research results have confirmed that tourism does not affect significantly urban hierarchy. Moreover, this may be the case in only pre-existing centralities like Lisbon, Porto, Funchal and conurbations around provincial capitals in the Algarve, Oeste and Minho, that have been pushed forward owing to new, tourism generated brand images (Edwards et al. 2000).
- The robustness of the coefficient associated with the spatial interaction potential shows the relevance of backwash effects related to tourism and spread effects associated with the activities displaced by tourism to the surrounding areas. This may be important for nature conservation since growing demand increases the value of natural resources. On the other hand, backwash effects from nearby central places may prevent the degradation of natural attractions. For instance, tourism in Funchal protects and values the Madeira landscape, tourism in Lisboa helps to conserve the value of the Forests of Sintra and Arrábida, while tourism in Porto rescues the Douro Valley.
- Concerning the impact of tourism on regional sustainability, the model generated the following conclusions: if a place remains at an infancy stage, no major sustainability problems are evident, because activities displaced by tourism can readjust thus enabling a place to hold its position in the urban hierarchy. Notwithstanding this, maturity should be used to redeem the 'prostituted' place by keeping it competitive in the global market and, by using backwash effects from the surrounding areas, avoid further decline.

19.5 Conclusion

Most places are 'serious', built and adapted to respond to the public needs with their institutional, economic and environmental contexts. Such places have less than 4.000 tourists per 100 residents and are capable of developing tourism sustainably since they do not depend on tourism development exclusively.

After exceeding the number of 4.000 tourists per 100 residents, spaces tend to adapt to tourists' needs dominantly, thus becoming a product for the external market. Such places deserve to be named as 'prostituted' places. However, it cannot be neglected that, by serving outsiders, tourism activities in a prostituted place stimulate the economic growth for the insiders. On the other hand, by adapting themselves to the needs of the external visitors, such places can destroy the very reasons that made them attractive and liveable.

The lesson for serious, non-touristic places is to follow other development paths while, at the same time, unobtrusively interacting with visitors and benefiting locals (Castilho 2015). The strategy for prostituted places is to specialise in a sector they are the most competitive (i.e. tourism) but also to continuously 'redeem' the place with adequate tax policy measures affecting adequate space usage. What we have learned is that once a place reaches such a scale, it has to stay competitive and try to develop sustainably within that specific sector. Yet, it must be noted that this goal would be much easier fulfilled if a place is located closer to centralities because spatial interaction enhances its resilience to external shocks.

References

Asmelash AG, Kumar S (2019) Assessing the progress of tourism sustainability: developing and validating sustainability indicators. Tourism Manage 71:67–83

Blancas FJ, Lozano-Oyola M, González M, Caballero R (2018) A dynamic sustainable tourism evaluation using multiple benchmarks. J Clean Prod 174:1190–1203

Butler RW (1980) The concept of tourism area cycle of evolution: implications for management of resources. Can Geogr 24:5–12

Butler RW (1999) Sustainable tourism: a state-of-the-art review. Tourism Geographies 1(1):7–25

Castilho VJ (2015) Tourism, the end of illusion: a social practice that still has its spatial importance but does not deceive people anymore. Alma Tourism. J Tourism Culture Territorial Dev 12:1–17

Edgell, D, L. (2016). Managing sustainable tourism: a legacy for the future. Routledge

Edwards J, Fernandes C, Fox J, Vaughan R (2000) Tourism brand attributes of the Alto Minho, Portugal. In: Richards G, Hall DR (eds) Tourism and sustainable community development. Routledge, London and New York, pp 285–296

Kappert J (2000) Community and rural development in Northern Portugal. In Richards G, Hall D (eds) Tourism and sustainable community development. Routledge Advances in-Tourism Series. Routledge, London, pp 258–267

Lee SH, Jan FH (2018) Can community-based tourism contribute to sustainable development? Evidence from residents' perceptions of sustainability. Tour Manage 70:368–380. https://doi.org/10.1016/j.tourman.2018.09.003

Modica P, Capocchi A, Foroni I, Zenga M (2018) An assessment of the implementation of the European tourism indicator system for sustainable destinations in Italy. Sustainability 10(9):3160. https://doi.org/10.3390/su10093160

Navarro JL, Martínez ME, Jiménez JA (2019) An approach to measuring sustainable tourism at the local level in Europe. Curr Issues Tourism 23(4):423–437. https://doi.org/10.1080/13683500.2019.1579174

Pan SY, Gao M, Kim H, Shah KJ, Pei SL, Chiang PC (2018) Advances and challenges in sustainable tourism toward a green economy. Sci Total Environ 635:452–469. https://doi.org/10.1016/j.scitotenv.2018.04.134

Richards G, Derek R (2000) Tourism and sustainable community development. Volume 7 de Routledge advances in tourism. Editors Greg Richards, Derek R. Hall, Routledge

Taylor P (1995) Measuring changes in the relative competitiveness of package tour destinations. Tourism Econ 1(2):169–182. https://doi.org/10.1177/135481669500100204

Torres-Delgado A, Palomeque FL (2018) The ISOST index: a tool for studying sustainable tourism. J Destination Mark Manage 8:281–289. https://doi.org/10.1016/j.jdmm.2017.05.005

Woo E, Uysal M, Sirgy MJ (2016) Tourism impact and stakeholders quality of life. J Hospitality Tourism Res 42(2):260–286. https://doi.org/10.1177/1096348016654971

Chapter 20
Summary and Outlook

Ante Mandić and Lidija Petrić

Tourism is, without doubt, one of the most critical drivers of economic growth, both globally and locally. Driven by a relatively strong global economy, the growing middle class in emerging economies, technological progress, new business models, affordable travel costs and visa facilitation, tourism grew 5% in 2018 and recorded visits reached 1.4 billion (World Travel Organization 2019). Although the COVID-19 pandemic suppressed the positive numbers, tourism to no small extent demonstrated its resilience, as well as desperate need to reconsider the current development trajectory, particularly concerning growth focus policies, failure to implement sustainability at the destination level, lack of public control over tourism development and negative impacts associated with both local community and ecosystems. The pandemic revisited the debate as to whether tourism and tourist flows can be controlled on a destination scale, and provoked more critical thinking on the sustainability of tourism development. More radical scholars, concerning criticism of the current tourism development mainstream, for example, Hall (2019), Higgins-Desbiolles (2018), Higgins-Desbiolles et al. (2019), Liu (2003), and Sharpley (2020), are gaining more attention. In one of the most recent articles published in *Journal of Sustainable Tourism*, referring to the "war about tourism", Higgins-Desbiolles (2020) concludes that advocates of industry rapid recovery stand opposed to broader efforts to reform tourism so that it should be more ethical, responsible and sustainable. Following publication, the article additionally spurred an online academic debate first on TRINET Tourism Information Network via email and subsequently via various online statements and position papers. One thing is for sure; the article demonstrated the existence of a profound divide between tourism academics about the success of the sustainability of tourism development.

A. Mandić (✉) · L. Petrić
Faculty of Economics, Business and Tourism, University of Split, Split, Croatia
e-mail: ante.mandic@efst.hr

L. Petrić
e-mail: lidija.petric@efst.hr

A. Mandić and L. Petrić (eds.), *Mediterranean Protected Areas in the Era of Overtourism*, https://doi.org/10.1007/978-3-030-69193-6_20

No matter what one stands for, it is clear that local-scale problems related to tourism development, particularly those related to the deterioration of ecosystems and quality of life, are present and are gaining growing attention. Most recently, tourism academics accepted the term overtourism, to describe these established challenges in a different context, and to discuss the need for careful planning and management of tourism and the respect for the well-being of the local community. The studies addressing overtourism initially drew the most attention in urban destinations (UNWTO 2018); however, recently, the debate moved to include nature-based destinations (Wall 2019, 2020). Due to escalating demand and increasing concentrations of visitors within pristine natural areas, Weaver and Lawton (2017) proposed the transition to a new visitation paradigm for PAs that sees visitors as an inherent opportunity and employs management and monitoring accordingly with visitor motivation for mass participation in on-site park activities. The pandemic additionally emphasised the need for this transition, i.e. a more holistic approach to visitor management and sustainable experience design, provision and monitoring, as, during the lockdown, many PAs have actually witnessed increasing visitation.

The academic research presented here reinforces the need for collaboration across and beyond disciplines among researchers, practitioners and communities in a search for plausible solutions to improve governance frameworks and management of PAs in the Mediterranean region in an era of overtourism. The contributors explore and discuss pertinent mechanisms, taking into account the rapid evolution of the tourism economy, as well as the critical long-term trends, such as demographics, evolving demand, digitalisation, climate change, transportation, everlasting tourism growth and overcrowding. With this book, we acknowledge that nature-based tourism, when built upon broad stakeholder engagement, sustainable development principles, and with the support of relevant policies, tools and measures can contribute to more inclusive growth through the provision of economic development in communities within and adjoining natural environments. It can also raise awareness of cultural and environmental values, and help finance the protection and management of PA, and the preservation of biological diversity. Holistic policies are needed to meet these targets. From the perspective of excessive visitation, we encourage coordinated actions across governments, communities and relevant institutions and invoke more responsible business practices within PAs through the integration of environmental and social criteria and resilience thinking into tourism policies and commercial visitor programmes. It is, therefore, essential that all stakeholders collaborate to take the work forward in meeting these goals and ultimately optimising nature-based tourism development.

20.1 Governance and Management

Governance and management of PAs are closely related but nevertheless distinct phenomena. While governance is about who decides what the objectives are, what to do to pursue theme, how decisions are taken and who holds power, management is about what is done in pursuit of the given objectives and the measures and actions needed to achieve such objectives (Worboys et al. 2015). Management of the PA is related to the type of governance it applies, which in a context of nature-based tourism encompasses laws, policies, regulations and management plans. Ultimately, PAs will aim to reconcile conservation with the provision of ecosystem services (ES), particularly recreation, ensure fair benefit sharing and build resilient ecosystems. PAs are experiencing a diversity of pressures (Dudley and Stolton 2018), requiring adequate responses to balance environment, economy and society. Within this section of the book, we were particularly eager to broaden the understanding of the way Mediterranean PAs adapt so as to mitigate the pressures associated with overtourism.

The challenges related to overtourism in the Mediterranean region comprise two parallel stories. One of them is related to almost six decades of continuous growth of the tourism industry, and the consequent tourism-dependence of many economies. The other one is related to inherent strengths and weaknesses associated with PA governance and management and visitor use management. Many tourism destinations have long pursued the goal of tourism growth, which in many cases resulted in ecosystem degradation. However, sustainability does not have to be an unattainable goal, if we manage to reframe the collective mindset both about our perceptions of what sustainability means and our worldview and value systems more generally. However, the progress in this area has lagged behind political and economic developments, suggesting there is significant scope for radical change (Chap. 16). An excessive dependence on tourism development leads to vulnerability both to overtourism and no-tourism (Chap. 15). For nature-based destinations, that means that high visitor concentrations threaten the sensitive environment, while local communities do not face significant long-term economic benefits (added value) arising from the provision of mass tourism products and services. This over-specialisation in tourism relates to lack of resilience, suggesting there is a need to support the development of clusters of activities not related to tourism, and interactions between tourism and related sectors to stimulate smart specialisation and innovation strategies in the Mediterranean area.

The conclusions drawn from the studies presented here are that there is a need for efficient leadership of PAs underpinned by professional management expertise, the experiences and tools to achieve such goals. The adoption of fair and effective governance, adaptive co-management (ACM) focused on the promotion of collaborative learning among all parties involved, the cooperation between institutions acting at different governmental levels, and different sectors will be essential for PAs to mitigate current and future challenges (Chap. 2). The efficient participatory approach advocated by ACM could foster the creation of the new institutional arrangements through various co-management initiatives enabling better interaction among different stakeholder groups, including the community as well as the tourism industry and the PA (Chap. 4). This holistic approach to the PA system could foster the reconciliation between hard-edge single objective conservation and those seeking much broader outcomes, such as the inclusion of the public enjoyment, community well-being, and changing the relationship between public goods and land use in terms of changes in tourism and recreation. To mitigate the impacts of tourism on PA ecosystems, established visitor management frameworks should be advanced by consideration of new theoretical and practical advances employing a system approach in which PAs are seen in the interrelation with other ecosystems (Chap. 3). Along with that, PA managers require innovative and functioning tools and resources, new approaches to data collection, and the active involvement of various stakeholder groups coexisting within a PA ecosystem so as to be proactive and forestall pressures.

20.2 The Local Community and Well-Being

This section of the book is focused on the influences of excessive tourism on local communities, their well-being, and the ambitions of hosts to have early warning of the development of overtourism.

Communities are central to nature-based tourism development. They are affected by the development of tourism, so, in addition to the knowledge they have about the reality and conditions of the place, they can collaborate in assessing the impacts and defining the strategies for the governance, management and planning of nature-based tourism within a PA (Chap. 10). Communities' attitudes towards visitors and tourism are primarily associated with the perception of the impacts involved. Excessive tourism refers to a community feeling aversion or to social rejection, which are in many Mediterranean destinations discussed as tourism-phobia and tourist-phobia. Such attitudes undoubtedly affect the way tourism products and services are shaped and tourism experiences designed and delivered.

Contemporary challenges emphasise the interdependencies of multiple stakeholder groups (Chap. 5). Excessive tourism can disrupt the essence of the local, as a PA is the setting for communal rites, customs and traditions associated with spiritual, physical, emotional and mental health. Enabling communities to reap benefits from visitor activities while keeping their core values intact, and facilitating the balance between the right to travel and residents' rights, should be the strategy to overcome the challenges posed by nature-based tourism development. From the perspective of the local community, overtourism requires a holistic approach and the inclusion of stakeholders to stimulate intrinsic change that will lead to equilibrium. The involvement of the stakeholders should not be a mere aspiration or designed to meet legal requirements. Instead, it should stem from a mutual commitment to improve the resilience of the PA and establish ecosystem stewardship. The imbalances in power among stakeholder groups, diverging attitudes towards tourism development, as well as the prominent core-periphery relations, will lead to widespread dissatisfaction (Chaps. 6 and 9). PA management needs to include equitable and meaningful participation of the community, promotion of collaborative culture and better monitoring of sites, i.e. destination carrying capacities. Genuine participation should entail empowerment for engagement, calling for education and capacity building of local communities to get involved in the process of PA management and visitor use planning. A relationship between conservation, recreation and education should be established to facilitate an integrated approach to the territory, promoting knowledge, awareness of ecosystems and sustainable development (Chap. 7). However, converting ideology into action, and bridging the gap between theory and practice is a long-term process (Chap. 8). With this in mind, particular attention should be given to initiatives focused on the young concerning their pro-environmental behaviour, accountability and education regarding what seems to be acceptable from the perspective of the society.

20.3 Visitor Experience Design and Management

This section of the book aimed to discuss the influence of excessive visitation on visitor experiences, visitor behaviour and the way PA managing agencies influence the setting to facilitate the construction of sustainable experiences.

Visitor experiences are central to nature-based tourism. They do not just happen but are built with the help of destination managers who influence the social, attraction and management setting attributes that provide the opportunity for visitors to construct an on-site experience. Nature-based tourism is characterised with intense experiences derived from various activities in nature. Providing visitors with engaging experiences in nature is recognised as a way to build societal support to deliver conservation goals, as well as benefits for communities and societies. As nature-based tourism and activities within exceptional scenic areas are becoming increasingly popular, the relationship between recreational activities and natural resources require further attention.

The management of visitors is central to the provision and upgrading of transformative experiences (Chap. 11). In a vibrant PA with high, seasonal visitation, management for the improvement of experiences requires effective policy measures related, among other things, to setting up limits for hourly peak use levels, redistribution of these peak levels throughout the day, changing of the supply opportunities offered, and tailored marketing activities. Such carefully planned policies to manage tourism volumes and associated impacts to maintain service quality and significant environmental values should be part of broader planning efforts, ultimately resulting in the purposive limiting of tourist growth. PA managers should consider creating a multi-experience site catering to different needs (Moyle et al. 2017) while broadening the scope beyond the current to include potential visitors too. Successful management of visitor numbers in a PA presumes cognition of the causes and consequences of overtourism, as well as an enabling governance framework, efficient management and inclusion of PA goals into regional (tourism) development agenda. The lack of consensus on how tourism should be developed and monitored, limits, among others, the potential of communities to harvest benefits associated with tourism development, as well as to implement responses to address its adverse impacts.

Overtourism relates to crowding, which is, in many cases, negatively related to visitor satisfaction. Satisfaction tends to induce visitor loyalty, such as a visitor's intention to revisit, to recommend the PA, to pay the entry fees, and volunteer time to work in it. The perception of crowding is closely related to visitor use. Thus, the optimisation of visitor flows and numbers requires a profound understanding of these phenomena. The mismanagement of a PA can lead to loss of authenticity, causing disappointment on the part of visitors, as well as conflicts and tensions within the community and between residents and visitors (Chaps. 12 and 13). Such findings provide one additional argument that self-governance arrangements are not optimal for PAs experiencing the symptoms of overtourism, as a lack of monitoring mechanisms, and policy interventions may harm the integrity of the ecosystem. Instead, public authorities should take responsibility for implementing adaptive co-management tools as consensus-building, persuasion and negotiation, supplemented with increased sharing of knowledge and collective learning (Chap. 14).

20.4 Intelligence in Nature-Based Tourism Development

Intelligence could be defined as the act of understanding or the ability to learn and deal with new situations.[1] Recently, within smart destination-related literature, intelligence has become a synonym for the knowledge associated with decision-making through structured processes (SEGITTUR 2015). This section of the book contains several perspective articles discussing the causes and consequences of overtourism, as well as collective adaptation to changes to build up the resilience of nature-based tourism destinations.

Since many PAs are located near urban destinations often subject to overtourism, they are not to blame for the problems of excessive visitation because they simply attract daily visitors who are already located in an adjoining urban destination. On the other hand, these backwash effects, i.e. concentration of tourism development in adjoining urban areas are favourable for a PA as they minimise the pressure on the protected features and may prevent their degradation (Chap. 19).

Areas with a higher endowment of natural resources attract large volumes of tourists. At the same time, they are significantly exposed to adverse shocks, implying severe socio-economic problems for local communities (Chap. 15). The most sustainable destinations are those with high levels of resilience; however, resilience alone may not be sufficient for sustainability as it might include responses that are not aligned with sustainability principles (Espiner et al. 2017). Resilience is a constituting dimension of sustainable nature-based destinations, as for a destination to remain viable over time, the capacity to adapt to a dynamic environment is crucial. This underlines the importance of incorporating, among other approaches, resilience measures, adaptive co-management, training and education into PA management and planning, as well as a shift away from reactive strategies towards long-term goals (Chap. 18). Enhancement of PA resilience requires a system approach (PA within regional development agenda), improvement of governance, crisis management and strengthening of the relations between communities and PA mainly through cognition and enhancement of its positive impacts on well-being.

[1]Definition of intelligence: Merriam-Webster dictionary. Available at https://www.merriam-webster.com/dictionary/intelligence.

Traditional approaches towards tourism development have their costs, including low added value, and in many cases social, environmental and economic fragmentation. However, expecting that a change in the way the various stakeholders think about tourism will happen overnight is naïve. It is more likely that in time when sustainable choices become the social norm, policy-makers will have to raise the threshold of what is acceptable and reflect the values of the majority. In other words, social change should pre-empt policy; as in democratic nations, governance should be a reflection of the collective mindset (Chap. 16).

PAs call for cost-effective, efficient and innovative solutions to handle a wide variety of threats; thus, PAs must use the digital revolution in their advantage. Although technologies are one of the enablers of overtourism, as increasing visibility of a PA on social media tends to contribute to an increase in visitors, they are also identified as a means to combat overtourism, in addition to or replacing traditional forms of sustainable management and de-marketing (Chap. 17). PAs are becoming smarter as managing agencies employ different technological solutions (for example for automated collection of ecological data via cameras, bio-tracking, drones; automated surveillance of the parks; security of staff and tourists; improvement of visitor experience) to address the diverse everyday pressures related to conservation and recreation. However, to combat the potentially detrimental effects of overtourism, comprehensive and holistic strategies based on a combination of technological and governance-related solutions are needed (Chap. 17). Information technologies enable the storage, analysis and integration of various data, allowing PAs to be viewed as integrated parts of an overall landscape, and enabling more efficient, accurate and informed decision-making. The implementation of technological solutions is vital to improve innovativeness and sustainability within protected ecosystem, and to capture the data related to tourism and tourists visiting a PA, to provide realtime, personalised services, and to optimise management to improve the visitor experience. Technological innovations lay the foundations for new tourism experiences (Hjalager 2015), but, to take full advantage of the current possibilities provided by smartness, destination and PA managers must integrate the entire range of smartness components and ensure interoperability and interconnectivity of both soft (innovation, social capital, human capital and leadership) and hard (presence of technology) smartness (Boes et al. 2016).

References

Boes K, Buhalis D, Inversini A (2016) Smart tourism destinations: ecosystems for tourism destination competitiveness. Int J Tour Cities 2(2):108–124. https://doi.org/10.1108/IJTC-12-2015-0032

Dudley N, Stolton S (2018) PROTECTED AREAS: challenges and responses for the coming decade. February

Espiner S, Orchiston C, Higham J (2017) Resilience and sustainability: a complementary relationship? Towards a practical conceptual model for the sustainability–resilience nexus in tourism. J Sustain Tour 25(10):1385–1400. https://doi.org/10.1080/09669582.2017.1281929

Hall CM (2019) Constructing sustainable tourism development: the 2030 agenda and the managerial ecology of sustainable tourism. J Sustain Tour 27(7):1044–1060. https://doi.org/10.1080/09669582.2018.1560456
Higgins-Desbiolles F (2018) Sustainable tourism: sustaining tourism or something more? Tour Manag Perspect 25(October):157–160. https://doi.org/10.1016/j.tmp.2017.11.017
Higgins-Desbiolles F (2020) The “war over tourism”: challenges to sustainable tourism in the tourism academy after COVID-19. J Sustain Tour 0(0):1–19. https://doi.org/10.1080/09669582.2020.1803334
Higgins-Desbiolles F, Carnicelli S, Krolikowski C, Wijesinghe G, Boluk K (2019) Degrowing tourism: rethinking tourism. J Sustain Tour 27(12):1926–1944. https://doi.org/10.1080/09669582.2019.1601732
Hjalager AM (2015) 100 innovations that transformed tourism. J Travel Res 54(1):3–21. https://doi.org/10.1177/0047287513516390
Liu Z (2003) Sustainable tourism development: a critique. J Sustain Tour 11(6):459–475. https://doi.org/10.1080/09669580308667216
Moyle BD, Scherrer P, Weiler B, Wilson E, Caldicott R, Nielsen N (2017) Assessing preferences of potential visitors for nature-based experiences in protected areas. Tour Manag 62:29–41. https://doi.org/10.1016/j.tourman.2017.03.010
SEGITTUR (2015) Smart destinations report: building the future. SEGITTUR. https://www.segittur.es/opencms/export/sites/segitur/.content/galerias/descargas/documentos/Libro-Destinos-Inteligentes-en-Ingls.pdf
Sharpley R (2020) Tourism, sustainable development and the theoretical divide: 20 years on. J Sustain Tour 0(0):1–15. https://doi.org/10.1080/09669582.2020.1779732
UNWTO (2018) ‘Overtourism’?—understanding and managing urban tourism growth beyond perceptions, executive summary. World Tourism Organization (UNWTO); Centre of Expertise Leisure, Tourism & Hospitality; NHTV Breda University of Applied Sciences; and NHL Stenden University of Applied Sciences. https://doi.org/10.18111/9789284420070
Wall G (2019) Perspectives on the environment and overtourism. In Dodds WR, Butler R (ed) Overtourism, issues, realities and solutions. De Gruyter Studies in Tourism, pp 3–27
Wall G (2020) From carrying capacity to overtourism: a perspective article. Tour Rev 75(1):212–215. https://doi.org/10.1108/TR-08-2019-0356
Weaver DB, Lawton LJ (2017) A new visitation paradigm for protected areas. Tour Manag 60:140–146. https://doi.org/10.1016/j.tourman.2016.11.018
Worboys GL, Lockwood M, Kothari A, Feary S, Pulsford I (eds) (2015) Protected area governance and management. ANU Press, Canberra
World Travel Organization (2019) International tourism highlights. Unwto, 1–24

GPSR Compliance

The European Union's (EU) General Product Safety Regulation (GPSR) is a set of rules that requires consumer products to be safe and our obligations to ensure this.

If you have any concerns about our products, you can contact us on ProductSafety@springernature.com

In case Publisher is established outside the EU, the EU authorized representative is:

Springer Nature Customer Service Center GmbH
Europaplatz 3
69115 Heidelberg, Germany

Batch number: 10732076

Printed by Printforce, the Netherlands